Building Your Own Home

POINTS OF INTEREST - P32 - SQ FT TO HOUSE SIZE
BUDGETING

Building Your Own Home

Murray Armor

Fifteenth edition

EBURY PRESS

London

First published in the United Kingdom in 1978 by Prism Press

3 5 7 9 10 8 6 4 2

Updated editions 1980, 1981, 1982, 1984, 1985, 1986, 1987.
Revised editions 1983, 1988, 1989, 1990, 1991, 1993

This fifteenth edition published in 1996 by Ebury Press
Random House
20 Vauxhall Bridge Road
London SW1V 2SA

Random House Australia (Pty) Limited
20 Alfred Street, Milsons Point, Sydney
New South Wales 2061, Australia

Random House New Zealand Limited
18 Poland Road, Glenfield
Auckland 10, New Zealand

Random House South Africa (Pty) Limited
PO BOX 337, Bervlei, South Africa

Random House UK Limited Reg. No. 954009

A CIP catalogue record for this book is available from the British Library.

ISBN: 0 09 180920 7

This book contains only general advice, and neither the author nor the publishers will accept responsibility of any sort for the consequence of the application of this advice to specific situations. In particular, planning and other legislation is described in outline only, and professional help should always be obtained when dealing with these matters.

Typeset by Colin Spooner
DTP by Imprimés
Printed in Singapore by Tien Wah Press.

Foreword

This 15th edition of *Building Your Own Home* is published 18 years after the first edition, and has since been completely rewritten many times. Looking back at the earlier editions it is interesting to see how the selfbuild scene has changed. In 1978 the first chapter was about how it really was possible for a family with no relevant experience to build for themselves. To do so was unusual to the point of eccentricity. The first chapter in this edition presents the statistics behind today's booming selfbuild phenomena, with about one in ten new homes being owner built. The proportion rises to at least one in four for houses with a value over £150,000.

What has caused this change? Simply the day for selfbuild has come. New homes have moved from a seller's market to a buyer's market. Twenty years ago prospective home buyers queued up to put their names down for a house on a speculative development, and were able to decide whether they wanted a blue or pink bathroom suite. Today they can make their choices at leisure, and if the house is not yet completed they can usually have it finished just as they wish. Consumer choice in housing has arrived, and selfbuild is its cutting edge. In this we are not being innovative, but merely catching up with how the rest of the western world buys new homes. As described later, 80% of Austrians, 60% of Germans, and 25% of American owner occupiers build their own homes. We have a long way to go yet.

The word *selfbuild* does not help, evoking an image of a heroic DIY enthusiast doing everything himself. There are very few of these. The definition of a selfbuilder is someone who buys a piece of land and who arranges for a new home for himself to be built on it. This is a complex business, and he has to make it happen himself. The management is the key to everything, and is the theme of this book.

The market for land, materials and services for individual home builders is currently worth two billion pounds each year. This market is now recognised as being a distinct sector of the building industry, and in recent years a large number of specialist firms have been established to supply it. While they serve to make the whole business of building for yourself a lot easier, they do require their customers to commit themselves to a particular way of building at an early stage, so that the first decisions are the most important. It is essential that potential home builders learn all that they can about all of their options before making choices, and this book is written as their consumer guide.

Building Your Own Home is completely independent of all the firms which are mentioned in it, and although I am grateful to them for photographs, the descriptions and appraisals of their services are wholly my own. The addresses of all the companies featured, together with details of various statutory and other organisations mentioned in the book will be found in an appendix.

This book would not have been possible without the co-operation of many families who built their own homes and then chose to write to the author. Some explain that reading an earlier edition led directly to them building for themselves, and it is always a worry that a letter might go on to say that all has been a disaster. Happily this has not happened yet, and the theme of the following pages is that building for yourself can be a low risk and very rewarding enterprise if approached in the right way. I hope the book will continue to be a real help to those who read it and decide to go on to build their own homes.

Finally, two matters may attract criticism. There are both male and female selfbuilders, architects, solicitors, planning officers and others. In the first draft the use of 'he or she' to recognise this was so tedious and repetitive that much of it was edited out. If any are offended by this they have the author's apologies. The second matter, for which no apologies are offered, concerns the use of metric and imperial terms in the text. In this, the day-to-day language of the building site has been followed, where the selfbuilder will refer to using a four inch nail to fix a 38mm rafter, and to laying fifty yards of 100mm drains. Neither selfbuilders nor authors have to be consistent.

Contents

1 Introduction

In 1994-95 at least eighteen thousand families moved into new homes which they had built for themselves. They all got exactly the home that they wanted, and invariably there was a significant margin between the cost and the value of the new property. All of them were involved in a great deal of decision making, most of them managed the building work themselves, some did some of the work with their own hands, and a few, a very few, did everything themselves. On average the whole process took them between a year and eighteen months, with six to nine months spent finding a site and setting everything up, and the same amount of time spent doing the actual building work. They built homes that were well constructed and finished, and, whether they had an architect to handle everything for them or whether they managed the whole project themselves, they invariably found it immensely satisfying.

Building Your Own Home looks at this scene in detail, explains what is involved in building for yourself, and describes exactly how it can be done in this age of regulations and restraint. It is full of options, schedules of requirements, programmes and financial plans. Somewhere there is mention of the need for hard work. Involvement in over two thousand individual homes in more than twenty years has taught me that this order of priorities is correct, and that if one is building a home for oneself the project planning is everything, and, compared with it, managing the construction work is all fun.

It is important to differentiate between the different ways of building for oneself. The situation is confused by the various ways in which the terms 'selfbuilder', 'individual builder', and 'owner builder' are sometimes used in the media and by advertisers.

Individual builders and selfbuilders

Individual builders is a term used to describe all those who arrange for a new home to be built for themselves on their own land, however it is constructed. They may have owned the land for years, or they may have bought it to build on it immediately: the essential factor is that the choice of design and of how it is built is theirs.

Selfbuilders are the individual builders who provide all the management which is involved in building a new home themselves, and who do not place a single contract for the whole project with a builder. They make use of sub-contractors who they employ separately for all the different building trades, and are easily identified by arrangements which they make to reclaim VAT. Provided that the home is for their own occupation they can reclaim most of the VAT which they have paid for materials and services without being VAT registered. In 1994/5, the last year for which the VAT statistics are available as this book goes to the printers, there were 10,529 claimants. However, at least 10% of all selfbuilders are estimated to reclaim their VAT through their business accounts when they built a home in connection with their businesses, and a further 200 families moved into new homes built by members of selfbuild housing associations, which are not included in the statistics. The whole business of VAT is dealt with at length in chapter 22: here it serves to demonstrate that the total number of homes completed by people in these three categories in the financial year 1994/95 was

---

Selfbuilders and individual builders annual statistics

18,000 individual builders build on their own land

of which

over 11,000 build as selfbuilders.
* They use sub-contractors without formal contracts.
* They provide all the management.
* They reclaim their VAT.

and

approximately 7,000 are *not* selfbuilders.
* They have formal contracts with builders.
* They make all the choices but do not provide site management.
* They do not pay VAT.

H M CUSTOMS AND EXCISE
VAT Policy Directorate
Construction & Utilities
4th Floor West, New Kings Beam House
22 Upper Ground, London, SE1 9PJ

Dear Mr. Armor,

Listed below are the numbers of DIY claims from selfbuilders under Customs and Excise Notice 719 that this Department has processed during the financial year 1994/95. I have listed them out in our collection areas which does not quite match those areas listed in one of your previous books.

BIRMINGHAM	330	SOUTH WEST	661
S.WALES & BORDERS	1297	DOVER	229
E. ANGLIA	662	N.E. ENGLAND	571
N.W. ENGLAND	549	LONDON PORT	41
LONDON NORTH	26	MANCHESTER	210
NORTHERN ENGLAND	1300	NORTHAMPTON	372
EAST MIDLANDS	670	THAMES VALLEY	267
SOUTHAMPTON	339	EDINBURGH	639
GLASGOW	938	NORTHERN IRELAND	1428

This gives a total of 10529 claims processed.

Your sincerely,

for VAT Policy Directorate

STATISTICS FOR SELFBUILD

The VAT figures quoted in the letter above enable the total of selfbuild homes completed to be estimated as follows:

10,529 families reclaimed the VAT which they paid in building their new homes in the financial year 1994/5.

1000 homes are estimated as built by the 10% of all individual selfbuilders who reclaim their VAT through their business accounts (mainly rural enterprises).

200 families are estimated to have moved into new homes built by members of selfbuild groups during the year.

This gives a total of 11,700 selfbuilt homes completed during the financial year 1994/5.

No figures are available for Individual Homebuilders who place a single contract with a builder, and who do not pay VAT on their builders invoices as they are zero rated. However, experience from the large selfbuild developments and other sources suggests that they total at least 60% of the number of selfbuilders. This gives another 7000 homes completed, which brings the total of individually built homes to at least 18,000, or 11.47% of all private sector homes built in the UK in that period.

11,700. As they completed in 1994/5 they started their selfbuild projects in 1993/4 in the middle of the housing recession — more of this in due course.

The other individual builders are those who place a single contract with a builder to construct their new home. They do not have a generally used identifying name, and are not involved with VAT, as a builder does not charge VAT to anyone who places a single contract for a complete dwelling. Their number is variously estimated at between 60% and 150% of the VAT total — a huge variation! In the author's experience 60% can certainly be demonstrated.

This gives approximately 18,000 new homes built in 1994/95 by all the individual builders in both categories. This is 11.47% of all private sector homes built in the UK in the period. An interesting comparison can be made with the latest published sales figures for leading commercial house builders, which are Wimpey 8078 and Barratt 5782.

The term *owner builder* is not generally used in the U.K., but will be found to describe self-builders in American and Australian selfbuild books. In these countries, as in most European countries, it is far more usual for individuals to be involved in building their own new homes than in Britain: the most recently available figures are in a paper published by Sussex University in 1992. Extracts from it are reproduced on page 13.

Effect of the recession

What has been the effect of the recession on the popularity of building for yourself? Statistics published since 1985 show a steady growth in the number of individual homes built each year up to those quoted here. As a result the individual homes market is considered to be booming, and as it is worth at least two billion pounds in land, materials and services, manufacturers and others are promoting themselves energetically in it at this time when the overall housing market is depressed. The general view is that the current availability of building plots, because developers and builders are not buying land, and the ease with which building tradesmen can be

found to work for selfbuilders, outweigh the financial problems experienced by so many, and that the number of people building for themselves will continue to increase.

One reason for this is that new homes have moved from being part of a seller's market to being in a buyer's market, with consumer choice dictating what the developers build. Until the end of the seventies those buying a new house or bungalow queued to get on a builder's waiting list, and they were offered a choice of a pink or blue bathroom suite in a house built to a very bland standard design. By the end of the eighties new homes were being offered in a wide variety of styles and the builders would incorporate all the internal fittings and fixtures that their customers wanted. It is certain that we are now moving to a situation where developers will build whatever purchasers require on their developments, and this will follow standard practice abroad. Building for yourself can be seen as the cutting edge of this move to individual choice, and will grow in popularity. How far can it grow? Informed

Owner built homes in regional styles. *Above* in Devon and *below* in Cheshire.

guesses suggest that within a few years up to 25% of all new homes may be individual homes, which will still be short of the proportion in other countries.

In the last three years all this enthusiasm for individual homes has led to the establishment of many new companies offering special services to those building on their own, to the appearance of two specialist monthly magazines, two national exhibitions a year and even residential training courses. As with all fast developing markets, the quality of the services offered varies enormously, and the proliferation of those wanting to provide help and assistance to the individual builder requires that he or she must be able to evaluate the options and make informed choices. The way to set about this will be a constant theme in this book.

Associations

Besides the proliferation of commercial undertakings, there are two non-commercial organisations which are important. The Individual House Builders Association is a trade body run by the established companies which sell into the selfbuild market and which offers advice on services available. The membership requirements and the high level of affiliation fees serve to make the IHBA logo an indication that a member company is well established, and the booklets and other publications of the Association contain useful advice and are a handy directory of services available.

The Association of Selfbuilders is a grass roots organisation which is run by the selfbuilder members themselves through an elected committee, offering seminars, visits and a newsletter for a very modest subscription. It attracts sponsorship that enables the association to have stands at selfbuild exhibitions and roadshows, and it performs a most valuable service in enabling prospective selfbuilders to meet those who are actually building.

Group selfbuild

One area of confusion in the individual homes scene is the position regarding selfbuild housing associations, which are also known as selfbuild groups, and which have two chapters of their own later in the book. They involve a group of families obtaining loan finance to be able to work as a team to build as many houses as there are members of the group. When the building work is completed each family buys their home from their association at cost, and the association is then wound up. There are two sorts of these groups: mainstream associations which are normally composed of existing owner occupiers who want to trade up to a larger home, and community selfbuild groups which are concerned with low cost homes for those in housing need.

Both of these type of selfbuild groups are in a state of change. Mainstream groups built about two thousand homes a year until 1989, but in the current economic climate they are unable to get the loan funding that they require and currently only a handful of houses are completed by those building in this way. On the other hand, finance for community selfbuild groups is now more widely available than ever before, as is their level of support from both government and local government.

These team activities by groups, both mainstream and community, are financed and operated in a completely different way to the activities of the individual homebuilders, and to avoid confusion they are ignored in this book until chapters 24 and 25 which are wholly concerned with them. Nothing in the intervening pages is directly relevant to them, but an understanding of how they work is useful to anyone wanting an over-view of the whole selfbuild scene. One reason for this is that selfbuild activity attracts media attention in reverse order to the size of the operation. Not unexpectedly the community groups draw the headlines first, followed by the team efforts of the associations. The individual selfbuilders tend to be ignored, except when a heroic mum busy laying bricks is featured in the tabloids or women's magazines. Anything which you are likely to read in the press about individual home building, or programmes which you may watch on television, are featured because they make an interesting story. They are also likely to be rather unusual stories and may not be relevant to the real world of individual homes.

Owner-built housing abroad

Self-help housing 1980-89
Estimated percentage completions

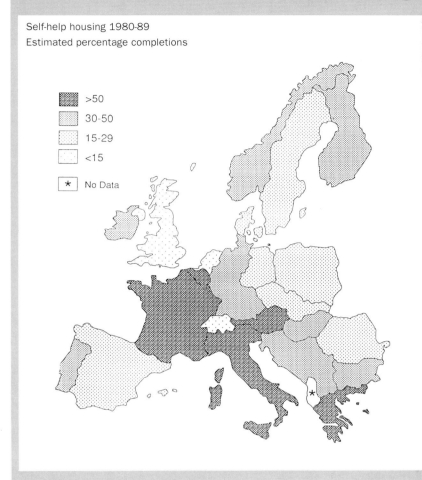

▨	>50
▨	30-50
▨	15-29
□	<15
★	No Data

Self-help housing as a proportion of new
owner-occupied housing, 1980-89

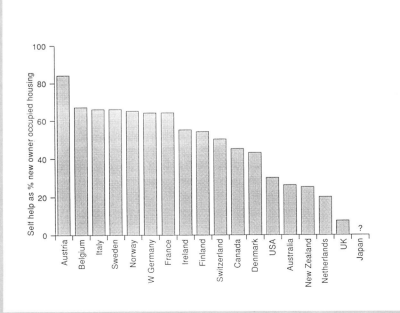

Sadly, the British lag behind the rest of the world in building their own homes, as demonstrated by the figures below. In Belgium, Germany, Austria, Italy and France more than half of all houses built in the 1980s were constructed by their owners in this way, compared with six per cent in the UK. As we move to a single Europe it is surely inevitable that we shall catch up.

The table and map are taken from a recent publication from Sussex University with the very appropriate title 'Self-Help Housing - The First World's Hidden Housing Arm', from which the following is reproduced with the permission of the authors.

"Self-Help Housing is a major form of housing provision in nearly all the developed countries of Western Europe, North America and Australasia. In many, like France or Germany, it accounted for the major part of housing output during the 1980s. Nor is self-help housing associated with backwardness, peripherality, or lack of market development. Self-help provision lowers the money cost of housing and usually ensures higher quality, and in this way enlarges the housing choices of middle-income nuclear families.

Self-help housing is defined as being when a household (alone or collectively) finds finance, buys land, manages the project and owns the finished product. In one form of self-help households do not engage in much of the actual building work, but rather accept a tender from commercial builders to build the dwelling. Alternatively they take on the role of general contractor in organising various builders and suppliers, including situations where they use kit homes from catalogue builders. Some carry out the bulk of the building work themselves.

In practice there is a grey area between the categories, and a rule of thumb is to consider housing as self built if the household is involved in all stages of the production process and contractors account for less than 50% of construction labour time.

Different mixes occur. In France, for instance, only about 20% of self-help is selfbuild, with the remaining self promoted housing being split fairly equally between that commissioned from individual builders and housing bought from catalogues. In Canada and Norway in contrast, about two thirds of self-help housing is self built while in Ireland about 15% of self-help housing produced in the 1980s was completely selfbuild, with 40% partly selfbuild and 45% contracted out."

Reproduced from *Self-Help Housing* by S. Duncan and A. Rowe, published by the Centre for Urban and Regional Research, Arts Building B, University of Sussex, Brighton, BN1 9QN.

But enough of this setting the scene: now for the actors. For most house owners the purchase of their home is the largest transaction of their lives, and the decision to buy a house is a step only rivalled in importance by changing jobs or getting married. A publicity machine promoted by builders, estate agents and building societies helps most people to their decision, and guides and assists them in the details. To reject all this professionalism by building for yourself is to flout the conventional, and before looking at how it is done it is important to understand what motivates those who participate.

Cost savings and motivation

Building for yourself offers significant savings to those involved and this often leads to an assumption that it is the cost savings that motivate selfbuilders and that they build cut price homes. This is very wide of the mark. The average individual home is a four bedroom, two garage detached house, and although there are plenty of small starter homes and small retirement bungalows, there are just as many very large properties. They are all built to a high standard, both in engineering specifications and in the detailing and finishes. The case histories on later pages will demonstrate this. Building for yourself attracts far more salaried and professional people than weekly wage earners, but has its enthusiasts at every level in the socio-economic spectrum. Selfbuilders range from senior managers to market traders, TV personalities to postmen, doctors to dockers. (Doctors often go on to selfbuild a group practice surgery, as the quirks of NHS funding make this a very advantageous thing for them to do.) Airline pilots are great selfbuilders, as it suits their pattern of work, as is also the case with firemen and taxi drivers. With all these people the common link is a capacity to get things done, and to enjoy doing it. This is the essential selfbuild common denominator.

There is virtually no economy self-build housing, and, except for community projects, there are few situations in which people decide to become selfbuilders because they cannot afford to house themselves in any other way. Invariably the budget for a selfbuilt home could be used to buy a smaller and nastier house that is immediately

on the market. However, Individual Builders do not see their situations in this light. Time after time one is told that 'this is the only way I can afford to build', when the house that is being built will make the self-builder the best housed individual in his family or among his work colleagues. Farmers are masters at this form of self-deception, persuading themselves that by building on a direct labour basis, they are just managing to afford to put a roof over their heads, when the roof is of a size and standard more usually seen in the stockbroker belt.

The real motivation in selfbuild is far deeper than costs per square foot. We were cave men for many thousands of years before estate agents arrived, and the urge to find one's own cave is deep rooted. In a world where opportunities for self-expression are diminishing, building one's own home both satisfies primaeval urges and exorcizes modern frustrations, with enough obstacles to be overcome to provide a sense of achievement. The rationale may be economic: the motivation is more basic. The new home is an expression of the individual himself, and it will be built to the highest standard to which he can aspire. At the very least the standard will be several levels above the property developer's lowest common denominator.

The level of individual involvement in building a home varies enormously. At one end of the scale we have the man who may contribute two thousand hours of manual labour, in the middle of the scale we have the person who contributes only the management of the project, employing self-employed workmen for all the building trades on a direct labour basis, while there are some who find the right land, the right architect to handle everything, and then go off for a winters holiday (or to work in the Middle East) and come back to collect their new front door key. They invariably enjoy a saving of anything from 15% to 30% on the market value of the property. This saving is not simply a builder's profit, or the value of a selfbuilder's own labour, although both contribute to it. It really comes from a different cost breakdown for the building, shown opposite.

Another misconception about selfbuild is the idea that those concerned are making a political

The selfbuilder's savings — a comparison with developers' costs

Developers' costs	Selfbuilders' costs
1 Land cost	1 Land cost
2 Interest on land cost over a long period	2 Interest on land cost over a very short period
3 Design and planning fees	3 Design and planning fees
4 Site labour costs	4 Site labour costs, which may be as low as the developers' or could be at premium rates
5 Labour overheads — cost of labour between profitable jobs, in periods of bad weather, holidays, Training Board levy, N.I. etc.	5 No labour overheads
6 Materials at trade prices	6 Materials probably at trade prices
7 Up to 10% of materials damaged/ wasted/stolen on site	7 No site losses
8 Office overheads	8 No office overheads
9 Staff costs and staff overheads	9 No staff
10 Expensive general contractor's insurance, NHBC warranties, Trade Association levies	10 Cheap simple site insurance and Zurich Custom Build premiums
11 Sales costs	11 Nil
12 Provision for bad debts	12 Nil
13 Interest on building finance assuming worst sales situation	13 Interest on building finance kept to a minimum
14 Corporation Tax or other revenue involvement	14 Nil
15 Return required on capital	15 Nil

statement. During the 1890s, and between the wars, a number of groups of people in Britain with a common social philosophy built villages in which to live together to put their theories into practice, sometimes as part of the Arts and Crafts Movement. Today they would be described as communes. The best known are extensively documented in histories of housing trends, and there are some who see selfbuilders as their successors. This is rarely so. Individual homebuilders want to do their own thing within the system, not to escape from it. Their boast is of what their houses are worth, and they usually build marketable homes of conventional appearance and layout. There are a few exceptions: those who

build ecologically friendly houses at Machynlleth or the members of the Findhorn Community who build beautifully crafted homes out of disused whisky vats are certainly selfbuilders, but they are a tiny minority. Ordinary selfbuilders are members of ordinary families who want to do their own thing, and invariably do it very well indeed.

Learning about selfbuild

How much technical knowledge do you need if you are going to manage a selfbuild project? The answer is, all the general information that you can acquire, together with as much technical information, or as little, as is appropriate to your style of management. All selfbuilders have to make sure everyone contributing to their new home does the right job, and they need to learn as much as they can about the organisation of the project and the different ways of arranging and co-ordinating the work. Some decide to go on and learn about technical matters, from drain laying to wall tiling. Others decide that as they can never become an expert at everything, they will concentrate on hiring the best experts and rely on them to do the best job. It is all a question of personal management style and it is doubtful if those who have read everything they can find about building skills end up with better homes than those who have not had the time for this. It makes sense to learn as much as you can, but success will not be guaranteed by familiarity with technicalities, but only by providing good project planning and effective management.

Before going on to examine the various ways of learning as much as you decide is appropriate for you to learn, it is important to recognise the limitations to whatever technical knowledge you acquire. Firstly, building techniques vary in different parts of the country, and there are very few guides to regional building practice. For example, in some parts of Britain moulded bricks are laid frog up, in others frog down. There are local ways of setting windows, and the use of quarry stone is always a local matter. Secondly, you cannot possibly hope to know more than the sub-contractors who you will employ, and there is a danger that if you pretend to a little knowledge they may assume that you know more than you do, and not use their own initiatives to the extent that they will if you emphasise that you rely on them.

The purpose of learning about the work of the different building trades is to know what questions you should be asking the experts who you have hired, to understand what they suggest, and, if you are not entirely happy with what they propose, to know where to obtain a second opinion. Reading about bricklaying will prompt you to ask your bricklayers about using additives in the mortar and covering up work in the winter, but it should not lead to you issuing instructions as if you were a site foreman. If you do have to

give instructions of this sort, you have got the wrong bricklayers.

Having made this clear, most selfbuilders want to know all that they can, knowing they must use their skimpy knowledge with caution. How do you set about it?

Books
There are other consumer guides besides this one, and it is worth reading all of them which are of general interest. Others are only concerned with special subjects, like community selfbuild or special types of timber frame housing. There is a bibliography on page 294.

Useful books in a different category are the Building Regulations and the NHBC and Zurich Custom Build handbooks, both of which are considered invaluable by some selfbuilders and ignored by others. The Building Regulations are published in sections, so you can buy only those which are of interest to you. You may find the various illustrated commentaries on them easier to read than the regulations themselves. All are on display at the Building Bookshop (see below).

The NHBC have a catalogue listing a wide range of publications concerning every aspect of their services, and the building standards which they promote. They are a good buy.

There are literally hundreds of technical books on single subjects like bricklaying, plumbing or laying drains, and they vary from DIY manuals for the layman to text books for students. It is unlikely that the ones that suit you will be in your local bookshop, and the only place to find all of them is in the Building Bookshop at the London Building Centre. If you cannot get there in person send for their catalogue — details on page 294.

If you are using a library to obtain books of this sort, look at the publication date of any book that you find there and ignore anything written more than five years ago. Also ignore American books unless you propose to build in America.

Magazines
Two specialist magazines for selfbuilders are *Build It* and *Individual Homes*, both available at newsagents or by subscription. They are essential

reading, and the address for subscriptions is on page 294. There are a host of magazines for DIY enthusiasts which sometimes have features relevant to those building a new home and they occasionally have separate selfbuild supplements. All magazines depend on their advertisers, and it is sometimes difficult to distinguish between the advertising and the features, which does not matter if you use your critical faculties. The advertisements themselves, with details of materials, special services, exhibitions and other opportunities are invaluable.

Courses

Organisations which run courses for selfbuilders are listed on page 294. All have their own particular flavour. Do not hesitate to telephone the organisers and cross examine them about how relevant their courses are to your own requirements. They will not mind this: they want to make sure you get on the right course just as much as you do.

Exhibitions

There are two major exhibitions, at London in the autumn sponsored by *Build It* magazine, and at the National Exhibition Centre in the early summer sponsored by *Individual Homes* magazine. Details will be found in these magazines. These exhibitions feature hundreds of companies providing all sorts of services for individual home builders, from manufacturers to financial agencies. There are show houses to visit, lectures and seminars to attend, bureaus at which you can obtain advice, and they are an essential day out for the serious selfbuilder.

Roadshows

Selfbuild roadshows are held at large hotels with conference centres throughout the year. Many of them are organised by *Built It* magazine, others by package companies or occasionally by builders merchants. Usually there are between twenty and thirty exhibitors, many of them local, and there may be seminars or advice services offered. Always useful if you live locally, and a good opportunity to look at other selfbuilders and perhaps make useful contacts.

Show houses

Some of the package companies have show houses, and Potton has a 'show village' of three homes, all available for inspection seven days a week, with regular seminars. A must if you are thinking of using the company concerned, but also arrange to see a house built by one of their ordinary clients.

Conferences

Community selfbuild is usually on the agenda at the various local authority and housing association conferences and the individual home scene is discussed at timber industry conferences concerned with timber frame housing. Details on page 294. The CSBA and the Segal Trust will arrange special conferences if there is sufficient interest in any area, and some of the package companies organise local exhibitions which may be called conferences.

The Building Centre

The Building Centre at 26 Store Street, London, a minute's walk from the Goodge Street tube station, has six floors of displays of everything to do with house building. Leaflets on the materials on display are available at a special counter, and there is both a general advisory service available and various agencies dispensing advice to enquirers. All of this is aimed at the architect and builder, but enquiries from the public are welcome.

Advisory services

Both trade organisations and some large companies offer advisory services and first class publications of different sorts, varying from British Gypsum's residential courses where selfbuilders can learn plastering skills to Pilkington's attractive publications about the use of glass. A few are listed later in this book, but opportunities for help of this sort are always changing. The magazines are the best place to find details of them.

BUILDING ON YOUR OWN:

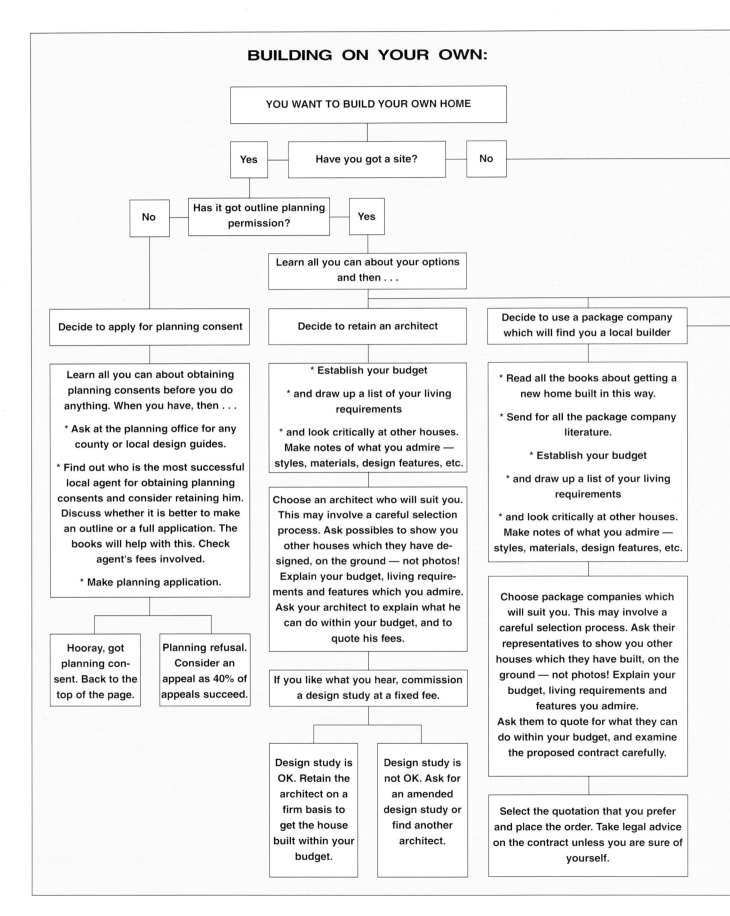

YOU WANT TO BUILD YOUR OWN HOME

Yes ← Have you got a site? → No

No ← Has it got outline planning permission? → Yes

Learn all you can about your options and then . . .

Decide to apply for planning consent

Decide to retain an architect

Decide to use a package company which will find you a local builder

Learn all you can about obtaining planning consents before you do anything. When you have, then . . .

* Ask at the planning office for any county or local design guides.

* Find out who is the most successful local agent for obtaining planning consents and consider retaining him. Discuss whether it is better to make an outline or a full application. The books will help with this. Check agent's fees involved.

* Make planning application.

Hooray, got planning consent. Back to the top of the page.

Planning refusal. Consider an appeal as 40% of appeals succeed.

* Establish your budget

* and draw up a list of your living requirements

* and look critically at other houses. Make notes of what you admire — styles, materials, design features, etc.

Choose an architect who will suit you. This may involve a careful selection process. Ask possibles to show you other houses which they have designed, on the ground — not photos! Explain your budget, living requirements and features which you admire. Ask your architect to explain what he can do within your budget, and to quote his fees.

If you like what you hear, commission a design study at a fixed fee.

Design study is OK. Retain the architect on a firm basis to get the house built within your budget.

Design study is not OK. Ask for an amended design study or find another architect.

* Read all the books about getting a new home built in this way.

* Send for all the package company literature.

* Establish your budget

* and draw up a list of your living requirements

* and look critically at other houses. Make notes of what you admire — styles, materials, design features, etc.

Choose package companies which will suit you. This may involve a careful selection process. Ask their representatives to show you other houses which they have built, on the ground — not photos! Explain your budget, living requirements and features you admire. Ask them to quote for what they can do within your budget, and examine the proposed contract carefully.

Select the quotation that you prefer and place the order. Take legal advice on the contract unless you are sure of yourself.

THE CRITICAL PATHS TO MAKING A START

```
                        ┌─────────────────┐
                        │ Choose between  │
                        └─────────────────┘
```

Choose between

Joining a selfbuild group

Finding and buying a site

Become an individual selfbuilder

Learn all you can about selfbuilding on your own.

Choose whether to build entirely on your own or to use a package company that provides a special service to selfbuilders.

Building entirely on your own

Learn all that you can:

* Read all the books.

* Find other selfbuilders and learn from their experiences.

* Go to conferences and exhibitions.

Follow the 50 stage critical path on pages 176 - 178.

Learn all that you can about group selfbuild:

* Read books and contact all the relevant agencies listed in this book.

* Attend selfbuild conferences.

* Send for brochures from Selfbuild Management Consultants operating in your area.

Decide either to form a group yourself or apply to join a group being formed by others. Whatever you do, do not enter into any commitment without being sure that everything follows the guidelines of the Federation of Housing Associations.

* establish a budget for the new home and how much you can spend on the site (books will tell you this).

* Assess whether you are likely to be able to build on it within your budget, and draw up a checklist of things to consider when looking at possibilities.

* explain to your solicitor that you are land hunting and that you will want some fast legal action when you find what you are looking for.

Start looking for a site:

* Contact estate agents in the area and consider appointing an estate agent as a buying agent to find you a site.

* Place 'Building Plot Wanted' ads in local papers.

* Cultivate personal contacts. Most selfbuilders buy plots this way.

* Study magazines with relevant adverts.

* Join a national land-finding agency?

Found your plot? Don't commit yourself without legal advice. While your solicitor is making his searches you investigate any snags over building on the plot. Read and follow the books.

Bought it! Back to the top of the page!

2 Finance

Building for yourself is about spending money, spending it so that you get exactly what you want, and also spending it so effectively that you will end up with a bigger or better house than the money would buy on the open market. The management of this money is as important as the management of the building work, and far more hazardous. The Building Inspector and others will do all they can to prevent construction problems, but no one is going to jump up and cry halt if you are heading for a financial disaster. In this chapter we shall look at the plans that have to be made, and the safeguards that must be built into these plans. Of course, no selfbuild venture can be entirely safe, but if you follow the basic rules you are most unlikely to have any major problems. Selfbuild disaster stories are very few and far between.

If you want to finance a new home, you have to know the answers to the following questions.

1. If a mortgage is going to be involved, what can be arranged, and how much is it going to cost when the property is finished?

2. Where is the money coming from to pay for the site before a start is made?

3. Where is the money coming from to pay for the materials, sub-contract labour and services as work progresses?

4. How can this be managed so that I pay as little interest as possible?

5. What warranties will be required by those who are financing the mortgage and lending the building finance?

6. How can insurances cushion the major risks?

7. How can all this be done without placing an unacceptable strain on the family's pattern of living?

First of all, not all those who build an individual home require a mortgage. If you read the magazines for individual home builders, or attend the exhibitions, the very high profile of the building societies may lead you to assume that all selfbuilders take out the largest possible mort-gages. The experience of the package companies is quite different, and suggests that up to 50% of those who build for themselves do so using family funds and the proceeds of selling an existing house. The figures given by some building societies for the number of selfbuild mortgages which they approve compared with the known number of selfbuilders tends to confirm this. Let us look at the position of these fortunate people, most of whom had parents who bought modest homes in the 1950s for perhaps £1,500 and who have now inherited properties which may be worth £150,000. Families inheriting two such homes are dealing with serious money, although they may think of themselves as ordinary people.

People with the problem of how to best deploy family wealth have an interesting situation when planning to build a new home, and if it is a new situation for them, it may not be easy to handle. Wealthy families which have guarded inherited money down the generations have their own attitude to it. They tend to consider their funds as family assets, and use trusts and other devices to ensure that it passes smoothly down the genera-tions with the minimum interference from inheritance tax. Often this is their favourite topic of conversation, and very boring they can be too! In contrast, those whose modest capital has always been kept in the Post Office until they inherited a couple of houses with telephone number values, tend to keep their affairs very much to themselves. They may refuse to face up to the fact that they cannot take it with them and that the taxman is lurking.

Inheritance tax is often described as a voluntary tax, because if someone makes appro-priate arrangements at the right time, nothing at all needs to be paid. However, this involves giving money away in one way or the other, and just try discussing this with Aunt Lucy who, after a lifetime of being careful with money, is now worth a cool £200,000. If she is able to take a realistic view of things, she should be able to continue to enjoy her present standard of living indefinitely and ensure that her favourite niece and nephew use her money for a new home so as to avoid inheritance tax. This all depends on her seeing her capital as a family asset.

If she does, there are various ways of going about it. If sums of six figures are involved, it is appropriate to make specialist arrangements and consider setting up a trust, which will cost money but which is cheap at the price. Finding the right professionals to handle this may not be easy, since they should be totally independent and not people who are wanting to sell you some sort of financial service. A large firm of accountants is probably best.

If the funds involved are more modest, there are a number of straightforward approaches. The simple way of obtaining help with the cost of your new house from an elderly person is simply for them to give you the money that he or she intends will be yours 'one day', and then to live for seven years. After that time the gift escapes inheritance tax. So do annual gifts of £3,000, and outright gifts which are considered to be 'normal expenditure out of income'. This 'normal expenditure' can usually cover the cost of a fitted kitchen as a birthday present. Other exemptions can be claimed for gifts made by parents on the marriage of a son or daughter.

A more complicated way of going about this, but one that may appeal to an elderly person who has capital invested to earn interest, is for them to give you a mortgage. Anyone can provide a mortgage, not just a bank or building society, and tax relief is available for the mortgagee in the ordinary way providing that the paperwork has been done properly. Any solicitor should be able to arrange this.

This really is a most useful approach. Suppose Aunt Lucy has £30,000 invested, currently earning her interest on which she pays tax. If she uses it to give you a £30,000 mortgage at 1% under the current mortgage rate she will get a better return (on which she will still pay tax), you will have a cheap mortgage on which you will get tax relief, and the money stays in the family. Aunt Lucy's will may have to be rewritten to avoid complications in the future, but this is not a complicated undertaking.

If both parties are interested in Aunt Lucy living in a granny flat in the new home, a whole new range of possibilities is opened up, as any contribution which she makes to the cost of the house can escape inheritance tax. This is discussed in a later chapter, and, if it is a possibility, it is essential to seek out disinterested professional help. Never hesitate to check out advice from one professional with another, and a final okay from your bank manager or solicitor is prudent.

Turning now to bank or building society finance, it is first essential to understand what happened to various building society schemes to finance selfbuilders when the housing boom ended in 1989, as this affects the current attitude which the financial institutions take towards individual home builders.

Prior to 1987, banks and building societies normally required anyone building for themselves to find their own money to buy the site, following which they would lend the money for a house to be built on it, with the land itself as the security. When the house was built this loan was repaid by taking out a mortgage in the usual way. It was very common for a bank to lend the building finance, and for a building society to issue the mortgage on the finished property, with the two of them working closely together. This meant that a very high proportion of borrowers had to sell an existing house in order to buy the site for a new one, moving into temporary accommodation until the new home could be occupied.

In 1987, when it seemed that the rise in property values would go on for ever, the National & Provincial Building Society launched a '100% finance' scheme for selfbuilders who had an existing home. This involved transferring their current mortgage to the N & P, who, after a careful assessment of their proposals, would lend them a further sum to cover the entire cost of both buying the site and building a home on it. The interest on this further loan was rolled up with the capital, so that no additional repayments were required until the new house was occupied, the old house was sold, and the special selfbuilders mortgage could be converted to a standard mortgage. This scheme attracted a four figure total of borrowers in a very short while, and everything went well until the housing market collapsed in 1989, when borrowers who completed their new homes found that they could not sell their old properties.

This caused serious problems for both the society and its borrowers and the situation attracted widespread publicity. The National & Provincial withdrew its scheme, much to the disappointment of a large number of people who were proposing to use this particular method to finance a new home. Other building societies, and particularly the Birmingham Midshires, had similar schemes although they were not so heavily committed, and avoided losses. Selfbuild lending generally became very unpopular with the financial institutions, particularly as those which had lent to selfbuild housing associations were experiencing even more serious losses as will be explained in a later chapter.

The memory of this lingers on, and it is now virtually impossible to borrow 100% of both the cost of a site, and the cost of building on it, from a mainstream financial institution. However, building societies promote themselves in the selfbuild market, most of them preferring to state that they 'deal with all applications for housing finance on their merits'. Some of the larger societies have leaflets which describe their arrangements for financing those who are building on their own. The offers which they

make are changed frequently, like mortgage rates. In general, they all require that a borrower should sell his existing house and repay any old mortgage, following which they will lend a proportion of the site cost and a further proportion of the building finance required. An example of the arrangements as this book is published is set out on pages 24 to 27, where we give the details of the current Bradford and Bingley's arrangements.

Other building societies which do not advertise a special service for those building on their own, and almost all of the banks, will always discuss making appropriate arrangements with established customers, and often these are marginally more advantageous than those offered by the building societies which have formal arrangements for selfbuilders. Their enthusiasm for this and the terms which can be arranged will usually depend on an applicant's standing as an existing customer.

All lending which eventually leads to a conventional mortgage on the finished property is limited by normal building society considerations of maximum overall loans, income multipliers, valuation fees and administration fees. The only unusual feature is that some lenders charge a

A realistic appraisal of your finances and careful financial planning is the first stage in building the home of your dreams.

premium or surcharge on top of the standard rate of interest during the time that the house which they are financing is actually being built. This is supposed to cover the cost of their 'management' of your scheme. Most selfbuilders are under the impression that they are providing the management themselves but it will not help you to point this out to the branch manager.

Another issue here is the choice that has to be made between endowment and repayment mortgages, the early redemption arrangements and of course the interest rates. The pros and cons are always changing, and there are magazines available on the bookstalls which provide useful up-to-date information. If your finances are complicated because you are self-employed, work abroad, or have a significant investment income then you may decide to use the services of a mortgage broker. A number of these specialise in finding finance for selfbuilders, and advertise in selfbuild magazines. They can be very useful, although if the arrangements that they suggest are in any way unusual it is important to have them checked out by a solicitor or accountant.

Two cautions. First of all, do not be tempted to borrow further from any other source without disclosing it to the principal lender. Secondly, do not be tempted to exaggerate the cost of the land so as to gain an advantage. In other words, if you are buying a plot for £60,000, do not tell the bank it is costing £80,000 in the hope that you will get a 75% loan which will be equal to the whole of the real cost. This is sometimes advocated by those who should know better or who will even offer to arrange it for you. It is dishonest, and a potential recipe for disaster.

Whichever route you decide to take when arranging finance for an individual home, it is important to give careful consideration to how to make your approach to the lender in the best possible way. Imagine yourself in the chair of the branch manager concerned. He or she earns their reputation with their head office by arranging mortgages that are simple, straightforward, and generate no problems of any sort at all. Normally these are ordinary mortgages on ordinary homes bought on the open market. A proposal linked to a house which is yet to be built, and which may

be being built on what they regard as a DIY basis is not in the category that they really like. You have to persuade them that your proposals carry no risk of any sort, and that you are very well able to handle everything in a completely risk free way. How do you do this?

Do not casually drop into a building society office and start asking about how they would help you if you decided to build for yourself. Instead, it is better to write and ask if they have any leaflets regarding the services that they offer to those who are having a house built on their own land. Study these leaflets carefully. At the right time, when you think you have learned all that is necessary to make it obvious that you know what you are talking about, make an appointment to call on the manager and go along equipped to demonstrate how well-organised a person you are. If you are married, or building with a partner, it is better that both of you should go along. Take with you a concise note of your budget for your lifestyle, your income and your outgoings, and also your rough budget for the project. If you know that you will only be able to borrow 75% of the project cost, explain where the other 25% is coming from. Produce a tentative programme for the work, and above all, make sure that you have allowed in your costings for the interest which you will have to pay on the loan — remember, as far as the manager is concerned, this is what it is all about!

One of the most important aspects of your loan will be the arrangements that are made for stage payments, which you will need to pay subcontractors and suppliers. Prompt payments to them are essential to your operations, but the bank or building society may not see the release of the funds that you will need for this as a priority. Typically you may be offered your building loan in five stages, at the completion of the foundations, at wall plate, at roofed in, at plastered out and on completion. At these stages the building will have to be inspected, usually by a surveyor appointed by the lender, and his certificate is necessary before funds will be released. There may be a further delay before the money actually arrives in your bank account. Two aspects of managing this are to ensure that the

A selfbuild mortgage - how it works

The key to trouble free arrangements for a selfbuild mortgage is to understand how they work, and to see that your bank or building society gets the information it requires to be able to arrange things to suit you. Both the rates and the terms for these progress mortgages change all the time, and the selfbuild magazines carry details, usually as tables which compare rates, terms and the extent to which the lender will finance your land purchase. Currently the Bradford & Bingley Building Society have the highest profile of all the selfbuild lenders and are to be found at the exhibitions and roadshows, but there are lots more. In choosing between them remember the relationship you establish with the local manager may be as important as the small print.

21 The Maltings
Hatfields Hill
Anytown
AT3 7GS

The Manager,
Bradford & Bingley Building Society,
High Street,
Anytown.

Dear Sir/Madam,

I would like to know if you can offer my husband and myself a progress mortgage for a house that we intend to build ourselves, and would like to call at your office to discuss the scheme described in your progress mortgages leaflet.

At this stage we have not found a building plot, but require to know if we can get a loan from you when we do. I add that we have read widely about selfbuild and about the way that progress mortgages work.

Yours faithfully,

Sonia Travers

Sonia Travers

BRADFORD & BINGLEY
BUILDING SOCIETY

1 HIGH STREET
ANYTOWN
AT1 1AA
Phone 01279 300333

Mrs. J. Travers,
21 The Maltings
Hatfields Hill
Anytown

Dear Mrs. Travers,

Thank you for contacting the Bradford & Bingley Building Society.

As requested I am pleased to enclose full details of the services offered to customers who are preparing to build their own home.

Up to 75% of the value of the land can be obtained initially and then up to 75% of the value of the property as your project reaches further stages of construction.

You must also take out insurance to cover the building both during its construction and once it is completed and the Society will be happy to arrange cover for you.

I hope to reach you by telephone later today to arrange for you to call, and in the meanwhile hope that the leaflets enclosed will be useful to you.

I trust this is satisfactory.

Yours sincerely,

C. Richards

C. Richards
Bradford & Bingley Building Society.

TELEPHONE MESSAGE

B & B rang.

Both of us to call Saturday 10.45

Remember to take

* Figures for the present mortgage
* Value of the house — talk to estate agent
* Evidence both our salaries etc.
* Balance in the new house savings account.

---

21 The Maltings
Hatfields Hill
Anytown
AT3 7GS

Bradford & Bingley Building Society.
High Street,
Anytown.

Dear Mr. Richards,

I refer to our meeting at your office yesterday regarding a possible building mortgage for an owner built house. We have decided to buy one of the plots at Meadowside at £48,000. Our existing house is on the market, with a potential buyer who is very keen but in a chain, but we anticipate a sale soon which will free about £35,000 as we discussed. In the meanwhile we can put £15,000 of our savings towards buying the plot and legal fees, and wish to borrow £35,000 from you for the land on the basis which you offered (i.e. - 75% of the cost). Our solicitor is Mr. Jones of Heaton & Letwell, who says he knows you, and wants confirmation that the money will be available. Can you please ring him to confirm, and let me know urgently when we should call in to sign the paperwork.

Yours sincerely,

Sonia Travers

Sonia Travers

BRADFORD & BINGLEY
B U I L D I N G S O C I E T Y

1 HIGH STREET
ANYTOWN
AT1 1AA
Phone 01279 300333

Mr. & Mrs. B. Travers,
21 The Maltings
Hatfields Hill
Anytown
AT3 7GS

Dear Mr. & Mrs. Travers,

OFFER OF ADVANCE REFERENCE 2719-DL-49162

Your application for a Mortgage Advance has been approved upon the terms of this Offer of Advance, which includes the enclosed special conditions and Additional Terms. Please read these carefully together with the enclosed copy of the Mortgage Deed, Mortgage Conditions and the leaflet 'Information for Prospective Borrowers'.

Your acceptance of this offer, and the resulting agreement, only takes effect when you have executed the Mortgage Deed and it has been dated by the Solicitor on your behalf. The Mortgage Deed incorporates your agreement to be bound by these terms.

Your Solicitor has received a copy of the Offer, the Mortgage Deed and the Society's Rules relating to Mortgages and will contact you to complete the Mortgage Deed.

The Society reserves the right not to proceed with the Advance or to modify these terms at any time before the Advance has been made. In addition the Society will cancel any Offers outstanding after six months, where contracts have not been exchanged, and are no longer needed.

Purchase Price	Advance	Interest Rate	Repayment Period	Monthly Amount	Reduced Monthly Payment	The property will be insured for
£155,000	£60,000	8.45%	25 years	£486.60	£456.60	£138,000

MONTHLY PAYMENTS

* This is the sum required to repay instalments of principal and interest over the repayment period in the case of a loan which does not qualify for inclusion in the MIRAS scheme.

** This is the sum payable because £30,000 of the loan qualifies for inclusion in the MIRAS scheme

CONTINUED

21 The Maltings
Hatfields Hill
Anytown
AT3 7GS

The Manager,
Bradford & Bingley Building Society,
High Street,
Anytown.

Dear Mr. Richards,

Progress mortgage 2719-DL-49162

After a lot of shopping around we engaged David Gainsborough, the architect in Park Street, to design our house and I enclose his drawing reference C2519-D. He tells us that this has now been agreed in principle with the planners, and got the quantity surveyors costings for us which you require. This is also enclosed.

We have now sold the old house, but had to drop the price so that it only made £32,000 after repaying the mortgage, which is safely in our deposit account with you. Together with what is left of our savings after paying 25% of the cost of the land and the legal fees the balance in this is now £39,000, and we hope to use £24,000 of your promised progress mortgage for the building work.

Our business plan is attached, and you will see we will want a first stage payment from you in October. Please note that the costings do not include anything for the kitchen: my parents are giving us this, and if for any reason they cannot, then the contingencies allowance in our budget will cover a modest kitchen anyway.

I will be grateful if you will get your surveyor to deal with all of this as soon as possible, as we want to move quickly. Can you deduct the fees involved from our deposit account, or shall I send you a cheque? Finally, we will want to open an interest bearing cheque account with you for meeting building costs: please send details.

Yours sincerely,

Sonia Travers

Sonia Travers

---

21 The Maltings
Hatfields Hill
Anytown
AT3 7GS

Dear Mr. Richards,

Progress Mortgage 2719-DL-49162

Our planning consent has arrived and I enclose a photostat. Does the council realise what this sort of delay costs people: it is three weeks since it was approved at the planning committee meeting.

At any rate, this has given us time to make final arrangements and I enclose

A copy of the Zurich Custombuild Warranty agreement
A copy of the Norwich Union Selfbuild Insurance schedule showing your interest in the policy
Details of the short term life insurance and permanent health insurance which you suggested that John should take out for the year that we are building: I do not know if this is a condition of the mortgage or just helpful advice that you were giving to us!

The latest version of our programme is enclosed, and you will see we plan sending the first Architects Progress Certificate in late January. We shall then need the first draw from you straight away, and relying on your promise that it will come promptly.

With best wishes for Christmas.

Yours sincerely,

Sonia Travers

Sonia Travers

BRADFORD & BINGLEY
BUILDING SOCIETY

1 HIGH STREET
ANYTOWN
AT1 1AA
Phone 01279 300333

Mr and Mrs J. Travers
21 The Maltings
Hatfields Hill
Anytown
AT3 7GS

Dear Mrs Travers

Thank you for your letter with all the enclosures about the new house. These have all been inspected and everything is in order.

As soon as we hear from Heaton and Letwell that they are in a position to complete on the purchase of the land we will arrange to release the 75% of the land cost as agreed. Please remember that you will need to give a months notice to withdraw the other 25% from your savings account if you are not to lose any interest.

As you know, you will have to use your own funds to get the foundations built. When the floor slab is finished let us know and we will get the surveyor round to see it just as soon as possible. The more time you can give us for this the better, but we all know that getting your first progress cheque will be very important to you and will move as quickly as we can on this.

If you have any further queries please do not hesitate to get in touch with me

Yours sincerely

C Richards

C Richards
For Bradford and Bingley Building Society

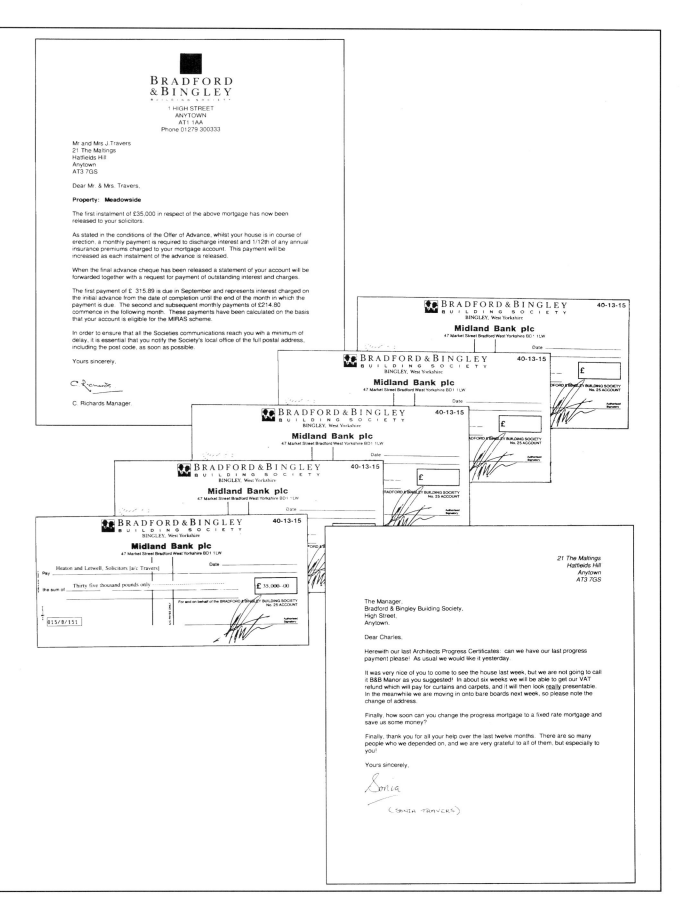

BRADFORD & BINGLEY
BUILDING SOCIETY

1 HIGH STREET
ANYTOWN
AT1 1AA
Phone 01279 300333

Mr and Mrs J.Travers
21 The Maltings
Hatfields Hill
Anytown
AT3 7GS

Dear Mr. & Mrs. Travers,

Property: Meadowside

The first instalment of £35,000 in respect of the above mortgage has now been released to your solicitors.

As stated in the conditions of the Offer of Advance, whilst your house is in course of erection, a monthly payment is required to discharge interest and 1/12th of any annual insurance premiums charged to your mortgage account. This payment will be increased as each instalment of the advance is released.

When the final advance cheque has been released a statement of your account will be forwarded together with a request for payment of outstanding interest and charges.

The first payment of £ 315.89 is due in September and represents interest charged on the initial advance from the date of completion until the end of the month in which the payment is due. The second and subsequent monthly payments of £214.80 commence in the following month. These payments have been calculated on the basis that your account is eligible for the MIRAS scheme.

In order to ensure that all the Societies communications reach you wih a minimum of delay, it is essential that you notify the Society's local office of the full postal address, including the post code, as soon as possible.

Yours sincerely,

C. Richards Manager.

Heaton and Letwell, Solicitors [a/c Travers]
Pay
the sum of   Thirty five thousand pounds only —————————
£ 35,000—00

For and on behalf of the BRADFORD & BINGLEY BUILDING SOCIETY
No. 25 ACCOUNT

015/8/151

21 The Maltings
Hatfields Hill
Anytown
AT3 7GS

The Manager,
Bradford & Bingley Building Society,
High Street,
Anytown.

Dear Charles,

Herewith our last Architects Progress Certificates: can we have our last progress payment please! As usual we would like it yesterday.

It was very nice of you to come to see the house last week, but we are not going to call it B&B Manor as you suggested! In about six weeks we will be able to get our VAT refund which will pay for curtains and carpets, and it will then look _really_ presentable. In the meanwhile we are moving in onto bare boards next week, so please note the change of address.

Finally, how soon can you change the progress mortgage to a fixed rate mortgage and save us some money?

Finally, thank you for all your help over the last twelve months. There are so many people who we depended on, and we are very grateful to all of them, but especially to you!

Yours sincerely,

Sonia

(SONIA TRAVERS)

lender knows how critical the timing of this is to you, and, if the payment of sub-contractors depends on you receiving a stage payment, tell them this when you engage them instead of leaving them to find out when their money is due.

There may be another problem here if you are using a package service, as most of the package companies require their money up front, or at least have forward loaded payment arrangements. This may involve you making advance payments into deposit accounts — see page 155.

Building warranties

All lenders will want you to prove that you have arrangements for building warranties. This is an important subject. It is usual for all new homes to have warranties of some sort to guarantee that they are properly built and to provide for the cost of putting right damage due to structural failures. Such warranties are required by banks and building societies as a condition of mortgages, and also by a bank or building society providing a mortgage for any future purchaser of the house. If houses do not have such cover, special surveys are usually required when they are sold, and the properties are not such a good resale proposition as those with appropriate warranties.

The three standard arrangements for this, and how they work, and how they are different, should be clearly understood. They are:

* The NHBC warranty, promoted as Buildmark, offered by the National House Building Council which is a building industry organisation.

* The Zurich Custom Build warranty, offered by the Zurich Municipal Insurance Company.

* Architects' certification.

The NHBC warranty

The NHBC warranty is only available to selfbuilders who are employing a NHBC registered builder. It is *not* available to selfbuilders who are using sub-contractors except in Northern Ireland and the Isle of Man, but it *can* be arranged for those who are using a registered builder to construct a shell which they will complete using sub-contractors.

Registered builders are obliged to provide NHBC certificates for the homes that they build as a condition of their NHBC membership. Before they start they have to apply to register the new building, and the NHBC will check out the drawings with care, particularly if there are any unusual ground conditions. In due course they allocate a registration number and the builder's customer is advised of this. They will usually need to produce this promise of a warranty to their building society or bank. The work in progress is checked by NHBC inspectors at regular intervals, and the inspector will be very involved in any problems with the design or workmanship. However, the inspector is not concerned with any difficulties in your relationship with your builder unless the builder dies or becomes bankrupt, when the NHBC will step in with limited assistance.

A very few selfbuilders and selfbuild housing associations register themselves as builders with the NHBC to obtain warranties on homes built with their own labour or with sub-contractors, but the procedures are cumbersome.

The NHBC warranty in Northern Ireland

The NHBC has a scheme for selfbuilders in Ulster and the Isle of Man. It is widely suspected, and hoped, that this is a pilot scheme for the whole of the UK but this is denied by the NHBC. Details are available from the NHBC Belfast office — see page 294.

The NHBC 'shell only' warranty

Sub-contractors cannot be NHBC registered, and cannot offer a warranty on the work that they do. Fortunately, a registered builder is able to get a warranty which is limited to 'structure only' from the NHBC if he is building you a shell which you then arrange to fit out for yourself. This is not usual, and is not advertised in the NHBC leaflets. However, regional directors are able to authorise this when approached by their registered builders, who have to obtain written confirmation from their clients that this is specifically required.

This NHBC warranty is only available in Northern Ireland.

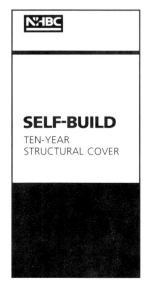

NHBC

SELF-BUILD
TEN-YEAR
STRUCTURAL COVER

The Zurich Custom Build warranty

Warranties which are in nearly every way comparable to the NHBC warranties are provided by the Zurich Municipal Insurance Company, and work in the same way as the NHBC arrangements with registration of buildings, inspections of work in progress and the issue of final certificates acceptable to building societies as a condition of mortgages. The difference is that selfbuilders are welcome, and have their own scheme known as *Custom Build*. This is the usual way for most selfbuilders to obtain the warranty that they require and is described on page 30.

There are separate arrangements for builders who take responsibility for a complete contract, and their scheme is called Zurich *Newbuild*.

Architect's certification

The third route to a warranty acceptable to banks or building societies is architects certification. This was very common in the past, when simple certificates from a registered architect said that he or she inspected the work at certain set stages, and certifies that the property was built in accordance with the approved drawings and good building practice. Today the value of such certificates is criticised. The only way that a major claim for structural failure can be established with such a warranty is to sue the architect, who will rely on his professional indemnity insurance. At the best this will be a lengthy process, probably taking years, while the claims procedures offered by NHBC and Zurich are claimed to be faster and user friendly. Also the warranty provided by architects progress certificates is usually specific to the architect's client, and cannot generally be transferred to a successor in title if the home is sold.

Anyway, architects progress certificates are still generally accepted by building societies, who normally require five of them at different stages of construction. They should be kept with the title documents of the house.

The problem lies in finding an architect who is prepared to issue these certificates. It must be emphasized that a selfbuilder normally wants 'architect's progress certificates' and not 'architect's supervision'. Architect's supervision involves the architect being responsible for supervising every detail of the work on site, and is very expensive, while progress certification refers only to five inspections made at specific stages. If you are retaining an architect to design your house for you, you will easily be able to persuade him or her to issue the progress certificates, but, if you have obtained the plans from another source, then you have an interesting job finding an architect who will inspect work on a design for which he or she was not personally responsible. This involves some shopping around and the best way of finding the right person is to find out who is certifying the work of other selfbuilders in your local area. If you are using a package service, the company will either arrange to have the work certified by their own architects, or will put you in touch with architects who will undertake this. In the former case the cost of the certification may be included in their contract price, while an independent architect will probably charge somewhere between £60 and £100 per visit.

A few building societies are not particularly interested in architect's progress certificates and will arrange for their own surveyors to make routine inspections and issue appropriate certificates. Sometimes these are staff surveyors of the society concerned, while others are independent surveyors appointed by the society, which will debit your account with the costs involved. Nothing comes free. If this is the way in which you are going to get appropriate certification for your own mortgage it is important to discuss with the building society if this certification can be passed on to a future purchaser.

The selfbuilder's warranty — Zurich Custom Build

The Zurich Custom Build structural guarantee is one of the three types of warranty available to selfbuilders, and is the only one which is designed specifically for them. It guards against structural problems by giving a full ten years of protection against major damage, including ground movement, provided of course that Zurich are happy with how the property is built. It is recognised by all major banks and building societies, but this is something to check out carefully at the appropriate stage in your project planning. If those providing your loan finance are less than familiar with this particular warranty, Zurich will deal with their queries, and obviously it is important to give them time for this.

To obtain a quotation for a warranty you submit full details of how you intend to build to Zurich together with a data sheet as reproduced opposite. This is required at least four weeks before the anticipated start date, together with a non-returnable premium deposit of £100. This enables Zurich to quote for the whole premium, which is payable before work starts.

Although Zurich assesses each proposal individually, and particularly concerns itself with any special ground situations, their general requirements are:

* The selfbuilder should have a source of technical advice of some sort — a 'professional friend'.
* The home must be designed by an architect or other professional.
* All skilled work — plumbing, electrical, etc. — must be carried out by professional contractors and tradesmen.
* Evidence must be provided of Building Regulation approval.
* Notice of the start on site must be given to the Custom Build surveyor not less than 48 hours in advance.
* All specifications and work must conform to the requirements of the Technical Manual, which is supplied to all applicants for a quotation.
* If there is any indication of hazardous ground a site investigation report and the foundation design details as submitted to the building control authority must be provided.

The first stage in all of this is to send to the address on page 294 for a Custom Build Information Pack, and to study it carefully and make sure that you understand how everything works. Get your mind around exactly what is offered by the insurers, the obligation which you will have to see that your new home is built to the required standards (which you are certainly going to do anyway), and the arrangements which will have to be made for the regular inspection of the building while it is under construction.

The next stage is to apply for the registration of your new home, and to pay the appropriate fees, following which you will be put in touch with your local Custom Build surveyor and will receive the Technical Manual. You will also get a letter giving formal confirmation that your home is being built on a Custom Build basis, and that subject to satisfactory routine inspections the guarantee will be issued. This will be required by your bank or building society as evidence that you are going to be able to provide them with the warranty certificate in due course.

Inspections of the work on the structural shell of the building are made at regular intervals from when the foundations are opened until it is roofed in. This is handled by the local authority building inspectors following arrangements made with councils in all parts of the country, and the visits are made at the same time that the authorities carry out its own statutory checks. In addition, quality control inspections are made by the Custom Build's own surveyors as work progresses.

When the warranty is issued it is fully transferable to subsequent owners, and this will be most important to you if you should sell the home within the structural guarantee period. The ten years cover usually can be extended for a further period of five years if you wish .

The cover includes the full cost of dealing with structural defects, paying for the work to be carried out by professional contractors and not involving the selfbuilder in any work himself. It also covers the cost of alternative accommodation, professional fees, debris removal etc., as well as the cost of complying with statutory requirements should these ever be necessary during the course of a claim.

The Custom Build warranty is only available to those who are *managing* the building of their own new home. This includes those who may be building a home with their own hands, those using subcontractors and all combinations of such arrangements that Zurich approves. This also includes using a builder to construct a shell, and then employing subcontractors to complete the building. However, if a full contract is arranged with a builder who will be responsible for the whole job the Custom Build warranty is not appropriate, but a similar scheme called the *Zurich Newbuild Guarantee* is available through the builder, who must make the necessary application himself.

Finally, it is important to understand that the Custom Build warranty does NOT provide cover in respect of:

* The failure of any builder or sub-contractor to do satisfactory work, or their business failure or simply their disappearance or if they walk off the job.
* Any losses on site due to theft, vandalism, etc., or any public liability or employer's liability cover. There are special selfbuilder's insurances for this — see page 206.
* Provide any warranty in respect of the failure of any of the installed services, such as the heating system, artificial ventilation arrangements, etc.

Custom Build Quotation Application

Policy Details

ZURICH MUNICIPAL

ZURICH MUNICIPAL

DECLARATION

I enclose with the Application the following documents and declare that the information given is true to the best of my knowledge and belief.

☐ Notification Data Sheet

☐ Site plans (1:500 or 1:1250)

☐ Full set of working drawings as submitted to the Building Control Authority (Preferred scale 1:50).

☐ Site investigation and foundation design (if applicable, see above).

☐ Copy of Building Control Authority approval notice or plans certificate.

☐ Outline of arrangement for site supervision and technical advice including "Professional Friend":

☐ Crossed cheque for £100 non-returnable premium deposit payable to Zurich Insurance Company.

Signed: _____

Dated: _____

...DATA SHEET

| | | Dev No: |
| | | Date Received: |

OFFICE USE ONLY	Scheme Code:
	Branch Code:
	BCA Code:

Name/Address (A) See 'Guidance'	Name:		Postcode:
	Usual Address:		Scheme Reg No: Z
		Fax No:	
	Tel No:		

Name/Address (B) See 'Guidance'	Name:		Postcode:
	Usual Address:		Scheme Reg No: Z
		Fax No:	
	Tel No:		

| Please give details of Building Control Authority | Name of B.C.A.: |

Please give details of the development	Dev Name:
	Address:	Postcode:
	Phase No:	Estimated Start Date:
	Previous 'Development Number' relating to this site, if any	

Ground Conditions - please tick appropriate box	Made-up ground	01		Mining or mineral extraction area	07
		02		Steep slopes	08
	Trees etc (existing) on clay soil	03		High water table	10
	Trees etc removed from clay soil	04		Sulphates	11
	Soft clays, silts etc	05		Peat soil	12
	Existing drains	06		Contaminated land	
	Old foundation			
	If the site investigation indicates any of the hazards listed above please attach the site investigation report and foundation design as submitted to the B.C.A.	13			14

Special information - please tick if appropriate	Thatched roof	13		Basement or semi-basement	05
				P.R.C. rebuild-timber frame	06
				Timber frame (Ecowall)	07

Construction type - please tick appropriate box	Brick and block	01		R.I.S.F. repair	08
	Timber frame	02		Others	99
	P.R.C. repair	03		
	P.R.C. rebuild-masonry	04		

The Budget

Having understood this framework, how much can you afford to pay for your new house and what can you hope to get for your money? Of course, everything depends on finding the right site, and on the design of the house that is appropriate to that site. However you have to start somewhere, and so make a start by looking at what you are going to be able to afford.

For most people the money available to them will be:

* The money that you will have after you have sold your existing home and have repaid your existing mortgage.

* Plus the finance that you can hope to borrow to fund the new project, which will eventually become your new mortgage. At this stage take this as two and a half times the family income.

* Plus any other capital which you intend to spend on the new home.

This will give you your total budget, and it has to be split between paying for your land, paying for the building work, paying fees of various sorts and paying interest on any loans which you take out until you can repay them from a new mortgage. What proportion of this do you allocate to buying the land? It depends wholly on where you live, and will involve some careful research. The cost of a plot for a four bedroom detached house can vary from 25% of the total project cost in Wales or the North of England, to 60% in the South East or the Home Counties. You have to arrive at the figure appropriate to the area in which you want to build.

To do this you must make a general assumption about the cost of building, and, in early 1996, take this as £40 per square foot if you hope to be a selfbuilder and £45 per square foot if you intend to have a contract with a builder. This includes a contingencies provision for anything unexpected. Next, make an assumption about the size of home that you want, using the information in the box below or consulting the various books of plans in the bookshops which give the areas of the designs featured. Supposing you are interested in a house of 1475 square feet. As a selfbuilder your guideline cost is £59,000, while an individual using a builder will allow

---

Areas of houses and bungalows

Up to 700 sq. ft.
Holiday chalets and one or two bedroom old people's bungalows only.

700 to 800 sq. ft.
Smallest possible three bedroom semi-detached houses. Small two bedroom bungalows.

800 to 900 sq. ft.
Small three bedroom bungalows with integral lounge/dining rooms and a compact kitchen.
Most estate-built three bedroom semi-detached houses.

Around 1000 sq. ft.
Large three bedroom semi-detached houses.
Three bedroom detached houses.
Small four bedroom houses.

Four bedroom bungalows with integral lounge/dining rooms.
Three bedroom bungalows with separate dining room or large kitchen.
Luxury two bedroom bungalows.

Around 1300 sq. ft.
Three or four bedroom detached houses or bungalows with the possibility of a small study, or second bathroom. or a utility room, or a very large lounge.

Around 1600 sq. ft.
Four bedroom houses or bungalows with two bathrooms, large lounges, small studies, utility rooms.

Around 2000 sq. ft.
Large four to five bedroom houses and bungalows.

Initial Budget

Total funds available £90,000

Land cost £30,000

Allow legal costs and interest £5,000

Available for building £55,000

Anticipated cost per sq.ft.= £40

Max size of new house = 1375 sq.ft.

Amended Budget

Total funds available £90,000

Land cost £30,000

Allow legal costs and interest £5,000

Available for building £55,000

Extra cost of special foundations £5,000

Available for building £50,000

Anticipated costs per sq.ft = £40

Max size of new house 1250 sq.ft.

£66,000. To keep things simple we shall stay with the selfbuilder.

If the selfbuilder's budget is £90,000 he will have roughly £30,000 to buy his land. Is he going to be able to find a plot to suit the sort of house he can afford for this sum? If not, then he will have to revise his ambitions. To deal with this our selfbuilder must get to know the figures at which plots change hands in his local area, making a start by consulting other selfbuilders, estate agents, and every other informed contact that he can make.

When a plot is found that is in the right price bracket, and which might suit the house which he would like to build on it, our selfbuilder will consider very carefully if there are any problems with it that will make it more expensive than usual to develop. This is discussed in detail in later chapters. If, for instance, there is a foundation problem of the sort that usually takes about £5,000 to solve, then either the size of house he can hope to build will be reduced to 1250 square feet, or else £5,000 must be knocked off the price of the land.

It is absolutely essential that everyone buying

a building plot is able to juggle figures in this way. Another way of showing the figures is in the box.

Will the 1996 costs of £40 per square foot quoted above need further consideration if you are reading this book in 1998? Of course they will. Every prospective selfbuilder has to find out his own probable costs, and has to keep updating the figure to match inflation as his plans progress. To do this he will find some sources of information better than others, and some of no use at all.

Firstly, disregard any analysis of the prices at which new homes are offered by speculative developers. Basically the developer sells a house for what he can get for it, and rarely sells any unit, or any estate of houses, at a price which is a set proportion of construction costs.

Other misleading sources of information are the quarterly statistics of housing costs published by the Government, local authority cost yardsticks for new Council housing, and the figures published by the Building Trade press and trade associations. All these figures are invaluable tools for the professional, but are of little use to the individual builder. Most houses are built on estates, and published figures relate to estates. They are built by large organisations enjoying economies of scale, but with enormous overheads. None of this is relevant to your new home on your plot of land.

Architects and builders will usually be hesitant about quoting prices, although local architects with experience of single housing units on individual plots should be able to quote guideline costs per square foot. Builders will tend to quote whatever costs they think the questioner would enjoy hearing, hoping it might lead to a contract. This is not because they are devious, merely that they have to survive in a harsh commercial environment, and without a site, drawings, consents and money in your pocket they regard you as making conversation rather than asking a question.

There are three sources of really useful information: personal contacts, magazines and the cost figures put out by the package companies. Individual selfbuilders will often quote their building costs with pride. Information of this sort is invaluable. Bear in mind that these costs only reflect one job, and assess their reliability. More readily available, and more reliable, are the construction costs quoted by the magazines in the case histories they publish. Finally, the commercial companies offering this sort of service quote average figures, and frequently justify them by giving details of the actual costs of recently completed projects.

In 1995 national average costs for bungalows and houses, built on single sites by individual clients, on straightforward foundations, including fittings and fixtures appropriate to the size of the property, central heating, double glazing, connection to drains or septic tank, garage, short length of drive, no landscaping were:

When built by a well established NHBC builder, *working from his offices, following a formal invitation to tender, formal contract.*
- £44 to £50 per square foot.

When built by a reputable small builder, *NHBC registered, working from his home, usually himself a tradesman, following an informal approach, contract established by exchange of letters.*
- £40 to £45 per square foot.

When built by a reliable small builder to erect the shell *with sub-contractors on a supply and fix basis completing the building.*
- £38 to £42 per square foot.

When built on a direct labour basis by a competent private individual, *using sub-contractors, without providing any labour himself.*
- £35 to £40 per square foot. Sometimes less.

The number of square feet of floor area in a property is the area enclosed by the internal faces of the external walls (which is very different from the plinth area).

The average figures quoted remain remarkably consistent irrespective of the size of building or whether it is a house or bungalow, with the more expensive fittings in larger properties being balanced by savings consequent on the economies of scale.

Taxes and the individual home builder

Any consideration of financing a new home involves looking at your tax situation, and there are a number of factors which you need to take into consideration.

Stamp duty

Buying a building plot at any figure over £30,000 involves you in stamp duty at a rate of 1% on the total sum paid. Your solicitor will pay this for you as part of the conveyancing cost, and will add it to his bill. As a result a person buying a plot for £40,000 must budget for £400 in stamp duty.

Income tax relief

If you are borrowing money to buy a plot, then as long as you borrow on the basis of an intention to build a new home, and not as an investment, you will be able to reclaim the interest on the first £30,000 of the loan against income tax. This is done through MIRAS. If you are borrowing from a building society and the loan takes the form of a mortgage which is paid to you in stages, then you will not actually have to make a claim for the refund. Instead, on taking out the mortgage loan you will fill out a MIRAS application form and then the tax relief will automatically be deducted from your mortgage payments at the appropriate rate. If you borrow from a bank, or are self-employed, you may make your tax claim at the end of the tax year when you make your tax return. This is something to discuss with the lender.

Claiming MIRAS on your selfbuild project will in no way affect any claim which you are making against interest payments on your existing home. Even unmarried couples who purchased their joint homes before the August 1988 deadline and are still enjoying MIRAS relief on two lots of £30,000, can claim allowances on a third £30,000 for a selfbuild project. They will however revert to interest relief on a single £30,000 when they sell their existing home and move into the new one.

Although MIRAS is only generally described as being available on your 'principle private residence', which is your main home and not a holiday home, double MIRAS on an existing property and a selfbuild project can go on simultaneously for up to three years until the selfbuild project is complete. This also applies to selfbuilders who are unable to sell their old home when the new one is complete, provided that ultimately they do move in to the new home. If instead they have to sell it, the relief given will have to be repaid.

If you are building your home with the intention that it shall also be part of a business activity, and financing part of it from the business, then the situation regarding tax relief will be very different. This is a matter about which you must consult your accountant, as the way in which you do things, and the way in which you describe the building can be critical. It may be important to describe a farmhouse with a study, an outside lavatory, and an outside store as a 'farm workers dwelling with integral farm office, including provision of lavatory facilities for employees and separate store for hazardous chemicals as required under EEC regulations'. This sort of matter is often important if part of the cost is to be reflected in the farm accounts.

Capital gains tax

Most selfbuilders achieve a 20% cost value differential by building their own home, and some do even better. Providing that the new property is your principal private residence, and that you live in it for at least twelve months, then you are normally at no risk of having to pay capital gains tax if it is simply a way in which you have moved from one home to another. Even if you are left with two properties because you are unable to sell your existing home, or if you are forced for some genuine reason to move on from the selfbuilt home in less than twelve months, you will usually find that the tax inspector will exercise his discretion on your behalf.

However, for some this is a tricky area. There are some selfbuilders who go on to build again and again, each time living in the latest home for a year while they are building the next one. The tax inspector will only tolerate this for a limited period, after which he will claim that you are in business as a developer and will ask for a tax return to be submitted on this basis. When you meet other selfbuilders at exhibitions and conferences you will hear many anecdotes of those who have built ten homes in ten years, often changing wives every three or four homes, and who have avoided paying tax on these activities. It is unlikely that they have really the successes that they claim!

Value Added Tax

As previously explained, those who have a new home built for them by a builder or contractor do not have to pay him any VAT, while selfbuilders can reclaim the VAT that they pay on materials and services. VAT has a chapter on its own which starts at page 216.

3 Finding a site

Any article on how to find a site on which to build your own home must be prefaced by a brief explanation of planning permission, which is dealt with at greater length in a later chapter. The Town and Country Planning Acts require that before any sort of dwelling is built on a piece of land, consent to the development must be given by the planning authority, which is the local council. This consent is usually given in two stages, as 'outline approval' and then as 'planning consent', and it is usual for land to be sold as a house site with outline approval for a house to be built on it. The purchaser then applies for the planning consent to build the particular house that he requires. Planning consents belong to the land, not the applicant, and consents remain valid when land is sold.

Land without outline planning approval is sold at only a fraction of the cost of land with such approval, and it is a site hunter's dream that he buys an unwanted corner of a field at agricultural land prices and then gets outline planning approval to build a house on it. Alas, these days such opportunities arise only rarely, and for the layman they should rank alongside winning the pools in a repertoire of daydreams. Land offered cheaply as a building site without planning permission, with the explanation that it will be to the advantage of both parties if the purchaser obtains outline consent after he has bought it, will invariably be found to abound in complications.

However, you may already own the land on which you would like to build, and if you are very lucky it may already have outline planning consent. If this is not the case, you will want to examine every possibility of getting planning consent on it. The difference in the value of your paddock or orchard with planning consent, compared with its value without consent, is going to be in the tens of thousands of pounds, so it is worthwhile paying for the most experienced agent you can find to act for you. Before you contact him or her, read all that you can about obtaining planning consent on individual plots so that you can discuss tactics with confidence. This is all dealt with in Chapter 10.

If, like most people, you do not already own a plot of land, you have to buy one, and you must find a plot with outline consent for the type of building that you want to erect and on which you are assured by a professional advisor (or the planning authority) that you can obtain full planning permission for the actual house that you want to build. Much of the business of buying a piece of land is handled on the layman's behalf by experts who can be relied upon to give the guidance required if they are asked to do so. Foremost among these is the purchaser's solicitor. If you do not already have a solicitor, the time to find one is before you find the land you wish to buy and not after. A visit to him, or a letter, saying that you are looking for a building site, and that when you find it you would like him to handle the transaction and ensure you do not make any mistakes, will encourage him to advise you more readily than a complicated letter announcing that you have arranged a deal and wish him to give it legal effect. When he hears of your plans your solicitor will tell you in no uncertain terms to make sure that you consult him before entering into any sort of obligation. It is essential that you are not committed to go ahead with the transaction if enquiries disclose any snags or difficulties. This is most important.

At this stage, armed with the knowledge that your solicitor is primed to safeguard your interests, all that remains to do is to find the site. How difficult this is will depend on the part of the country in which you want to live. In the South East, the Home Counties, favoured suburbs of large towns and attractive villages in rural areas, the competition is such that individual plots rarely come on the open market. In these areas you simply have to get out and buy a plot as it is quite unrealistic to sit back and wait until one is offered to you. This means putting your ear to the ground and chasing up every lead, however unlikely. Fortunately in the rest of the country things are a little easier, particularly at this time when the building industry is in recession.

As with most decisions in selfbuild, everything starts with a great deal of careful analysis of your requirements. When you are looking for a building plot, the first of these is your price bracket. This was discussed at length in the last chapter. Are you likely to find the plot that you have in mind at your figure? Discuss this with those who have more experience than you have, or with local estate agents. If they are encouraging, then you can go ahead with some confidence. If your budget for the plot is inadequate, then you will have to accept that you must plan for a smaller house. If it is more than adequate, then you can expect to be able to afford rather more luxuries in the new home than you anticipated!

One thing that you must do is avoid over developing or under developing your land. Most selfbuilders are very concerned that their new homes should be the best possible investment, and it will not be this if a large house is cramped on a very small plot in the wrong area, while a very small house or bungalow will not optimise the value of a large plot in the right area. This point has to be given very careful consideration. Selfbuilders are notorious over-developers of plots, and at Milton Keynes, where there are hundreds of selfbuilt homes, it is said by estate agents that they can be identified because the plots are so often 'built up to the boundaries'. It is not a good idea.

Now let us move on to finding your site. The first thing to be emphasized is that you are going to have to find and buy it yourself. The chances of finding the land on which you want to build advertised in your local newspaper, making a phone call to the advertiser when it suits you or taking a leisurely stroll down to the estate agents, making up your mind at your leisure and letting the vendor know you are going to do him a favour by buying it are very small. YOU have got to make it happen.

Building plots are in short supply in this country and have been ever since the advent of the Town and Country Planning Acts in 1948. But the plots are there. Thousands of families built their own new homes as individual selfbuilders last year, and that means that thousands of people found a piece of land on which to build. How did they do it?

First of all, 50% of all selfbuilders find their land through some form of local contact. They either buy from a member of their family, or from someone who they know, or they were told of land for sale by someone else, or they heard of it at work, or they knew of a piece of land and

made it their business to find out who owned it and then made them an offer that they could not refuse. If you think that the chance of you doing this is rather remote, then you have to go out and find the land yourself. Half of all of those who build on their own do this.

How do they do it? First of all estate agents sell land. It is better not to wait until an estate agents advertisement catches your eye. You need to put your name on their lists and ask them to send you advance details of individual plots before they are advertised. If possible, you want to call at their offices on a weekly basis. If the local paper comes out on Thursdays, it would be a fair guess that the estate agents have to get their advertising copy to the paper on Tuesday afternoons. If you call at the local estate agents every Tuesday morning, there is a possibility that one day you will hear 'yes we do have this plot which has not been advertised yet' and so on.

Next, while you are in touch with the estate agents, you should write to everyone who might have land for sale where you want to live. At this time when builders are feeling the pinch, the obvious choices for this are the local builders. Invest £20 in enough first class stamps to be able to write to all the small local builders in the Yellow Pages saying that you are looking for a serviced building plot, in such and such an area, in such and such a price bracket and that you can arrange to purchase very quickly. Many families looking for a building plot find one in this way.

It is also worthwhile advertising yourself as a purchaser. Take out small ads in all the local papers on a regular basis. These should say something like *'private purchaser seeks building plot for new house for own occupation, will accept that vendor must approve design, spot cash available'*. You will get all sorts of replies, and in this way you hope to hear from a little old lady who wants to sell the orchard at the bottom of her garden, but does not want a nasty builder to be involved because she is not sure what sort of a house he would put on it. All you have to do is persuade her how ideal a family you are to have living there. You may also hear from a builder who wants to raise money but does not want more people than necessary to know he has cashflow problems.

Another approach is to find out what your local authority is doing. Let us start with some history. Selfbuild in Britain started in its modern form in the late 1940s with selfbuild groups formed by ex-servicemen's associations building on council land with council mortgages. A lot has happened since then, but some local authorities still have a tradition of helping selfbuilders. A number have a policy of regularly selling off council land as serviced plots for those who want to build on their own, and some give priority to local residents.

Other councils do not have special schemes for selfbuilders but have decided that they receive a better return when selling land for housing by marketing plots instead of letting a developer buy a whole site.

Again, all councils have lots of small sites of all sorts such as gaps left in a row of houses for access roads that were never built, old swamps that are now drained, or redundant school playgrounds. Some dispose of these quietly through agents, but others arrange for planning consent and sell them as building plots.

The schemes that are specifically for

selfbuilders are often of 50 or more plots, and typically the plots are released in phases of ten. There are exceptions: as this book is being written there are 70 selfbuild plots on offer in Milton Keynes. There is usually a very comprehensive brochure available which includes a design brief, full particulars of services, details of the covenants and everything else that a selfbuilder wants to know. Until fairly recently

Local authorities sell land for homes of all sizes and values. This house is on a development of plots for individual builders sold by the local authority in Northampton.

This house by Medina-Gimson demonstrates the regional styles which are now very fashionable. They are not appropriate to every site, so that plots which will suit them are often very expensive.

SHARING A SITE WITH OTHER SELFBUILDERS

Individual selfbuild is a lonely business even with the most supportive family: the idea of being involved with others, perhaps developing a site together, sharing work on common services and co-operating in hiring plant and employing tradesmen, seems to make a lot of sense. And so, of course, it can, but it is certainly not as easy at it sounds and this description of how it can he made to work is going to be a catalogue of potential dangers and dire warnings. This does not mean that selfbuild in collaboration with others cannot be a huge success, but making sure that it is nothing less than a success requires very careful consideration of the legal position.

It is a general principle of law that when people act in concert to do something together, they are all equally responsible for what is going on, and from this it follows that all are responsible for what any one of them gets up to. If building materials are being bought by a group of friends who are building together, then any or all of the group are equally responsible for paying the bill. If the group enters into a legal agreement in respect of land, access, services, or responsibility to keep VAT records, then all of the group are 'jointly and severally' responsible for what has been agreed.

In business dealings people avoid this unlimited liability by forming a company or by using some other legal framework. Unfortunately, with selfbuild this tends to create as many difficulties as it solves.

If a group of people form a company to buy land and build themselves houses on it, they may or may not have any liability beyond the money that was put into the company, but the company will want to borrow, and almost certainly they will find that lenders want 'joint and several' guarantees from all of them. This means that any one of them is responsible for paying back all of money. The individuals will also have no control over what the company does beyond their vote as a shareholder at a board meeting. The individual or group with a majority of the votes has total control, and if they decide to do something which upsets the minority, the minority can do very little about it. For example, if Smith, Brown and Jones form a company to buy land and build three houses to sell to the shareholders at cost. Brown and Jones can decide to build their homes first and to leave Smith's house until the next century. Tough luck for Smith. Of course, there is probably some sort of agreement that all the houses are built at once, but for a minority shareholder to take action against majority shareholders to force a company to take action will mean huge legal fees and take years. And what sort of a house will Smith get if he goes to law?

You may think this far fetched, but suppose Smith loses his job and wants his money out, and Mr. Jones leaves his wife to set up house with someone else, leaving Mrs. Jones (who is legally aided) to claim that half his uncompleted selfbuild house is hers! These things do happen, and if they do, it is worse if all concerned are legally tied together in this venture.

If everything does go well, and the company sells the houses at cost to the shareholders, the sums transferred will hopefully be much less than the market value of the homes. Up will jump the tax inspector to claim that this cost/value discount is a benefit to the shareholder and taxable at his top rate of tax.

These two scenarios are just a start: there are many other things that can go wrong. So, if you have little or no experience in forming, running and winding up small development companies this is not for you.

A better approach is through a selfbuild housing association. These are a very respectable and well established way of building a home as a member of a team, and they have been going for over 40 years. They usually involve between a dozen and twenty families forming a Housing Association which they register as a Friendly Society, which has a very different legal structure to a limited company. Such an association is able to borrow finance for both land and building costs with no security beyond the land itself, and sells the finished houses to members without any tax implications. Provided the association follows the rules laid down by the Federation of Housing Associations, the members should be able to cope with problems that arise, with the rights of members safeguarded. Self Build Housing Associations are dealt with in chapter 24.

Another way of building together is to form a housing co-operative. These are popular in a few parts of the country, and are like associations in that you must set them up in the correct way and probably will require advice and help. Invariably schemes of this sort are launched by those who are enthusiasts for the cooperative movement.

Most selfbuilders who want to collaborate with others do not want to set up complicated arrangements of this sort. How can a smaller number of people collaborate to develop a site without any legal risks and avoiding tax demands? Easily the best approach is to devise some way of doing this as individuals. The ease or difficulty of this will depend on the nature of any road or drainage works involved.

In most local authority areas three or four houses can be built with access off a private drive. The drive will not be adopted by the local authority and the home owners will have to maintain it themselves. This makes it simple for each of them to buy a plot with the usual reciprocal liabilities to be responsible for the drive and drains, which are either built by the vendor of the land, or by one of the purchasers who buys it from the vendor more cheaply than the other purchasers in consideration of his entering into a bond to do this. Sometimes all the purchasers get together to share the common costs. If they adopt this course they will be wise to look at safeguards, explained later, but the cost of a joint drive is unlikely to be a major proportion of the development costs of any one plot, unlike the cost of a road.

If there are to be four or more homes requiring a new access it is likely that the authorities will require the short length of road and the drains under it to be constructed to highway standards, with a road bond to lead to formal adoption of the road by the authority. This costs a great deal more money and is complicated, as either a single person or a formal body of some sort has to take legal responsibility for the bond. This is not always easy to arrange, but it is far safer for individuals to buy the land that they are going to build on, and to make complicated arrangements for the road, than to let anyone else have an interest in their own plot.

All this supposes that a vendor of a site for two or more homes is willing to sell it to multiple purchasers. Ignoring situations where someone has set out to sell serviced plots, what happens when Joe Selfbuilder sees a site for three units and thinks how nice it would be to build one of them himself, with two other guys building the other two? He has two options: either he buys the whole of the site himself and immediately sells off two plots to people who he has found to buy them, or else he and his two friends persuade the vendor to sell the site in three parts. In either case all the contracts will be signed at once, either with everyone using the same solicitor or else with a gaggle of solicitors sitting round the same table. The contracts should either include appropriate arrangements for any shared drive or services, or take into account any requirement for an adoptable road, or else a separate agreement for this should be signed by all the purchasers at the same time.

Either way, the first selfbuilder who takes the initiative in this will probably have to obtain an option of some sort to 'hold' the land while he finds his fellow purchasers and sorts out the details with them. Sometimes this may take the form of conditional contract. Invariably he will either have some experience of this sort of thing or he will retain an estate agent to handle it. Occasionally a solicitor who is particularly interested in land may help set up the arrangements, (which is distinct from simply giving legal effect to arrangements). However, whoever helps with the details, it is Joe Selfbuilder who will have to provide the enthusiasm to make it all happen.

The actual arrangements to build any common drive, road or drains have to be made very carefully. Avoid any joint responsibility for this work: it is far preferable that it should be the responsibility of one person, who may be one of the selfbuilders or more often will be a contractor or civil engineer. This person or firm should provide a guarantee that the work will be done, preferably as a bank bond so that if he defaults the bank will step in and get the work done. In return he will look for a guarantee of payment from all of the plot owners, often with the money deposited with a solicitor, or with a second charge over their plots. In this way there are mutual obligations all round that will ensure the work is done, without any one participant's special circumstances threatening the whole job.

In spite of all of this hassle, building with others does have enormous advantages. If you have an opportunity to build in this way, go for it — but get the best disinterested advice on how to set it all up.

Found a site for three plots?

* See if you can find two friends so that you can offer to buy it as three plots.

* Or buy it yourself and sell on two plots to others who you have already found.

* Or buy it yourself and hope to sell two plots later.

* Beware buying it in conjunction with others.

Found a site for seven or more plots?

* Consider forming a Selfbuild Housing Association or Housing Co-operative, taking advice from those with wide relevant experience.

* Obtain impartial advice regarding any proposals to build in conjunction with others in any other sort of way.

Sharing access and services?

* A road, drive or drains to serve a number of houses should either be arranged by one selfbuilder who takes out a bond in favour of the others to guarantee that he finishes the work, and requires them to guarantee payment of agreed shares of the cost.

* Or all the selfbuilders jointly retain a contractor or engineer for the work, requiring a bond from him and offering him acceptable arrangements to guarantee payment.

* Every aspect of arrangements for the maintenance or management of common services or access should be a part of a contract for the purchase of the plots, or should be embodied in an agreement signed at the same time.

such plots were sold on a first come first served basis or by auction, but it is now usual for them to be offered using a sealed tender procedure.

The prices of plots bought from local authorities vary enormously. At the time of writing the cheapest cost around £20,000, while the Wellingborough Borough Council still has 15 plots available at its prestige Redhill Grange Individual Homes Development at prices up to £76,000. Council owned plots are sometimes considered to be downmarket: nothing could be further from the truth. They are usually large, in pleasant surroundings and specifically for detached homes. Sometimes the plots are for sale simply because the land value is such that statutory cost limits preclude its use for council housing.

A number of local authorities used to offer plots on a 'build now - pay later' basis, but the financial pressures on councils have now made this unusual. Currently only Daventry District Council offers this facility, with an arrangement that you obtain a licence to build on one of their 26 selfbuild plots at Lodge Lane on payment of 30% of the price, with the balance payable after 18 months when you have built the house and arranged a mortgage on it. Incidentally these plots are of a quarter to half an acre and cost £60,000 each. This sort of scheme would be even more appropriate to smaller plots for first time home owners.

How do you find out more about this? If you are hoping to build a home in one particular council area then write to the Town Hall, marking your letter for the attention of the Estates Department. Explain that you want to build a home for your own occupation, and that you are looking for a serviced building plot in such and such a price bracket. You will get a reply: it may be a brush off, but you may hit the jackpot.

Quite separately from selling plots, there are about 20 councils which will help you by sending you a list of all the sites for which they have recently granted Outline Planning Permission. Some are free, some cost up to £20. Some are nicely printed and bound, some are nearly illegible computer print out sheets. This service

is really an extension of every citizen's right to go the local planning office and demand to see the planning registers. This is always an interesting thing to do: sometimes you will be warmly greeted and asked if you want any further help, while other councils react as if you have made an indecent proposal! Of course, most of the plots that you will hear of in this way are privately owned, and the owners who have recently sought planning permission probably know exactly what they are going to do with their plots! Still, it is a way of investigating all your options.

Returning to other ways of finding somewhere to build, there are also commercial land finding companies. They invite you to subscribe to their service in some way and they keep you advised of opportunities to buy plots through them. Often this is linked to their own special financial or management arrangements, but, provided that they only require a small fee for you to register with them, their lists will help to give you a clear idea of what is on offer, and of prices. You are unlikely to get a bargain in this way, but you may find exactly the plot you want at the market value.

Some people find a plot by seeking out the owner of a possible site for a new home and

making a direct approach to them. Doing this requires a talent for sleuthing. One helpful factor is a fairly recent change in the law which obliges the Land Registry to disclose the ownership of any registered title. Start by writing to HM Land Registry, 32 Lincolns Inn Fields, London WC2A 3PH and ask for their leaflet number 15, which explains the procedures and gives the addresses of the various regional land registry offices. Modest fees are payable for this service.

In your search for somewhere to build you may come across two or more plots which are for sale as a single site, giving you the possibility of either buying together with another family or families who want to build, or buying the whole site yourself and then selling off the plot or plots that you do not require. There are no problems in this provided that you find an experienced solicitor to help you to plan how you are going to go about things. There are a number of different ways of handling the purchase and subsequent division of the land, and to some extent the best approach will depend upon whether there is to be a shared access drive or common services, or whether you intend to build in collaboration with the other plot owners. These options are shown on pages 40 and 41. When looking at the alternatives keep asking yourself how you and your family would be situated if the other people buying with you happen to lose their jobs, fall under a bus, or get divorced during the operation. If the Bloggs family are buying one third of a site for three homes at a price that reflects that they are paying for the bottom part of the access drive and the drain connection, then there should be some sort of a bond with a bank to ensure that this work gets done even if Mr. Bloggs suddenly disappears abroad closely pursued by the police, or is otherwise unable to discharge the obligation.

A house in a regional style by Potton.

4 Evaluating a site

Before you get into your car to go to look at a building plot, there are some things you must be sure about: such as how much you can afford to pay for the plot that you want, the size and character of the home that you want to build on it, and all the other factors that you must consider when you are looking at a potential site for the first time.

The sum that you can afford to pay for a suitable piece of land, and the probable size of the house that you will be able to afford to build, have been discussed in an earlier chapter. The whole business of design, and what the planners are likely to let you build, have chapters of their own later in this book. It is important that you have read them before you start plot hunting.

There are many other things to consider. Some of them, such as the slope of the land, the view, the orientation, drainage arrangements, etc., will affect the design of any new home that you build, and these factors are dealt with in the chapter on designing a home to suit your site which starts on page 80. They may, or may not, have a significant effect on building costs. The considerations which are dealt with in this chapter are those which may affect whether or not the site can really be developed at all, or which may detract from the value of the finished property. Remember that the fact that a piece of land is described as a building plot does not mean that it is a practicable place to build a house: it merely means that the vendor has used the words to attract prospective purchasers. Remember also that outline planning consent, or even detailed planning permission, is merely consent for a house to be built there by someone if they can. There is a legal term 'caveat emptor' - which means let the buyer beware. What follows is about what you must beware of.

Start by wondering just why no-one has built there before. Really first class individual building plots have been in very short supply for the last thirty years. Why is this still a building plot? Why did no one build there twenty years ago?

There may be a simple answer. The site may have been part of the garden of a large house and only split off from it recently, or planning consent may have been granted very recently following a change in the local structure plan, which has hitherto prevented any development. Or it may have been covered by old buildings which have recently been demolished. If this is so, all well and good. If not, try to find out why no one has built on it. The reason may not preclude you building on it, but it is something that you have to establish at an early stage.

Footpaths

First of all, there may be legal problems. In rural areas a common difficulty is that there is a public footpath running over the land, possibly coming over the fields from the next village and joining the village street through the land where you would like to build your new home. It does not matter that the footpath is disused: if it has been gazetted and appears on the local authority's footpaths map, then it is as firmly there as if it was a public road.

Now it is possible to have a footpath moved, but usually only if the realignment is going to be more convenient in every way for those using the path. This usually means that the new alignment has to be shorter, less muddy, more easily maintained and provide a nicer view and a more pleasant walk. It is absolutely no use thinking that the footpath can be shifted to run round the edge of your site, which will not bother anyone because hardly anyone uses the path anyway. Not only are footpaths protected in law and jealously guarded by local authorities, but there are also footpath preservation associations with members who make it their business to try to make sure that they are never, ever moved. Probably the best advice on moving a particular footpath will be obtained from a local solicitor, who may know what applications to do so have been made in the past, and why they failed.

Covenants

Another legal problem which can prevent the development of a site is the existence of a covenant on the land. Covenants are clauses in an earlier contract for the sale of the land which are binding on all future purchasers, and either require that something shall be done, or dictate that something shall not be done when the land is developed. Such covenants may be very old indeed, and may have passed with the title to the land since the middle of the last century. The reason for them is often lost in the mists of time. Quite often one comes across covenants preventing any development on a piece of land in case it should spoil the view from a manor house which has long since been demolished. The manor house may be no more, but the covenant marches on.

Positive covenants typically require that any property to be constructed shall be of only one storey or shall be set back a certain distance from one of the boundaries, often to preserve a distant view. Another common covenant gives a right to an adjacent landowner to pass through the plot to gain access to a distant field, or, in urban areas, gives a neighbour who once owned the land access through it to his own back garden. In these cases the land was probably sold off as a paddock and the covenants are appropriate to that use. Although it may now have planning permission as a building site, the covenants remain.

Negative covenants are more usual, and often preclude any building at all. Sometimes they forbid any development unless the plans are submitted to the original vendor, who can approve them or disallow them as he or she wishes.

Covenants can sometimes be removed and this is a job for your solicitor. In cases where the covenants were imposed when a large estate was split up, with standard covenants on all the land sold, a local solicitor may have local knowledge of the situation which will not be obtainable from a lawyer at a distance. A way of dealing with them is to make an application under Section 84 of the Property Act 1925 which gives the Lands Tribunal discretionary power to modify or remove covenants, but the outcome of such applications can be uncertain and the procedures take a long time. It helps if the character of the whole neighbourhood has changed. It is usually quicker and easier to take out insurance to guard against anyone claiming rights under the covenant, and this is discussed on page 46. Typically this is done when very old covenants in restraint of trade preclude "any wheeled vehicle being kept on the premises", and no one knows who are the heirs and successors of the person who imposed the covenant.

Unknown owners

There are other reasons why a dream site may seem to be impossible to develop. Sometimes this is because no-one knows who owns the land, although in fact the owner may live close at hand and simply does not want anyone to know that he has the title. A quick way to force him to surface is to borrow a caravan and park it on the land. This will usually bring a furious owner into view very quickly indeed, although it is really a most unfair thing to do. However, the situation may be that an owner went abroad, or was killed during the war, or simply that an elderly owner has forgotten about it. Sometimes there is an unresolved question over probate in a will and the title deeds were never registered and are lying in a deed box in a solicitor's office, probably never to be seen again.

A problem can also arise when a businessman buys a piece of land in his wife's name for tax reasons but does not bother to mention this to her. Subsequently they may split up and the resulting problems are enormously complex, particularly if the reason why the purchaser put the land in his wife's name in the first place was to avoid the attentions of the Inland Revenue. The situation is by no means unusual.

If you find a plot without an owner it is possible in theory simply to squat on it and after a period of twelve years occupation 'without let or hindrance' you may register a possessory title. After a further three years you may register an absolute title. One hears of this being done, but it is neither ethical nor easy. However, if you have a small piece of land which has been abandoned adjacent to your own property which is a nuisance, since it is full of weeds and spoils your enjoyment of your own property, you may decide to incorporate it into your own garden until such time as the owner arrives to claim it from you. If he does not appear for twelve years you can arrange to call it yours, and arrange insurance against an owner appearing to claim it from you.

Legal contingency insurance

Defects in titles can take many forms, and many can be dealt with through special single insurance policies. As far as restrictive covenants are concerned, the policy protects the insured and their successors in title aganst enforcement or attempted enforcement of the covenant. It includes the costs, expenses and any damages in connection with a court action or lands tribunal action, the cost of alteration or demolition following an injunction, the loss of market value of land as a result of development being prevented and abortive capital expenditure.

Legal contingency insurance can also cover situations where title deeds are lost, problems over uncertain rights of way, services indemnities where the rights to use drains or other services is uncertain or unknown, and even the validity of possessory titles.

This is a very technical area of insurance, and each proposal is evaluated separately. DMS Services Limited, who handle the standard selfbuilder's policies are able to help. Ring them on 01909 591652.

Ransom strips

This leads to the interesting matter of ransom strips, which are pieces of land, owned by somebody else, which prevent the site in which you are interested being developed. They are not there by accident: someone has arranged for them, and the purpose is either to stop anyone building on the plot at all, or to make him pay a huge sum to buy the strip. This may sound like some sort of sharp practice, but in fact it is rarely anything of the sort. What has happened is best understood by looking at the background history to the situation.

The usual situation is that the ransom strip is in place to stop anyone building on the site at all, and the story probably starts when all the land, including both the strip and the plot, were part of the garden of a house or possibly a smallholding or farm. This may well have been a hundred years ago. The land owner was probably approached by someone who wanted to buy part of the land for some purpose or other, but who did not intend to build on it. The buyer probably knew the owner did not want anyone building there anyway, perhaps because he wanted to preserve his privacy or the view from the house.

Now the land could have been sold with a covenant to the title to prevent anyone ever building there. However, to make it even more certain that no one could do so, it was sold with just a pedestrian access and no way of getting a vehicle to it at all. A refinement would have been to sell the land with the access uphill from it, so that in the days before electric sewage pumps there was no way of running a drain from the plot. As the purchaser did not want to build on it anyway, everyone was happy with this arrangement.

A hundred years later the big house has been demolished and the site where it stood is covered with modern homes at ten to an acre. At the time when they were built, it would have been sensible for our plot to have been developed as part of the same estate, but, maybe because the owner could not be traced, the plot lay dormant. Perhaps for many years.

Then, possibly by inheritance, the plot passes to someone who bothers to go to look at it. It is a gap in a built up area and as such likely to get planning consent. The new owner applies for outline planning permission, not knowing or not bothering to mention that his boundary does not go quite to the road. Consent is granted.

What does our plot owner do now? He can trace the owner of the ransom strip, who will probably now be the hard hearted builder who developed the site of the big house, and ask him to sell the strip. Alternatively he can build anyway and hope that no one else remembers about the ransom strip.

If he does the former, the owner of the ransom strip may ask him for a sum just below the value of the whole plot as a building site, reasoning that it is worth nothing at all to the owner unless he can buy the strip. If the owner takes the other course and simply annexes the strip and builds a house, he then runs the risk of being detected and ending up with a house without an access.

One way out of this problem is to sell the plot, now with planning permission, as a building plot and to let the purchaser deal with this situation. The vendor can claim that he knew nothing about the ransom strip. In every sense of the words, the unsuspecting purchaser will then have 'bought it'. You must make sure you are not such a purchaser. There are thousands and thousands of plots like this in Britain, and architects and surveyors who deal regularly with one-off houses for private individuals come across them quite frequently.

A similar situation arises when land without planning permission is sold and the vendor wants to ensure that he will receive a share of the increased value if it ever does get planning permission. To do this he creates the ransom strip. Both parties will agree to this, but a subsequent purchaser may not be told about it.

So what can you do to guard against finding you have a ransom strip situation? First of all, your solicitor will make sure you have a good title, but as far as access is concerned he can only rely on the vendors' solicitor's reply to a standard question in his enquiries. If the vendors' solicitor has not been told of the ransom strip, he may not be able to detect it from an old deed plan to an unregistered title. The planners may also simply not know it exists.

Does this frighten you? No need. If you follow a few simple rules you can spot a ransom strip very easily.

Rule one is that whenever you buy an isolated building plot you must first ask yourself 'why has no one built here before?'. The vendor may give you a good reason, which you should treat with suspicion until it is confirmed. By following this simple rule you can avoid buying plots which are village duck ponds that have been filled in, or have a war time air raid shelter below the ground, or suffer a horrid smell from a tannery across the valley whenever the wind is in the east, — or which have a ransom strip.

Rule two is to measure every part of the boundary on the ground and compare it with the plan on the title deeds — and this means the title deeds and NOT the plan in the estate agents particulars! Why is the plot the shape that it is? If there seems to be a mystery go to the public library and look at the nineteenth century

ordnance survey maps of the area. They often provide clues as to what has happened.

Rule three is to drop in at the local pub, strike up an acquaintance with the local busybody and ask his opinion of your prospective purchase. You may get told the soil is full of eelworm and won't grow brassicas, or that it was once the scene of a horrible murder, or that 'it has a funny title you know'.

Finally, if you do find a ransom strip situation remember that it may be possible to sort it all out at a practicable cost if you go about things in the right way. Do not tell the owner of the strip that he is a despicable rogue. It is unlikely that he is. He did not create the ransom strip. It is often better not to approach him at all, but to employ an estate agent as an intermediary. And certainly do nothing at all until you have taken professional advice.

A similar problem can arise when the access is off a private road. The title to the land may include the right to use the private road, but some other user of the road may have made a practice of blocking the end of it by parking his family cars there for many years. He may be advised by his solicitor that has now established a right to do so. Again, you have a special situation, and resolving it may cost you a lot of money. The usual answer to the problem described, which is not uncommon, is to build a garage for the car owners on your own land and give it to them! This may cost a great deal, but it may be worth it in the long run.

Once again, in some situations insurance is available to deal with old ransom strip or access title problems.

Roads and access

This leads to consideration of other aspects of how you can actually enter your building plot. Few aspects of building for yourself can generate as many challenges as the question of the access. This is the term used to describe arrangements to take a drive across a boundary on to some sort of road. This can cause very real difficulties for selfbuilders who have bought a piece of land on which to build after carefully checking the title and planning situation, and who then find that there are legal or physical problems involved in driving a car onto the land. If you are plot hunting, it is a subject you should know a little about.

Let us start by considering the road which leads to your proposed new front gate. It will fall into one of three categories: it is either a private road on a land owners' land or it is on common land, or else it is a private or 'unadopted' road belonging jointly to all the people who have properties leading from it, or else it is an 'adopted' road belonging to the council and maintained by them. If there is any doubt, your solicitor will establish exactly which category it is in.

If you are fortunate enough to be building in the country on a site reached by a road in the first category, then you are involved with special circumstances. You should make sure that you are getting the best advice on the arrangements which the land owner will have suggested when he offered you the land. If your solicitor has a local rural practice then he should be an expert: if he is not, discuss with him where he is going to get you the best local advice.

Access onto a private unadopted road is a different manner, and again you need a solicitor who has, or will get, relevant local experience. There are many sorts of unadopted roads, but the ones most usually found by selfbuilders were built in the first forty years of the century. At that time developers often sold plots leading off a private road with no intention that it would ever be taken over - or adopted - by the local authority. This was considered to be a way of guaranteeing that the development would always be suitably exclusive in the days before the planning acts. The plots and the houses built on them were often very large, although sometimes this way of doing things was used in low cost 'bungalow towns' like some of those on the Sussex coast and in Essex.

Fifty years later these unadopted roads are often in a sorry condition. Planning restraints have now largely removed the fear of unacceptable further development, and the residents owning the road would usually dearly like the authority to take it over and put it in good order.

The council will not do this unless all the residents pay 'road charges', which may amount to tens of thousands of pounds for each property. As it is most unlikely that they all can (or want to) do this, the road remains unadopted with only the most urgent repair work paid for by some form of residents association, usually only when they find of cost of filling the potholes is less than the cost of replacing the exhausts scraped off their cars!

Houses on these developments have large gardens, and high land values means that they will be subdivided and sold as building plots if planning consent to do this can be obtained. Plots like this are very popular with those wanting to build an individual home, who should consider certain aspects of such opportunities very carefully.

What are the arrangements for the ownership and maintenance of the road? What road charges are likely to be levied if it is ever adopted? What are the chances of it being adopted? Above all, what is the long term future for property values when the 1930 vintage road in terminal decay deteriorates further? Local solicitors and estate agents will be well aware of the local politics in such a situation, which is why a vendor may be hoping for a prospective purchaser from a distance. Also consider that road charges are levied at a rate per yard of road frontage, so that residents may wish to sell off a plot with a disproportionately long road frontage. You have been warned.

There is one interesting possibility in all of this, a legal loophole so arcane that few solicitors will pursue it, but which really works. If the road predates the original Local Government Acts of the first half of the nineteenth century, no road charges are payable. All you have to do is prove that pedestrians, horsemen, carts, cattle, gentry, clergy, the lower orders, etc, etc passed and re-passed along it regularly at the time of Waterloo. Witnesses to this will be hard to find, and land title plans and even tithe maps will be scoffed at by those in authority. However, find your local historian and with his help look at the report of your local Enclosure Acts Commissioners and the local Enclosure Act itself. If, in 1794, Commissioner Pecksniff reported that he travelled along Common Lane to arrange to terminate the villagers' rights of turbary and commonage, and marked it as a road on his map, and it is on this road that you now live, you may ultimately cause much discomfort in the council roads department. This can be good fun, and will take years.

Adopted roads, looked after lovingly by the council, would appear to present far less trouble. However, councils make rules about anything and everything, and their roads are no exception. Their rules for roads on new housing developments are demanding and precise, with thought given to the shape of cul de sacs to enable a fire engine to turn round easily, etc, etc. Arrangements for the junctions between new estate roads and highways are very carefully considered, and this is all very right and proper. However, unfortunately for those who buy infill plots fronting a highway, the same careful consideration is given to their own modest drive access.

In general terms, the higher the category of the road — A, B, unclassified, etc. — the stricter the requirements, and whatever the classification there will be a concern for road safety. At the very least this will involve an absolute ban on vehicles joining or leaving the road at a sharp corner or at a blind spot. More usually the requirements will also involve setting back any gate a fixed distance from the carriageway, and providing a visibility splay. The actual junction may be required to be level or at a gentle gradient for a certain distance inside your property, with a further maximum stipulated gradient for the next few yards of your drive. It is no use saying that you and your friends have Landrovers: the councils concern is the brakes on their dustcart. You will be expected to make sure that surface water from your drive cannot spill onto the road, and you will also have to meet specific standards for turning radii, which themselves may depend on the width of the highway at your proposed entrance.

None of this is relevant to the selfbuilder who is taking access off an existing private drive to another house, and whose planning consent shows this arrangement, or to the individual who is buying a serviced plot in a residential development where the road is not a highway. However,

Highway Department requirements for access to an infill plot in a village where the road is also a highway are absolutely crucial to every aspect of the project, and must be given very careful consideration. The fact that there is a planning consent is not enough: the writer has seen planning consents on such plots with a standard condition that 'the access shall be formed to the requirements of the Highways Authority', where these requirements involve 60% of the plot to be excavated to achieve required levels.

Highway requirements vary in different parts of the country, and the rules in Suffolk are obviously very different from those in Powys or Cumbria. If you are considering a plot with access from a highway, you must get hold of the local highway department requirements at a very early stage and consider whether the development that you propose for the plot will be practicable. You will be able to get details at the Town Hall, and bear in mind that it is most unlikely that you will be able to negotiate any relaxation of the requirements.

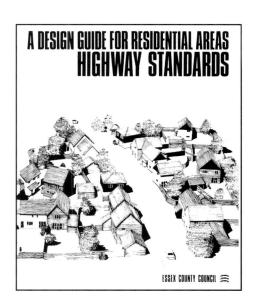

Visibility splays

One of the biggest difficulties is with visibility splays. Standard highway requirements may well require that access onto a particular category of road should be formed as shown in the illustration. As will be seen, a visibility splay is required

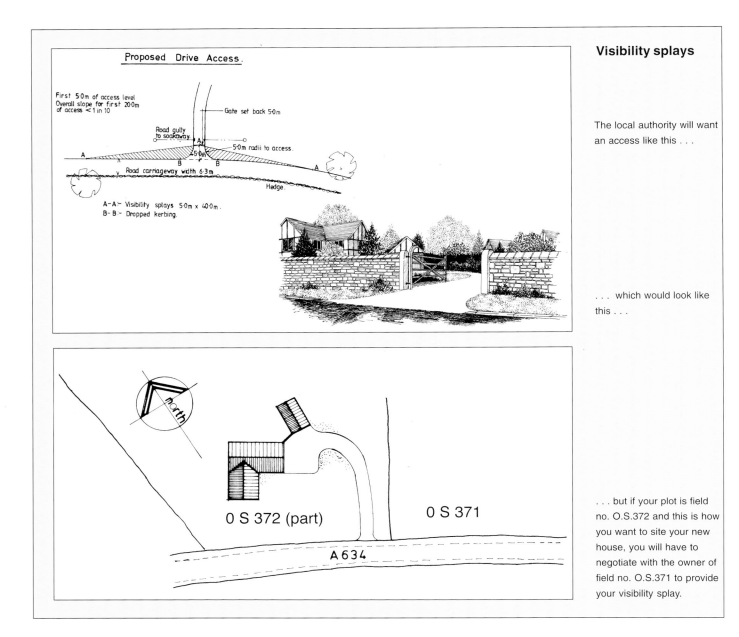

Proposed Drive Access.

First 5·0m of access level
Overall slope for first 20·0m
of access < 1 in 10

Gate set back 5·0m

Road gully to soakaway.

5·0m radii to access.

5·0m

Road carriageway width 6·3m

Hedge.

A-A :- Visibility splays 5·0m x 40·0m.
B-B :- Dropped kerbing.

O S 372 (part)

O S 371

A 634

Visibility splays

The local authority will want an access like this . . .

. . . which would look like this . . .

. . . but if your plot is field no. O.S.372 and this is how you want to site your new house, you will have to negotiate with the owner of field no. O.S.371 to provide your visibility splay.

of 80 metres by 5 metres. This is so that the driver of a car which is joining the main road will have an uninterrupted view of approaching traffic at a distance. It is irrelevant to the Highways Department that your frontage is not wide enough for the required visibility splay. They will expect you to make some suitable requirements with the neighbouring land owner. Needless to say, this may not be what the neighbouring land owner has in mind at all!

Shared drives

Because selfbuilders are often concerned with plots that have been carved out of a large garden,

they are often involved with arrangements for shared drives. If planning consent has been given on this basis, then the existing access can continue to be used. However, if you alter it in any way, the local authority may take the view that you are forming a new access, and will insist that it meets all of their requirements. This requires careful thought. Equally careful consideration has to be given to the arrangements which you make with the established owner of the drive for its future maintenance, for its use by vehicles delivering materials to build your new home, and for parking on it. Local authorities will usually permit up to four houses to share one private drive, beyond which they are likely to consider it to be a road and subject to all sorts of regulations. The basis on which you may be sharing a drive with three other people requires very careful consideration. A very basic requirement is an arrangement that there is no parking of any sort under any circumstances on the shared part of the drive. It is also very desirable that the shared length should be constructed to a very high standard, and should be properly kerbed, so that it is as indestructible as possible to avoid difficulties over sharing the cost of rebuilding it at some time in the future. And when that time arrives, are you going to be able to insist that it is rebuilt, and that your neighbours pay their share of the cost?

Shared drives have short term advantages and long term problems, and if there is a shared drive some building societies will restrict the percentage of the value of the property on which they will grant a mortgage. Solicitors will elaborate on their disadvantages to prospective purchasers. However, this is frequently the only way in which some individual plots can be developed, and they are a necessary evil.

So much for a few possible legal problems. Now let us look at the engineering difficulties which come with some sites.

Foundations

First of all, foundation problems, which can sound very worrying. Sometimes a site is 'made ground', which in most English villages means a pit from which brick clay was dug in the eighteenth century or earlier which has been filled in with rubbish, or the site of a village pond which has been filled. There can be local geological problems, such as the site being on a small pocket of greensand, or a 'slip plane' where an outcrop of rock presents possible movement problems. In some old mining areas, such as in parts of Derbyshire, there may be old mine shafts, sometimes dating back to Roman times, which are a very good reason why you find a gap in a row of village houses. Springs that can appear on the land after heavy rain are another potential source of trouble and so are underground streams which may run across the site to

surface in the next field.

The good news is that none of this need be a serious problem these days. The cost of special foundation work as a proportion of the value of a finished property has dropped in recent years, and new techniques for dealing with all of the problems which have been mentioned are now very well established and do not cost a great deal of money. The important thing is to get the very best advice, and this will come from a soils engineer who is a specialist in these affairs. Do not assume that you have necessarily got to build off piled foundations, or use an expensive raft foundation without having sought out advice on the excellent modern alternatives available for compacting bad ground.

Drainage

Another hidden problem may concern the drainage. There are two sorts of drains required for any dwelling, the foul drains which take the waste water from the kitchen and bathrooms, and the surface water drainage system which takes the water from the gutters. Surface water is usually discharged into a soakaway, which is a large hole dug at least five metres away from the new building and filled with rocks. However, in some circumstances, if the water will not disperse into the ground from the soakaway, other arrangements have to be made. Usually this involves piping the surface water into an existing surface water drain in the road, or, if the Water Board will agree, taking the surface water into the foul drain. If there is any doubt about whether or not a soakaway is going to work you will have to arrange for a percolation test, of which more anon.

Foul drainage is rarely a topic for polite conversation, but it is a fascinating subject. It is only the drains running beneath all our cities and below many rural roads that stand between us and the cholera epidemics of the eighteenth and nineteenth centuries. However, useful as drains are, nothing will flow along them if they go up hill. If you want an ordinary drainage connection into the council sewers, it is absolutely necessary that your new house should be higher than the main drain or sewer. If this is not the case, you have a problem site. However, there are ways of dealing with drainage problems, and they are discussed in chapter 7.

Trees

Most prospective selfbuilders dream of a building plot with mature trees on it, but if they find one they will find that the trees that they so admire will complicate their plans, and may limit the design of the house that they want to build. There are many angles to this, and they are best considered under four separate heads.

Trees and neighbours

If your neighbour has a tree which overhangs your boundary and which will interfere with the house that you want to build, you must clear the situation with your solicitor before you commit yourself to buy the plot. In most circumstances you can cut the branches which overhang your fence, and you can cut roots that will affect the foundations of the house that you want to build, but this can be a complex issue. It is particularly difficult if the result of your activities would result

in the tree becoming unsafe. Perhaps it is already unsafe? If it falls onto your house, you hope that your neighbour is properly insured or else very rich. He may be neither, and almost certainly he is unlikely to tell you, which is why you have to take out a selfbuilders policy to cover you until the house is built and your domestic insurance takes over. Also remember that a neighbour may choose to fell his own trees and seriously affect your view. There is little you can do about this but it is something you should consider.

Everything is simpler if the trees are on the land that you want to buy, but you must still consider the implications at an early stage. First of all, your tree may fall across the boundary and damage your neighbour's property. This risk is covered by your selfbuilders insurance policy while you are building and by your householders insurance when you have finished, but the policies will require that you exercise normal prudence in this matter. This definitely includes being aware of dangerous trees and trees that present a special hazard, and doing something about it. If there is a potential problem here you

A timber frame from Hedlunds.

should get the tree inspected by a qualified tree surgeon who can either give you a report saying that there is no hazard, or can advise on sorting things out. If you are buying a plot with potentially dangerous trees you might be able to negotiate for such a report to be a condition of your purchase.

Trees and planners

This is where the trouble starts. Planners can place a Tree Preservation Order on any tree or group of trees that they think worth protecting, except for fruit trees and orchards. They often do this with trees that are obviously dying or not worth keeping just so they can have a say in what happens to them. An application for planning consent that involves felling TPO trees always receives special consideration, although a planning consent showing that trees have to be removed overrides the TPO.

The first problem is knowing whether the trees on the plot have a TPO on them, and if you write or ask in a direct way at the Town Hall you may trigger a TPO being issued. You will have to make general enquiries — 'can I look at the TPO map for the village?' and proceed with caution. If there is not a TPO you may decide to fell any tree which is in the way straight away. If this is likely to upset anyone, do it very quickly, so that a TPO is not issued in a hurry while you are sharpening your chain saw. It does happen. Local advice is essential in this sort of thing.

If there is a TPO, you will have to consider the situation very carefully. If there is outline planning consent, then a condition of the consent will probably be that the tree has to be retained, even though this may seriously limit the size of the property that can be built on the plot. You will have to show the protected trees on your site plan for the detailed application.

At this stage it is probably worthwhile getting a tree specialist to make an inspection, and to provide you with a report. Hopefully he or she will write *'the specimen is over mature and should be replaced'* or at least *'an unremarkable specimen which can be replaced without affecting the character of the local arboreal environment'*. For every tree that you show on your plan to be removed you should show at least three replacement trees elsewhere in the garden, and make sure that the species are from any recommended list in the local design guide.

This will usually do the trick, especially if there is already outline planning. However, you may find that you get involved in negotiating about it with the planners, often with the suspicion that they care far more for trees than they do for people. As a last resort you will appeal. Whatever you do, this is one area where the guidance of someone with local experience is invaluable.

Trees and foundations

Trees can affect the foundations of houses in three ways: by the roots simply pushing into the foundation brickwork and cracking it, or by roots rotting and leaving voids under the foundations when the tree is felled, or, in clay soils, when a tree is felled and, no longer taking hundreds of gallons of water out of the subsoil, the ground may heave and rise by two or three inches. All these potential difficulties are dealt with by a combination of special foundation design and a limit on the distance that a new building can be constructed from a tree.

Foundation design to meet problems due to trees is a matter for your architect, and your plans will not receive Building Regulation approval unless they are satisfactory. One thing to watch for is a bad ground situation due to a tree having been felled before you bought the plot, often because the vendor wanted it out of the way before he advertised the land for sale. Sometimes he will have carefully filled in the hole. If the subsoil is clay the work involved in dealing with this can be considerable.

The minimum distance which a house can be built from a tree depends on the species, its potential height and the nature of the ground. There are strict rules for this, and these are most conveniently set out in the NHBC Handbook. Again this is something to be dealt with by whoever is designing your foundations. As a rule of thumb, on any soil it is unwise to build nearer to a tree than four metres or one third of its mature height, whichever is the greater. On some soils this distance may be increased. This applies only to forest species and not to fruit trees or ornamental species. Some trees are notoriously more of a problem than others, particularly popular, elm and willow.

Tree roots can also cause problems with drives, which will be a nuisance but not significantly expensive to deal with. What is more likely is that the planners will require you to align a drive away from protected trees to avoid compacting the ground underneath them, which may kill them. Drains can also be affected by tree roots, and it is usual to case them in concrete if there is any risk of this. If you forget, the problems can be unpleasant, and may surface — literally — on Christmas day.

Trees and your design

A large tree or a group of trees within a hundred feet of any house or bungalow is going to enhance the appearance of the property, but must be taken into consideration at the design stage. Will part of the building be in shade at certain times of the day? What about views from key rooms? Should rooms be arranged to suit this? Above all, remember that an attractive green woodland scene in June can be a dank, dark miserable outlook in January. Big trees need their own sort of garden, bearing in mind that they affect everything growing beneath them, produce huge quantities of leaves, and usually rule out having a swimming pool. Living with trees means letting them dominate your garden: most people like that very much indeed, but do not let it come as a surprise.

If you are planting trees, possibly to replace ones cleared to make way for a house, keep in mind that they will be there for a century at least. If your garden is big enough to split off a building plot in the future, do not cause problems some day by putting a tree in the middle of the potential plot. If you are planting a tree at the corner of your vegetable garden, make sure that it is on the north corner so that it does not shade your vegetables in the years to come. Choose a species carefully, and consider whether you want to make a modest investment in something really worthwhile. A ten foot Catalpa, the Indian bean tree, or a similarly sized Tulip tree, may cost about £100. This may be more than you intended to spend, but what other present for your partner will grow to 80 feet tall!

Decisions on problem sites

The important point to remember about any problem site is that someone has probably overcome similar difficulties before in the local area, and that if you set about things carefully, and get the best advice, you will probably find that the snags can be overcome at a cost which is acceptable. Unfortunately, making the arrangements will probably take a great deal of time and it may be necessary to buy an option to purchase the site, which has to be exercised within a set period to give yourself time to make sure that everything can be arranged in the way that you hope. Options have to be arranged by a solicitor, and to be legally binding must be bought for money, although this can be as little as £5. However, the expert who you will engage may be quite expensive. Whatever the outcome, you will find it very interesting, and even if you are not able to go ahead, you may be able to sell the technical or legal reports that you have commissioned to some other prospective purchaser who has a deeper pocket than yours.

A new home surrounded by mature trees like this is a delight, but the trees themselves will present you with problems which will have to be dealt with in your project planning.

Looking at a building plot

What are the questions you ask yourself when you get out of your car and walk over to a hedge with a 'For Sale' board planted in it?

1 Is it where you want to live? (Surroundings, outlook, travel to work, shops, schools, etc.). Do not let your enthusiasm for selfbuild override a critical consideration of whether an ideal plot for an individual home is really where you want to live.

2 Is it in an area where property values can be expected to appreciate ahead of the market, or at least keep in step with it?

3 Is the price appropriate to the total budget?

4 Has it got planning consent? Are you likely to get approval for the sort of house that you hope to build? This is likely to be so if there are other similar new homes in the area.

5 Is it in an area where the planners are likely to require expensive traditional materials like natural stone, handmade bricks or clay tiles? If so, the budget will have to be adjusted.

6 Will the dimensions at the building line permit you to build a house of the size and style that you want without it appearing cramped or out of place?

7 Does the shape of the plot and its boundaries match up with the plans in the estate agent's particulars? It is surprising how often they do not, and this is the time to investigate discrepancies.

8 Is the access off an adopted road without any obvious requirements for visibility splays or expensive earth moving? If not, then consideration has to be given to the legal position regarding access and any effect that this problem will have on the budget.

9 What about drains? Is there a sewer in the road to which access can be gained at a reasonable cost? If not, there is a special situation to be evaluated and costed which may affect the budget.

10 What about surface water drainage? Does the soil look well drained enough for soakaways? If not, are there surface water drains in the road - look for road drainage grids.

11 Do the ground conditions look normal? Is there any sign of wet ground as evidenced by unusual vegetation? If so the cost of special foundations will have to be allowed for in the budget.

12 Are there any signs of old buildings or of the site being a filled quarry or clay pit? Either might involve special foundations.

13 Are there any mature trees on the site? Might you ever want to cut them down? If so, might they be subject to Tree Preservation Orders? Are they within 30 feet of the house site, thus requiring root barriers?

14 Are there any signs of trees being recently removed — often done by a vendor to make a site appear larger? This may involve special foundations.

15 Services. Do water, gas, electricity and telephone services appear to be available? If they are not close at hand, then their provision and the cost involved become major factors.

16 Easements, etc. Are there any signs of underground services running through the plot or of overhead power lines? This may be a reason why the plot has never been developed.

17 Public footpaths and rights of way. Are there any signs of others using the land, especially any gates or stiles which give access to land beyond the boundary?

18 Is access adequate for deliveries of building materials? Many infill sites with access along an existing drive cannot be reached by a lorry, thus adding difficulty and expense to a building operation.

19 Why has no one built on this site before? There has to be a reason in these days when building plots are in short supply. Make sure that you find out, and that it is not because of a hidden problem.

20 Finally, what can the locals tell you about the site in the local pub? Is it worth having a word with the village builder before taking things further? Remember the estate agent is working for the vendor, not for you. It is up to you to find out all that you can about it. Your solicitor will only concern himself with the title. Everything else is up to you.

5 Buying a site

Owning a piece of land involves far more than simply paying for it and moving onto the site. The business of buying or selling a building plot has its own vocabulary, and its own set of procedures. Solicitors, land agents, and estate agents make their living by operating this system, and they do not go out of their way to explain what they are doing to their clients. If you are to buy land it is important to realise that you are going to have to accept the system, and learn the vocabulary. Property titles *can* be conveyed by private individuals. Money *can* be borrowed with land as a security without a legal charge or other documentation. Easements, access rights and other matters *can* be dealt with by those who are not solicitors. Depending on your viewpoint, this can be worthwhile, or it may be a recipe for a legal disaster. For the individual home builder the savings obtained are simply not cost effective. The time and effort spent in transferring land without a solicitor is disproportionate to any saving, and the suspicion with which your title will be regarded when you use the land in some way, as security for building finance for example, can be a nuisance. If you are looking for finance you are loading the odds against yourself with a home-made title to your principal security.

Solicitors

If you already have, or know, a solicitor who you wish to act for you, that is fine. Otherwise you will probably entrust your affairs to the man whose office you first contact. If you are buying a plot of land with planning consent, it is unlikely that there will be anything to choose between the competence with which different firms of solicitors will deal with the purchase, but there may be a big difference in the speed with which they will do the job and the advice which they will give to you as someone building on your own. If time is important, it can speed things up to use a local solicitor as he will be familiar with local titles.

The best way to find the right solicitor is by a personal recommendation from another self-builder who tells you what a super legal service they had, but you may have to start from scratch with the yellow pages. Start the ball rolling by writing a letter which might read:

I hope to buy a building plot at Harewood and require a solicitor to act for me. Will it be possible for me to meet one of your staff for ten minutes on Thursday afternoon to discuss whether you wish to handle this for me?

At this meeting you should explain that you are building for yourself and that you will be looking for advice on various matters involved with this. Ask if he or she has come across selfbuilders before. Explain your timetable and ask if it is practicable. Ask about fees. With luck the solicitor will take a special interest in an unusual client and be of considerable help. If you do not establish a rapport you can ask outright if he or she thinks you are the right client for them.

If you decide to go ahead, solicitors now have an obligation to send you a 'Client Care Letter' which sets out exactly what they can do for you, who will do it, who you complain to if you are dissatisfied, and what it will all cost. This is all new, and very reassuring, but you should still point out that because you are going to self build you may have more questions to be answered than most clients.

Always have a word with your solicitor once you have decided to buy a piece of land, but before taking any action to buy it. He will advise you on how to make your offer, and will make sure that you are not irrevocably committed before he has a chance to examine the title. This normally involves signing papers only if they are qualified by the words 'subject to contract'. This means that you can back out at any time until you sign the contract. However, if you buy at auction, once the hammer has fallen you have 'bought it', snags and all, and there is no going back. If you are going to an auction, consult your solicitor beforehand.

If the land you want to buy is being sold by an estate agent, he will probably ask you to sign a note and pay a deposit. This is quite usual, provided that the words 'subject to contract' appear, and that the deposit is refundable if you do not proceed. If you are negotiating a private purchase, you may wish to make a written offer, or to accept an offer made by the vendor. Your solicitor will prefer to write these letters for you. Alternatively, you may simply shake hands with someone on an agreed deal, with no paperwork involved, in which case all you need to know is the name of the vendor's solicitor.

Your solicitor will be concerned with the title to your land. It is his job to see that it is transferred to you together with all the advantages of the title, and that you also know of all the disadvantages of the title. He will see that it is registered in your name when you finally complete the purchase. You can expect that he will be very punctilious in this, and that he or she will make sure that everything is explained carefully to you. So far so good, but beware. A solicitor will deal with the situation regarding your land on paper, and only on paper. He is unlikely ever to see the plot, and will certainly not get involved in peering down manholes or wondering why part of the ground looks so green in dry weather. It is your job to relate everything that he finds out about the title to the actual situation on the ground, and it is not really part of his job to make sure that you do this. You have to get cracking on it yourself, and, by and large, it is something that only you can do unless you are engaging an architect to handle the whole project for you.

In considering this matter, let us first look at exactly what your solicitor is going to do for you. First of all, he is going to make sure that there is a clear title. This means that the land is really owned by the fellow who is selling it, and that if he has mortgaged it or offered it as a security in some other way, the mortgagee has agreed to release his rights over it.

Your solicitor will also examine what covenants are attached to the land, together with any rights that you may have over adjacent land, or which owners of adjacent land may have over the plot that you want to buy. This will also include looking at easements, which are the legal rights that others have to make use of your land. Examples of this are when a statutory authority such as the water board have a pipe running under the land, and the easement giving permission for this gives them the right to enter your property and dig down to the pipe if they ever want to. Other easements may be for electricity

cables or gas pipes under the ground, telegraph and electricity wires above the ground, and rights which others hold to connect into any drains that you may decide to install. All these covenants and easements are normally in documents which form part of your title.

Both covenants and easements can sometimes be altered or extinguished, but this is a long winded business and it is unlikely that you will have the time to get involved with this, as most people buying a plot for a new home want to move quickly. It is often easier to deal with an old covenant or easement with Legal Contingency Insurance — see page 46.

Your solicitor will also be concerned with your legal situation in relation to the planning acts. Let us assume that you are buying a plot that has outline planning consent. If you are not, you are in a very special situation and should either be taking advice from someone who is very well qualified to help you, or else should be making yourself an expert on a very complicated subject. The planning consent document is normally kept with the title documents, and your solicitor will obtain it at an early stage and will confirm to you that it is valid, and let you know the date on which it will expire. In some parts of the country mining reports, or other advice of special factors affecting the development of the site, are attached to the planning consent document and form part of it. Sometimes they become separated from the planning document: make sure that you have them.

The draft of the text above was shown to the senior partner of a firm of solicitors. He made the interesting point that in a large practice the member of staff who deals with conveyancing normally depends on the value of the property, so that a transaction to buy a plot will normally be handled by a junior who handles the simple transactions for properties worth £50,000. However, buying and selling building plots tends to involve rather more than buying and selling bottom-of-the-market houses, and it is reasonable to hope that an experienced solicitor will handle the matter. You should not hesitate to ask for this.

Planning consent

Outline consent is approval for someone to

carry out the development which is detailed on the consent form. The consent attaches to the land, and not to any individual. Whoever buys the land has the consent. It is called outline consent because it is just that: it says that someone may build a house, often subject to certain conditions, and that the person wanting to build must submit fuller details to the local authority for their approval. The details will include the design, the access, the exact siting, and the materials. All this has been mentioned before, and has a chapter of its own later in the book. It is very, very important.

Obviously you want to make sure that an outline consent can be used to build the house that you particularly want, or at the very least a house that you are prepared to accept. You have therefore got to clear this particular point before you enter into the irrevocable contract to buy the land. This gives you something of a problem. If you cannot agree matters quickly at the planning office it is probable that you will find that the vendor of the land will be chasing you very hard to complete the purchase. If you do not complete, perhaps he will sell the land to someone else. On the other hand, if the planning officer insists that he is not going to approve the design of the house that you want, then you are obviously not going to sign a contract for the land until you have ascertained whether you can change his mind, either in discussion or through a formal appeal procedure.

If you meet with this situation, then you will have to discuss your whole position with your solicitor. One possible way out is to buy an option to purchase the land from the vendor, subject to you getting approval for the design that you want. Another possibility is to sign a conditional contract. However, most people manage to deal with this in the month or so available to them while their solicitor is dealing with his side of things.

Make a start by getting hold of a photostat of the actual planning consent document. This is the one signed by the planning officer, and not an alleged copy of it in an estate agent's particulars. The first thing that you will notice is that it is dated, and that it also has an expiry date. If this expiry date is fairly close at hand, perhaps giving the consent less than a year to

run, then you want to take professional advice on any problems that this may present. If the consent has at least a full year to run, and you intend to start building as soon as is practicable, you have no problems.

Next read exactly what the consent document says. If you are very lucky, it will simply refer to 'the erection of a detached dwelling' and the only condition will be that 'details of the design and materials shall be approved by the planning authority'. If this is so you should have little difficulty in obtaining approval for any design in a local style and local materials. However, such

Arranging to build a new home like this will probably involve more work before you start to build than the actual construction process itself. If you are good at making things happen you will not find it at all difficult, but you have to be in charge at all times and not simply let the arrangements drift.

open wording in the outline consent is unusual. It is much more likely to read 'consent for a detached two storey dwelling not to exceed 2000 square feet to be designed in sympathy with adjacent properties and to be constructed in local materials'.

It is quite possible that the consent will go

on to list more than a dozen conditions. These may define the position of the building on the site, the route to be taken by the access drive, the roof pitch, the style of windows, and many other matters which you probably think should not be the concern of the planning officer at all. Unfortunately, you are stuck with the planning laws as they are, and you have no choice but to conform.

Approval of the materials is no less important than the approval of the actual design, if only because a four bedroomed house in handmade bricks under a Westmorland slate roof will cost £20,000 more than an identical looking home built with machine made bricks under artificial slates. All of this is dealt with in chapter 10.

Building Regulations

Your Building Regulation approval is a different matter and you are not very likely to obtain it before you sign the contract. This need not matter, as it is not concerned with what you build, but with how you build it. However, there are two matters concerning this approval which you should clear at an early date, and these concern the foundations and the drains. It is worth finding out informally if you are going to be involved in any special foundations, or in any special drainage arrangements, as these can add significantly to the cost of the whole project. There is also an outside chance that you may find that there are serious drainage problems, as when a connection cannot be made into the council sewers and the water board is not likely to permit a septic tank.

Services

The same applies to services. You may not wish to make formal applications for your water, electricity, gas and telephone services before you have actually bought the land, if only because you wish to avoid paying the application fees, but you should satisfy yourself that they are readily available at a reasonable charge. In particular check the cost of a water service.

Buying the plot

To return to the actual arrangements to buy what you now regard as 'your' building plot, the actual purchase of the land is started by the vendor's solicitor sending your solicitor two copies of a 'draft contract' or agreement to buy the land. This is unsigned and comes with details of the title. Your solicitor will look at this with a critical eye and initiate various standard procedures to make sure that you really know what you are buying. Many of these procedures are called searches, and include asking the local authority to confirm that it is not aware of any orders relating to the land, that it is not about to be compulsorily purchased and confirming any existing planning consents.

Next your solicitor will ask the vendor to complete a questionnaire in which he should disclose everything that he knows about the site. In theory this will include such matters as the fact that it floods every winter, or that a neighbour has announced his intention of suing whoever happens to own the land for some spurious reason. Vendors are sometimes economical with the truth in these matters.

In this way your solicitor will satisfy himself that he has discovered all about your purchase from the title deeds, the local authority, and the vendor. On the basis of what he finds he will advise you on whether you may safely go ahead and sign the contract. When you have done this you are legally committed to buy. You will probably have to pay 10% of the value straight away, and the other 90% when the transaction is completed, which is usually 30 days later.

Now, diligent as your solicitor will be in dealing with all of this, he will handle the whole situation in his office. You have to be a great deal more practical, and must make your own investigations on the ground, at the planning office, and in the local community at the same time as he is dealing with matters in the abstract. What follows is a rough guide to the action you must take, although it is not possible to prepare an exhaustive list. There is nothing in what follows about Verderers Rights in the New Forest or about access to the foreshore on coastal plots in Norfolk. When you buy a piece of Great Britain you are buying a part of our landed heritage and not a pound of sugar. Rights and obligations attaching to land vary in different areas, and there can be no simple procedural guide.

Title deed plans

First of all, there is the question of the plan. Probably more than one plan. There will be one plan in the title deeds, another to which the planning consent refers, and there may well be a third plan which was attached to the estate agents particulars. All or none of these may bear some resemblance to a large scale Ordnance Survey map. None of them necessarily represent the site that has been pointed out to you by the vendor. Don't worry: this is not at all unusual.

Start off by getting yourself a 30 metre tape and measuring the plot on the ground, noting both the dimensions along the boundaries and of the diagonals. If what you find is different from the plans with which you are concerned you should discuss the matter with your solicitor. Sometimes the discrepancy may not matter. If it affects your road frontage, or the width of the plot at the building line, or the way in which you can meet a local authority requirement for a sight line at the access, you may find that you are involved in some very complicated negotiations with the vendor, or with the land registry or possibly with others. Needless to say, the time to sort this out is before you sign the contract.

Access

The next thing to look at is the access. Again you may find a discrepancy between the access on the ground, the access shown on the plan attached to the title deeds, and the access shown in the planning consent. Remember that the access shown on the planning consent is the one that has been approved by the highways authority, and they may not readily agree to it being moved. Any problems here should be referred to your solicitor as soon as detected. At this stage you should also give consideration to whether lorries with your building materials will be able to enter the site. It is possible that they will have to park in the road to unload, and in some cases this needs careful consideration. Even more important is to check whether there are any requirements for sight lines, otherwise called visibility splays, imposed as conditions in the planning

consent. These may be hidden in a phrase in the planning consent reading something like 'the access onto Mill Lane shall be formed in accordance with the County Highway Requirements'. You may find that these requirements involve a visibility splay across a neighbours land. Has he granted an easement for this? If not, you need to go back to the solicitors office for another chat.

Drains

Now is also the time to consider the drains. A sordid topic, but one that must be investigated. The foul drains that you will build will probably take sewage from your new property and discharge it into an adjacent council sewer. Is this simple and straightforward? Do you know that the council sewer is really there, and that you will be able to connect into it without any difficulty? Is it deep enough for a proper fall on your drain? Usually the investigation involves borrowing a lever called a 'key' and lifting heavy manhole covers in the road. The council does not allow ordinary citizens to do this, but people do. Remember it is considered good manners to replace the covers. Sewage is very bad at flowing uphill and you have to think about this. If it is not practicable to make a cost-effective connection, there are a number of alternatives, including a pumped system, a septic tank, or a cesspit. Here you need very special advice, and as anything unusual may be expensive, you have to settle it all before you sign the contract.

Surface water drains take rain water from your roof, and from your drive. They lead either into soakaways or into adjacent ditches, or into council surface water drains in the road. They are rarely a problem but it is essential to make sure that you can make appropriate arrangements, and in particular do not forget that the Highway Authority may want to ensure that your drive does not discharge rain water onto the road.

Diverting existing drains and services

This book repeatedly urges anyone considering buying a building plot to find out why it has not been built on in the past. One reason can be that there is a sewer, gas line or electricity cable across the site, and that the authority which installed it bought an easement to do so from the previous owner of the land, and the easement prohibits anything being built on top of the drain or service, or within a certain distance of it. Unless the drain or service can be diverted, the plot cannot be developed.

This is a situation which is regularly exploited by selfbuilders, because unlike builders and developers, they often have time to arrange the diversion, and the patience to cope with the huge correspondence that is likely to be involved.

If you meet such a situation it is important to arrange an option or conditional contract on the land before you research the problem, as otherwise you may get everything sorted out only to find that the vendor puts up the price or decides to build on the land himself! The price should reflect the cost of the diversion, which can be significant, and so you have to establish this before committing yourself in any way.

Make a start with obtaining a photostat of the actual easement granted to the service authority, which should be with the title deeds. If it is not in intelligible language you will have to seek the help of an expert to work out exactly what it says. If it is a very old easement, particularly a 19th century one, you may find that there is a clause which requires the authority to move their service at their own expense if the land owner wants to build his dwelling on top of it. This is often a feature of old electricity cable easements. Do not celebrate too soon: in the 1950's the electricity boards offered sums to all the land owners who owned such rights to extinguish this part of the easement, and many of them accepted. If they did, the documentation involved may never have been with the title deeds. However, the board will not have lost their copy!

Usually you can expect to be quoted a most unreasonable sum for moving a service, and it may take a lot of patience to get a quotation at all. However, once you have a quotation it is an admission that it is technically possible to move the service, and you can then start to negotiate. Try to reach the engineers concerned, rather than the legal department of the services authority. The engineer may permit you to seek tenders for this work to be done from a list of approved contrac-

tors, which will probably be cheaper. It may also be possible for you to open the trench involved, lay the gravel bed and back fill after the drains, pipe or cable have been laid by the contractor. This will probably save a lot of money.

If you are successful in negotiating some arrangement for the diversion, and the arrangement to buy the land reflects the cost of the diversion, make sure that the legal arrangement with the authority refers to you *or your successor in title*, so that if you buy the land but have to sell it on for any reason a potential purchaser does not have to re-negotiate the whole business.

Ground conditions

Now let us consider the ground itself. Ideally you are hoping to build on good bearing ground, which will support the weight of your new house using simple and cost effective foundations. Sadly this is not always possible and you must be conversant with other options. Incidentally, the cost of dealing with difficult foundation situations as a proportion of the total cost of the whole house has dropped dramatically in recent years as a result of new building techniques. It is unusual to find a site with foundation problems that cannot be overcome at an acceptable cost.

The first potential problem is that there may have been mining activities in the area and that there is consequently some danger of ground subsidence. This is normally detailed in a mining report attached to the planning consent, and at the very worst it means that you will be building on a reinforced raft instead of on orthodox strip footings. The additional cost is unlikely to be more than 5% of the total cost of the new home.

Geological problems are more complicated, and more difficult to detect. The principle hazards are that there is a spring on the land or a slip plane between two types of rock which outcrop on the land, or that you have a pocket of greensand under the turf. Fortunately all of these hazards are normally easily detected by someone with a practised eye, and in areas where they are likely to occur you will find plenty of people to point them out to you. The usual warning is an area of grass that is unnaturally green in the summer and marshy in the winter, perhaps with indicator sedge grasses. At the worst, this can involve you building on piled foundations at an additional cost of up to 20% on the building cost, but it is more likely that you will have to install a few hundred pounds worth of land drains.

A serious foundation problem can exist if you are building on filled land. This is not always readily detectable, but if there was an old building on the site you should make sure whether there are cellars below the ground. Again, for hundreds of years there were brickpits in villages where the locals fired their own bricks for their homes. When this ended the pits became rubbish tips, and now they are gaps between homes which appear to make ideal building plots. So they do — provided that you understand the need for special care in designing the foundations.

Final checks

If you deal with all of these matters, and make all the checks, you should avoid any unpleasant surprises. However, there is one other thing that you can do which may bring a special situation to light, and that is to find out just why the plot has not been build on in the past - something that is mentioned repeatedly in these pages. Our planning acts date from 1948, and ever since then individual building plots have been in short supply. Why has no-one built on it before? This may be very obvious. It may have been part of the garden of a larger property which is only now being subdivided, or the area may only recently have been taken out of the Green Belt. The vendor may tell you that he has only just inherited it, or that he was keeping it for a retirement bungalow but has now made other plans. All this would make sense. However, you might also find that it was the village duck pond until it was filled in with building rubble and then grassed over a couple of years ago, or that there is a problem with the access. One good way of finding out why it is still a building plot is to buy a pint or two in the local pub while chatting to the locals.

Undisclosed problems

And if you do find a problem that had not been disclosed by the vendor? First of all, happily you know about it before you sign the contract and commit yourself. If you still want to go ahead, it gives you an opportunity to reopen negotiations over the price, and to see if you can purchase the plot more cheaply in view of the difficulty that you have discovered. If you do not find any difficulties, then you can sleep more easily while you are building, knowing that you have already looked for all the potential problems that you could think of, and have not found any!

Signing a contract

Eventually you will reach a stage where your solicitor has satisfied himself that you are buying a valid title, and has explained to you exactly what rights and obligations are involved. By now you will have obtained your own planning consent, or you are satisfied that you are going to get it, or you are happy to purchase with the existing consent. It is important to differentiate between these situations, which are that either

1. *the land has an outline consent and you have obtained approval of the reserved matters in that consent to build your dream house.*

2. *OR the only issues left relating to the reserved matters concern details which you know will be resolved and you will go ahead anyway.*

3. *OR you are buying with outline consent and will submit your detailed design later, knowing that the planners will have to be convinced that it is the right design for the site.*

4. *OR you are buying with a full consent and will build to the design for which the consent was granted.*

5. *OR you are buying with a previous full consent and have obtained your own subsequent full consent or a letter agreeing to a variation of the original consent design.*

6. *OR you are buying with a full consent and are confident of obtaining consent for your own design which you have started to negotiate with the planners.*

7. OR you are buying with a full consent with the intention of getting the design varied after you have bought.

If your situation does not fall into one of these categories, you require special advice.

Having cleared all this you sign the contract, which has already been signed by the vendor, and pay a proportion of the agreed price. You are now committed and if you try to pull out, the consequences are expensive. You must have the money to buy the land before you sign a contract, and if it is not already in your bank account, you should discuss the way in which it is going to get there with your solicitor. He will take a jaundiced view of a client signing a contract with the backing of a mother-in-law who has promised to produce the money when the time comes.

When you have signed a contract you have only contracted to buy the land, but have not yet bought it. Technically you may not use it, but your solicitor will probably be able to arrange with the vendor for you to start building work on it, particularly if you have already given him the full purchase price and he is able to assure the vendor's solicitors that he has the money.

Insurance

At this stage take out the insurances appropriate to your new situation as a landowner. Even if you are not starting work at once, a selfbuilder's policy will protect you from any claims, however unlikely, such as from trespassers who come to harm on your land, or a passer-by on whom your tree happens to fall. The cost is part of the standard policy, which runs for two years or until you finish the new home, whichever is the sooner.

Completion

The contract to buy the land will require that the conveyance of the land shall be completed within a set period, usually a month. This is known as 'completion' and involves your solicitor drawing up an unintelligible document which is signed by the vendor and delivered to your solicitor in exchange for the balance of the purchase price. This final part of the transaction is something of an anticlimax. The actual title deeds, which essentially comprise all the conveyance documents relating to the land in date order, finishing with the one conveying it to you, are usually sent straight to the building society or bank as security for your building finance, and you never see them again until you pay off your mortgage.

Buying a building plot: checklist

Checks to be made on site before signing a contract. These are essentially the same as those made when appraising a site, listed on page 55, which you made to alert yourself to possible problems. This time you are making the checks to be sure you know all that you need to know before you sign the contract.

1 Boundary dimensions to be compared with the title plan, planning consent plan, etc.
2 Access to be compared with the access shown on the various plans.
3 Any visibility splays required to be checked on the ground, and especially if there is any risk that they cross the plot boundary.
4 Check access for building materials.
5 Foul drains - where is the existing sewer? What is the size, depth, etc.?
6 Surface water drains - any existing surface water drains, agricultural land drains, ditches? Is the land suitable for soakaways?
7 Ground conditions. Any sign of wet ground or unusual vegetation? Any sign of demolished buildings? Any sign of quarrying or spoil pits?
8 Trees. Tree preservation order situation. Species and heights of trees (which affect permissible distances from new buildings). Any sign of trees recently removed which will involve special foundations.
9 Services. Are water, gas, electricity, phone services really available? Are the costs of the services going to be acceptable?
10 Easements. Are any cables or pipes where they are supposed to be? At the correct depth?
11 Planning. Do you know exactly where you stand?

6 Designing a new home — the basics

Among many popular illusions concerning an individually built home is the idea that those building for themselves are able to express their own personalities by building homes which are uniquely their own, and in which they can indulge their wildest dreams. In fact all of them suffer many of the same constraints as the developer, and if they succeed in avoiding standardised features it will only be with considerable effort. All influences in modern society urge towards conformity. Many features of this conformity are desirable, and reflect the best in modern living standards, but others are imposed, directly or indirectly, by the planners, the dictates of finance, and fashion. This situation ensures that a new home is the best possible investment.

For all who build for themselves the budget has a huge effect on the design. Invariably the aim is to build as large a conventional building as possible within the budget, which means the lowest cost per square foot which can be achieved using conventional construction methods. These two qualifications are important, for although revolutionary building techniques and experimental designs are sometimes claimed to offer very low costs, they invariably present other difficulties, particularly over mortgage valuations. Unconventional design and new materials or building techniques make good media copy, but they are very, very unusual.

Achieving the low costs goal depends on the basic approach to the project. For instance, a level site is most important. A split level building or a ground floor built up above surrounding land on a sloping site can add significantly to the cost of the most simple structure. This does not rule out building on a sloping site, but to minimise costs the building must be constructed on a terrace cut into the slope, rather than built out from the slope. Imaginative landscaping can make the two approaches look much the same, but the costs are very different. This is not to say that houses built halfway up a hill should always be built into the slope, rather that the cost implications of doing otherwise should be understood at an early stage.

Similarly, the shape of the building above the ground should be as cost effective as possible.

The cheapest structure to build is a rectangle with two end gables. Buildings with complex shapes, complex roofs, gable projections and hipped roofs all increase unit cost. Unfortunately these are the very features which characterise today's styles, so that the challenge to a designer is to incorporate them into plans in the most cost effective way. Doing this is a very technical business. For instance, gable dormers with windows below the wallplate are cheap, while gable dormers with windows breaking through the wallplate are expensive. This sort of matter is beyond the scope of this book, but it is dealt with at length in the two companion books which are featured on page 299.

Unless one has experience in designing houses, it is certainly not a field for do-it-yourself. Virtually all who build for themselves use the services of architects, experienced commercial designers, or recognised design services. This chapter is written for readers who want to know as much as possible about their design options before they discuss them with the person who is actually going to draw the plans.

Any logical consideration of the design of a new home must start by finding out the likely size of the building that can be erected within the budget, and the nature of the site on which it is going to be built. The site determines all, for the size, shape, slope, orientation and much else about the plot on which you hope to build will determine what you build there. The appearance of neighbouring buildings and the general character of the area will determine what the planners will let you build there. However, if you are going to buy a particular plot you must be able to visualise what can be built on it . This means that all that follows should be read to explain what you should have in mind when making a quick decision about whether or not to give further thought to a plot for sale. If the plot does actually turn out to be really the site of your new house, what follows should help you to consider the design features of the house which will be built on it.

Let us start by assuming that you have a level plot which does not have any special site problems or any special advantages which dominate the situation - circumstances like an absence of drains on the one hand, or a marvellous view on the other. The first issue is the size of home which you are likely to be able to build, and how you determine this has already been discussed in an earlier chapter.

The next thing to consider is whether the home which you are going to build will be the ultimate version or whether it is likely that you will extend or adapt it in the future. If it is likely that at some future time you will add a granny flat, or turn an internal garage into a games room and build a new detached garage, or simply that you hope to add a conservatory when you can afford it, this ambition should be recognised and provided for in your design. Indeed, whether or not you intend to build a garage at all at an early stage, is an important decision and this is discussed in detail later.

The external appearance and the internal layout of a house or bungalow are dealt with separately. The appearance of the property, together with its size and position, is very much the concern of the planners, while your interest is that it should look the way that you want it to look, and that it should be a good investment. You will also be very concerned about the costs of all of the options. However, it is likely that both you and the planners will want to build in a well established and local style which is commonly used for new homes in the area: the planners because they are concerned to protect that style, and you because it is probably why you are interested in building in the area anyway. Remember that a house in the appropriate local style will optimise your investment.

The styles of houses built in Britain since the Industrial Revolution have varied enormously, and a little of the history of this may help to give a better understanding of the design scene today.

Most homes over 120 years old were constructed in local materials, and local design styles reflected the best way of using local materials. Often these materials were relatively expensive. Then the railways made most of the country a single market for building materials, so that cheap Fletton bricks and cheap Welsh slates lowered the real cost of building. Design reflected social attitudes and economic conditions, so that in urban areas we had acres of back to back housing designed to house workers as cheaply as possible. The more affluent, who then lived in the town centres, were concerned that their homes made a statement about their relative success in life, and this determined the appearance of Victorian town houses.

After the turn of the century the motor car and suburban commuter trains enabled the affluent to escape from the towns into the suburbs, where developers provided them with standardised housing that was designed to be affordable to buy and run. This housing was well promoted with image advertising. As a result, most houses reflected a supposed popular taste which had actually been established by the developers

By the 1930s there was a marked distinction between the avant garde homes considered by architects as being the influential designs of that decade, and the design of popular housing generally. The homes admired by the architects had flat roofs, architectural use of glass, rounded corners and a German name - the Bauhaus style. Meanwhile, in this era of no planning constraints, builders and developers were building mock Tudor homes which were as different from the architects homes as they could be. Then came the war, and after it the planning acts arrived, although as far as domestic architecture was concerned the new planners had little idea what they wanted.

The architects continued to experiment with what they thought was important, seeking to express the essentials of function in bricks and mortar, avoiding unnecessary decoration and contrived features. At the same time the developers were offering homes with simple (and cheap) shapes, and with low pitched roofs and picture windows (as in Hollywood films). Later in this period contrasting panels of timber and plastic became almost obligatory.

The next factor in all of this was the reorganisation of local government in 1974, which gave the new planning departments in the new local authorities an opportunity to spread their wings. County design guides were produced which set out clearly and unambiguously exactly what could be expected to be approved and what could not, and these had a far reaching effect on the design of low density housing.

The design guides were drawn up when there was no coherent establishment style for popular low density housing, and in the absence of anything better they looked back to the early nineteenth century and the essential elements of the local buildings of that era. In Nottinghamshire, farm houses used to have gable roofs with steep pitches, so hipped roofs and low pitched roofs were forbidden in Sherwood Forest. Essex cottages once had black boarded gables and dark stained joinery: these features became essential to speculators developments in Basildon. In 1830 no builder had stuck panels of contrasting materials below windows, and so this 1960s fashion came to an abrupt halt. An even more far reaching innovation was the complex shaped and

An understanding of recent fashions in house design is useful if you are choosing a design for your new home. The house above was the most popular D & M four bedroom design in the early seventies, while below is a typical house of the same size built by D & M clients in the nineties.

An individual home in Kent in the local regional style.

An individual home in Hampshire.

involved roof lines which were introduced to give interest to a design, in marked contrast to the simple shapes and functional structures of the previous decade.

Architects protested vigorously against the idea of design guides, but, facing commercial pressures, they hastily became heavily involved in the minutiae of the regional styles. However, their dislike of aesthetic controls resulted in the planners' views on design issues being tested by hundreds of planning appeals, and by the mid 1980s a mixture of appeal findings and ministerial directives established that the duty of the planners was to approve what was acceptable, and not what they considered desirable. The spirit of the county planning guides was to live on, but only as a guide, and not as a rule book as it had been when it was first published.

Meanwhile the public was delighted with the move to new homes that were prettier and more interesting than the stark homes of the 1960s and 1970s, and the building industry, which was by then facing a slump in house sales, was quick to give the public what they wanted. As a result we now have Tudor homes, Georgian homes, Victorian style homes, homes in regional styles and post-modernist homes. The variety and choice is in keeping with the spirit of the times, and it is hard to imagine that 'function' is ever likely to return as a key factor in house design. As

long as we have a consumer market the popular view of what looks nice will determine what is built. Something of the options open today can be seen in the box on the following pages.

Surprisingly, the costs involved in building in any of the design styles featured are likely to be remarkably similar. Differences in costs between modern homes in any style will depend mainly on the complexity of the roofs and the roofing materials chosen, and on the walling materials. A house to any design that has a roof which is based on trussed rafters and is tiled with popularly priced tiles will be many thousands of pounds cheaper than a home in a similar style with a purlin roof and clay tiles or stone slates. Handmade bricks or natural stone will add significantly to the appearance and value of a home, but will add even more significantly to the cost. Whether the building is in the distinctive East Anglian regional style, or is a half timbered home based on a Warwickshire original, or a twentieth century version of the Georgian style, will not affect the cost nearly as much as how the roofs are engineered and cost of the materials which are used for the envelope of the structure.

House or bungalow?

At this stage it is appropriate to look at the position regarding bungalows, and also the role of the garage in the design of a new home.

Georgian

The Georgian look covers everything from houses which are reproductions of real Georgian homes to bungalows with full bar windows and a hip roof. Rules for the real thing are:

- The proportions of the building, and the proportions and spacing of the windows are the key to the Georgian style.

- Sash windows, or windows made to look like sash windows, are essential. Eaves details are important.

- Georgian homes were restrained — an elaborate porch is appropriate to only the grandest home.

- Real Georgian homes had neither patio windows nor 'Georgian' garage doors. Use appropriate French windows, and side hung folding garage doors that mimic those on Georgian Coach Houses.

Tudor

The word 'Tudor Home' usually refers to a style established just before the First World War which is very loosely based on Elizabethan half timbered homes. As used today it is applied to any house with half timbering, a vertical emphasis to the windows, and a steeply pitched roof. The more complex the shape of the home, the better. Projecting first floors, known as jettied floors, simple solid doors made from vertical planks, and windows with heavy mullions are part of the style, with big external chimneys and inglenook fireplaces.

Design Types

The names describing these house styles are those commonly used by estate agents and almost everybody else except architects. The descriptions of the styles will horrify architectural purists, but are in general use.

Regional Design — Kent and Sussex

Tile hanging, complicated roofs with dormers, windows with a vertical emphasis. Brickwork in multi-coloured stock bricks, white render, or white-painted brickwork, often on a brick plinth.

Regional Design — Cotswolds

Stone houses, often in complex shapes, usually with dormers set in steeply pitched roofs which should properly be clad with stone slates. Leaded lights in small windows. Often built with a brick or stone plinth to knee height.

Regional Design — East Anglia

White or cream render, sometimes with a design in the plaster called pargetting. Woodwork often painted black. Projecting timber window bays.

Victorian

The Victorian house is coming back into fashion. Flamboyant use of ornamental brickwork, with steep pitched roofs clad in slate or in small plain tiles laid to patterns. Ornamental ridge tiles and bargeboards.

Rural Design — Northern England

Simple severe shapes in stone or brick, gable roofs with slates, often with tabling to the gables. Simple porches to front and back doors. Windows with a strong vertical emphasis.

Post-Modernist

The functional "modern" architects' approach of the 60s and 70s softened by using less glass and more natural materials, usually brick. Careful attention to detail shows what could have been done with the shapes of the modernist era.

Cottage Style

Very small homes in rural areas are once again being designed to look pretty, with any mix of styles as long as they are easy on the eye. Pundits may sneer at them as gingerbread cottages: most people find them charming.

Suburban Design

Planners considering applications to build in a suburb usually consider whether new properties enhance the 'street scene' rather than bothering about a local style or the merits of the individual building.

Regional design styles — Regional design requirements can be met in timber framed construction just as easily as with traditional construction.

A timber framed house by Prestoplan — this time in an appropriate style for a site in East Anglia.

This timber frame individual home by Prestoplan was built in a Northumberland village.

Typical bungalow which is not in any style, but is usually acceptable if it is not in a conservation area and there are other bungalows in the vicinity.

A major consideration in designing a new home is whether you wish to live in a house or a bungalow, using the word "bungalow" to describe a dwelling on a single floor. Dormer bungalows with two floor levels are really houses with low eaves lines. The word bungalow is Hindi and it was brought home from India by tea planters in the last century when single storey homes were a novelty except as gatehouses on ducal estates. Admittedly the Welsh longhouse and the Scottish bothy have a long history, but in most parts of the country the yeoman's home was two stories high just as soon as he could afford more than a hovel, and the bungalow was certainly not part of our architectural heritage. However, its convenience and utility have now given it its own place in our domestic architecture, and it is here to stay. In many areas, particularly in Wales and the South West, bungalows are now well established and accepted without question by the planners. Everywhere else the rule is that unless you are building among other bungalows you must assume that you will have to argue your case. The number of bungalows built every year shows that it is often a perfectly good case.

If you think that you may be able to get consent to build a single storey dwelling on your site, what are the pros and cons of bungalows versus houses? A major factor is that a large bungalow requires a plot with a very wide frontage, so that if you want four bedrooms you will be obliged to build a house unless you find a plot at least 60' wide. If you find one, then other considerations are set out on page 74.

Consideration of the house/bungalow option invariably leads to discussion of dormer bungalows and to the possibility of building a bungalow which is designed to allow extra rooms to be put in the roof at a later date. This is very rarely as simple as it may seem to be.

The rectangular dormer bungalow with first floor windows in the gable end and a flat roofed dormer window in one roof slope had an enormous vogue 20 years ago, and it is seen everywhere in suburban areas throughout the country. At that time the statutory requirements and lower labour costs combined to make this a particularly cheap way to provide four bedrooms on a narrow

plot. Circumstances have since changed, and this particular design style no longer offers significant advantages. Of course, there are very many homes which have the roofs sweeping down to the first floor level with first floor windows set in the roof but they are properly considered to be houses.

Garages

Moving on from the house/bungalow option, another basic element in the design of a new home is what you do about a garage.

Garages are expensive luxuries which are built to accommodate a car which is completely weather proof and will depreciate no more quickly if kept outside. Keeping a car inside is a hangover from the days when motor vehicles were less reliable and the chauffeur or proud owner wiped them down after every trip out in the rain. A garage does enable you to get into a car that is marginally warmer on a cold morning, and you can unload groceries under cover when it is raining, but the cost of these advantages is considerable. Nevertheless, having a garage is part of the general pattern of housing in this country, however illogical, and you are unlikely to be easily deterred from building one. If so, there is a lot more to think about than simply making sure that the floor slopes from front to back so that the rain water from the car will run out under the door.

The first decision you will have to make is whether you want a detached garage, or one that is integral with your house. If you are building on a plot with a narrow frontage you may not have any option in this, but if you do have a choice it requires considerable thought. There is no doubt that in many ways an integral garage is more convenient, with a connecting door into the house which ensures that you do not get wet when it is raining. For many people, particularly the elderly, this is an important consideration and one which is often given priority.

On the other hand, there is no doubt that a detached garage has many advantages. To start with the separate building gives complexity to the appearance of the whole homestead, and a house with a garage which is a subsidiary building, perhaps linked with it by a wall, looks much more imposing than a single building. This can often enable the drive to be kept well clear of the house, with a path through a lawn or flower beds leading to the front door. In addition to this, a detached garage is usually significantly cheaper than an integral garage, as it can be built on simple foundations with uninsulated solid walls, while an integral garage has generally to conform with the building regulations requirements for a dwelling.

The arguments for and against a detached garage are summarised on page 75.

If you decide to build a double garage you have an interesting opportunity to make it look much more interesting than the simple rectangular box that is usually provided by developers. You should also consider whether your detached garage should not have some other function. If you want to use it as a workshop, as a garden shed, or to provide office or guest accommodation this can be easily arranged, and an example is shown opposite. This is the Carsholme design, which is from the book *Home Plans*, and has been built hundreds of times up and down the country with the space in the roof providing guest or Bed and Breakfast accommodation, used as an office, as a studio, and as a games room.

Carports

The alternative to providing a garage is building a carport, or simply making provision in the turning area to your drive for a permanent parking area. The latter is by far the cheapest

arrangement, and if you give proper consideration to it before the drive is constructed it can look an attractive and integral part of your landscaping.

Carports are very popular abroad, particularly in the parts of Europe and the USA, where it is desirable that a car is parked under cover to be out of the sun in very hot weather. The only logical argument in favour of them in Britain that they enable people to get in and out of their cars under cover when it is raining. The essential point is to make sure that they are seen to be an integral part of the structure, and not an after thought. This is best achieved by making sure that the uprights are really massive, with a heavy fascia. Carports are usually roofed with translucent material, and while this adds significantly to their utility it is very difficult to use this material in a way which enhances the appearance of the property, and thus the value of the property. Ornamental screen walling blocks are sometimes used for carports, but the situations where this can be done successfully are few and far between.

Designing a garage

Turning to garages, start by considering the optimum size. Even if you are not going to use it as a store for garden equipment etc, and you have no use for it other than as a shelter for your car or cars, making the right decision about the dimensions is very important. This is where most speculative builders decide to economise and they build garages which will exactly accommodate a Ford Granada, with room for the doors to open just wide enough for a contortionist to get out of the drivers seat after having dropped his passenger outside in the rain. Typical internal dimensions are 8'6" x 17', permitting an 9" reveal on either side of a 7' up and over door which is set behind the door jambs. If the doors are set in a frame within the jambs then you have another 6" of width, and this is a good idea. If you can afford something better than this it is well worth giving the garage an internal length of 20 feet and considering an 8 foot garage door. These doors are now available as standard from many manufacturers, and the off-square proportions look very attractive.

The smallest double doors are 14 feet wide,

The garage on the right has an unusual and attractive appearance and is much more appropriate to a rural situation than the usual rectangular structure.

A double garage in an area where a steeply pitched roof is appropriate offers lots of opportunities. The upper floor of the design (right) is shown here fitted out as a study.

and make it inevitable that a carelessly opened door on one car will bang into the side of another car which is already in the garage. Two single doors with a substantial pillar between them look a great deal better, and give you room to move around the parked vehicles without transferring mud from them to your clothes in bad weather.

There is a very wide choice of styles of up and over garage doors these days, as well as an interesting trend back to side hung garage doors and newly arrived roller shutter doors. With the exception of the side hung and roller doors, all of the up and over doors have one of three different sort of 'up and over' track systems. The most expensive of these involve horizontal tracks at head height which are more robust, open the door more smoothly, and are also easy to mechanise with an electric garage door operator. If Britain is to follow overseas trends we will find that garage door openers will become commonplace within a decade, and even if you are not considering them now it makes sense to ensure that they can be fitted in the future. They cost about £200 each complete with radio control. If you are thinking about buying one, remember that a separate pedestrian door to the garage will prevent a malfunctioning operator trapping a child or disabled person in the garage until someone chances to hear their calls for help. You will also want to be sure that a power cut will not leave you on the other side of a door from your car!

It is usual to provide at least one window and pedestrian door in the garage, and this is where you should resist any temptation to economise. Garage joinery should match the joinery in the main house and be of the same quality and finished in the same way. Care should be given to the height of the window, as it is usual for the level of the window sill to match the level of the windows in the house, while the floor of the garage is invariably 6" lower than the house.

It is also worth making a decision at an early stage about how you intend to complete the garage. Will you plaster it out, with a ceiling, or is fair face blockwork and exposed rafters all that you want? Will you have a concrete floor or something more attractive? And if you are having an integral garage, how will you treat the six

House or Bungalow?

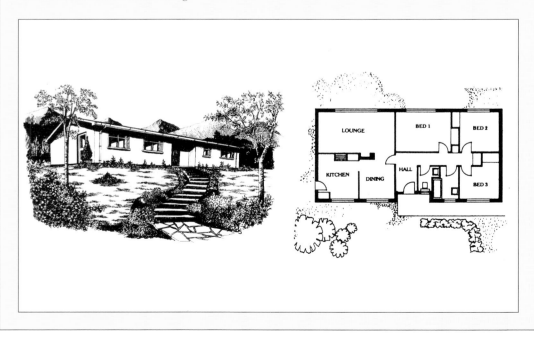

* A bungalow is likely to cost marginally more to build than a house of the same floor area.

* In most parts of the country a bungalow is usually worth rather more than a house of the same size. If this is an important factor, you should check it out with a local estate agent.

* A bungalow will take up more of a plot area than a house, and this can lead to it looking cramped between the boundaries. If your frontage is limited, a bungalow may not have the feel of gracious living that comes with a bigger garden around a house.

* Bungalows usually have less waste space than a house if you regard the staircase well and the landing as wasted space.

* Bungalows generally permit more flexible layouts as fewer walls have to be load bearing.

* Split level and multi level homes are more easily arranged as bungalows, and so are changes in ceiling height and sloping ceilings.

* If provision is to be made for a home to be extended at a later date, the ease with which this can be arranged, and how much room on the site is available for it, is often a key factor in deciding to build a house or bungalow.

* Routine maintenance work like painting and cleaning gutters is more easily dealt with by the average householder if he is working on a bungalow. This often appears to be a major consideration with those faced with making a choice.

* Finally the "no stairs" factor is enormously important to many elderly persons.

Integral or detached garage?

* A detached garage can be positioned in relation to a house or bungalow so that the group of two buildings look better than a single building with an integral garage. A wall linking the two structures can add to this effect.

* A detached garage can be sited in the most logical position in relation to the drive, leaving the house to be sited in the most logical position relative to the view or other considerations.

* A detached garage is cheaper to build than an integral garage, as it does not require to meet the building regulations for a dwelling.

* A detached garage can be built after the house or bungalow is constructed, or when it is approaching completion. If it makes financial sense to build it at some time in the future, or if you simply want to be sure that the house is built within its budget before you start the garage, this may be useful.

* It is more expensive to put a water tap, electricity or an outside w.c. to a detached garage than it is into an integral garage.

* An integral garage provides somewhere to put the central heating boiler, or an extra freezer, and in many ways tends to be useful simply because it is part of the main building.

* An overwhelming advantage of an integral garage for the elderly or infirm is the opportunity to get in and out of the car "out of the weather".

inch step down from the house floor level to the garage floor level which is required by the Building Regulations?

Building a detached garage also gives you an opportunity to incorporate all the other accommodation that your lifestyle requires. If you want to keep a lawn mower, camping equipment or children's toys in the garage, then consider whether it should simply be a little larger than usual to accommodate them, or should have a separate store. If an outside lavatory is required, or may be required in the future, then make provision for it in the garage building. If you want a workshop, then planning it properly will make working in it much more enjoyable.

Before you get carried away it is worth reflecting that a large garden shed, situated where it is seen to be part of the garden and integrated into your landscaping with a trellis or pergola, can provide storage space at a lower cost than a similar area in your garage. An extra 20 square feet of garage space is likely to cost you £500, which will buy you a very large good quality garden shed with a wooden floor at the bottom of the garden. The choice is yours.

Garage services

Services are also important. There should be plenty of lights, both internally and externally, including at least one which ensures that the interior of the boot is well illuminated when you are unloading at night. Switches should be situated at both the main door, the pedestrian door, and, if appropriate, inside the house. If a security light with one of the new automatic switching systems is required, or is likely to be required in the near future, then provision should be made for it to be wired separately. Power points should be provided where they are going to be convenient for power tools, for a vacuum cleaner to clean the car, and also where they can be used for electric garden tools if there are no other external sockets. A garage is often a convenient place for an outside tap, sometimes with arrangements for the piping to run in the garage where there is some measure of frost protection, terminating at a position where there can be one tap inside the garage and another

outside the garage for a garden hose. Finally, when a trench is open between a detached garage and the house, drop a length of multicore telephone cable in it: in the future it may be useful for an intercom, a burglar alarm or some other device that has not been invented yet!

External design

Having given thought to whether your new home should be a house or a bungalow, and the type of garage which you will probably want, let us return to the style in which you want to build. Some of the options were shown on pages 68 and 69.

Whichever style is adopted, there are a number of aspects of the appearance of the home which require careful attention. The first is the front entrance. Most people like their front door to be the focal point of the front of the building, usually accentuated with a porch of some sort. Further emphasis can be provided by using double doors. We are very bad at designing impressive but restrained front entrances in this country - the Americans are far better at it. Whenever you look at a house design, look at the front doorway and ask yourself if it does the building justice.

Another major feature in the front elevation of many new homes is a garage door, while at the back there is often a patio window. Both of these are wholly unnatural elements in a house in any of our traditional styles, but are often rightly considered essential to today's living pattern. Integrating them into a traditional design is a challenge, and success usually comes with the way in which they are designed into the shell.

Drawing up plans for a house in the style which you prefer, incorporating all the features that you want, and doing so in a way that will enable your house to be built within your budget is a job for your architect or designer. Your own task is to become aware of all of your options, and to decide on the design style and features which you want. There is a very simple way of doing this: start making a scrap book, and, if you don't already have one, buy a very cheap camera.

At one end of the scrap book start pasting in pictures and photographs of houses which you particularly like which are about the size of the home that you intend to build. You can get illustrations from books, magazines, and estate agents brochures. You can take photographs through your car windows, and however badly they come out, they will be clear enough to be seen and understood. Besides collecting illustrations of what you like, consider what you don't like. If you see a feature which you think spoils a house which you otherwise like, include it as well.

At the other end of the book start pasting in pictures of details: a front door, a way in which a garage is linked to a house with an interesting stone wall, a particular pattern of brickwork or the way in which roof tiles are kicked up at the eaves in the Spanish style. If you do this, you will both stimulate your own interest and also give yourself a good way of explaining exactly what you do and do not want to your architect or designer.

Books of house and bungalow plans can have a useful role in making your design choices. Many people buy them simply to make a choice from the hundreds of designs shown and they then order a full set of drawings from the publishers. A better use of them is as a source of design ideas, and as a way of deciding which of the design companies you will approach about getting something drawn to your own requirements.

However, useful as the plan books are, they are no substitute for spending afternoons driving round the part of the country where you want to live, snatching photographs of houses which you admire. A lot of bother perhaps? Certainly. And buying all the books of plans will cost a penny or two. But you are unlikely to build many homes, and you surely want to make sure that you know exactly what you want. When you do know exactly what you want, you can be firmly in the driving seat when you are talking to your own architect or the council's planning officer.

Internal design

Happily the planners will play no part in the internal design. Here you have far wider opportunities to construct rooms to suit your own living pattern. For example, it will cost little more to have a family living room, dining room, and a small TV room than to have a large open plan living space, and this may be the way in which you want to live. The choice can be wholly yours. If you do want something unusual, perhaps a drawing room with a sunken floor and an island fireplace, it is simply a matter of being able to afford it, and you should be able to get a clear idea of the costs at an early stage.

Once again, make a scrap book or keep a note book with details of every idea you come across that might be right for you. It will definitely be worth your while. Some of the details that you should make up your mind about are overleaf.

Design details checklist

* Start with the hall. There is a modern trend to make this as large as possible. If your guests come straight into a room that has a table in the middle with a flower arrangement on it, and doors off all the way round, it is a lot more impressive than stepping into the usual cramped space at the side of the stairs. In the USA, the dining hall is very popular and is now occasionally asked for in this country. You either like the idea or you don't.

be? Double doors and the ceiling 9" higher than usual to give a feeling of authority, or a cosy room with a fireplace set diagonally in the corner? Fireplaces are important: whether they are Inglenooks, classical Georgian, or built in a stone faced wall they will look their best if included in the original design concept.

* If you want patio doors leading out onto a terrace, give some thought to how you will arrange your furniture in relation to them.

This bungalow has a number of interesting design features. The carport is an integral part of the main building; there is a dining hall; the larger of the two bedrooms is en suite with a sunken bath; and there is a lot of storage accommodation. Building for yourself provides an opportunity to have a home that suits your own personal living pattern. Make sure you take advantage of this.

* Staircases give tremendous character to a house, and even the most expensive of them will cost less than your kitchen units. A splendid staircase is probably the cheapest way to make the definitive statement about your home to your guests!

* Moving to your living area. Do you want the largest possible space, with just an arch through to the dining area, or do you want a drawing room with a separate dining room? If you are building a house of over 1800 square feet you can probably have a T.V. room or study as well, and at over 2000 square feet you can have both.

* What do you want the feel of these rooms to

Consider the look of the terrace from your living room, because that is the view you will have of it for most of the time.

* How many people will be able to sit around the table in your dining room after you have put a sideboard in there? Does this suit your lifestyle? Some designs in the plan books have very small dining rooms, and do not indicate the limited number of people that can sit around a table in them.

* Move on to the kitchen. Fashions change: once the vogue was for smallish kitchens and a separate utility room. Now a larger kitchen with a table in the middle is popular, called a farmhouse kitchen. A combined kitchen/

living room or kitchen/dining room is all the rage across the Atlantic: it is supposed to promote family awareness that the cook is not a skivvy, and the trend will eventually come here. Take your pick.

* Consider a projecting back porch instead of a utility room inside the house: it will be a cheaper way of providing the space and can be built later if you wish. Projecting porches also add interest to the appearance of a

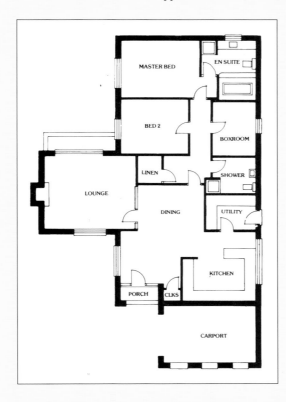

building, and are currently fashionable.

* Do you want a downstairs cloakroom with a lavatory? If so, should it be convenient for the front door, or for the back door? (If it is at the back door, remember that the lavatory must have a ventilated lobby between it and the kitchen.)

* Where are you providing a cloaks cupboard. Also remember somewhere to park a pram: if you have no plans in that direction yourself, a couple in the house at some time in the future may decide to reproduce themselves

* Where are you going to keep the vacuum cleaner and other household objects? No family ever has too much storage space.

* Moving up the stairs, do you want your designer to look at the possibility of keeping the master bedroom suite completely separate from the other bedrooms? For some people this is a valued feature: for others it is of no account at all. What about you?

* Staying with the master suite, how large do you want it to be compared with the other rooms? Some families building a three or four bedroom house want the master suite to be as large as possible, with the other bedrooms relatively modest in size. Others prefer a more egalitarian arrangement.

* The same decision has to be made if you are having two bathrooms. Until recently the norm was for a large family bathroom and a modest bathroom or shower room en-suite with the master bedroom. This is changing: the issue today is which bathroom will be used most, and who is paying for it anyway.

* The family and guest bedrooms. Need they all be the same size, or would you like one prestige double room with other single rooms relatively small?

* Dressing rooms to master bedrooms are popular, but need careful thought.

* Built in furniture or alcoves for free standing furniture? The former hold twice as much but the later are fashionable. On cost grounds built in furniture will win every time.

* Can you use the landing - for a desk or a sewing table for example?

* Many people who build for themselves want really huge walk-in airing cupboards which can hold all the family linen, etc. If it is big enough it can house a second vacuum cleaner as well, saving carrying one up the stairs thousands of times in a lifetime!

* Many designs in plan books show first floor balconies. Before enthusing about this ask yourself how many balconies are actually used in this country.

All this is applicable to any new home on any building plot, but of course every plot is different and every individual builder has their own individual requirements. How to deal with the special situations you may encounter is dealt with in the following chapters.

7 Designing a new home — to suit your site

As more and more of the countryside becomes covered with houses, and as concern grows to protect rural areas from further development, more and more homes are built on sites which have not been developed in the past because they were considered to have some sort of problem. Some are infill sites, typically a gap in the houses in a village street which was a duck pond until it was filled in during the war. It will require special foundations, and no-one has tackled this in the past because of the high costs involved. Now that almost any foundation problem can be overcome for £10,000 it becomes a very desirable building plot. Similar situations arise with steeply sloping sites, sites where there is no easy access to main drainage, sites with access problems, and sites covered by trees which carry preservation orders. If these sites are to be developed, these special factors will play a major part in determining the design of the home to be built on them.

Sloping sites

Consideration of the design of a house or bungalow on a sloping site depends largely on the part of the country in which it is situated. In parts of Wales, or in the West Riding of Yorkshire, homes are commonly built on sites with one in five slopes, and local styles and the local building practices are geared to this. In other parts of the country any slope at all is deemed to merit special consideration. Wherever you are going to build, the first thing to do is to have a careful look at how other people choose to build on slopes in the local area, and to try to analyse the basis for the regional practice.

This may depend on the ease with which excavations can be made. If there is rock just below the surface, it will probably determine that buildings are built out from the slope because of the high cost of quarrying into it. If the subsoil is easily excavated, there are many more options. It may be that local cottages nestle into the hillside because in earlier centuries they had to do so to escape strong winds which had an adverse effect on poor local building materials. This may have given local villages a particular style which the planners will expect you to accept.

Wherever you build, there are two approaches to be considered: should you arrange to remove the slope, or should you design a home to make use of the slope? If the site permits, it is invariably cheaper to excavate a level plinth for a new home, adjusting the levels and spreading the surplus soil as part of your landscaping. This involves either just digging into the slope, or else digging out part of the plinth and using the excavated material to raise the level of the other part. This latter approach is called "cut and fill" and the sketch on page 82 shows what is involved.

Digging out a level plinth is not always possible, sometimes because the site is too steep or the ground too rocky, but usually because the plot is too small to allow for the necessary changes of level. Remember that you cannot excavate close up to your neighbour's fence: in law his land is entitled to support from your land. In this case you have to consider a design to make use of the slope, which will usually mean a multi level home.

A property of this sort is more expensive to construct than one which provides the same living accommodation on a level plinth, and changes of levels in a home are more attractive to the younger generation than to the elderly. This may affect the resale potential, and it is generally true that while split level homes are often exceptionally attractive and lend themselves to exciting decor, they often have a limited resale market.

If a plot slopes to any extent this will also affect the design of the drive. It will be a much more complicated matter than you may at first imagine. If your drive will join an A or B class road the Local Authority will probably want a visibility splay at the point where the drive joins the road, and in hilly country this can involve expensive excavations. The authority will probably insist that your gates are set back a car's length from the road, and they will impose maximum gradients on the drive. They will also insist that rainwater from your drive should not cascade across the pavement, which is often difficult to arrange. There is obviously a maximum slope which your car can negotiate, but long before you reach this limit, problems will arise in designing the drive so that your car does not "hang up" when making the change of gradients as it gets onto the road, or into the garage.

The appearance of a drive is always particularly important with a sloping site. Any drive is almost always the largest feature in the front garden, and it has to be designed with the landscaping proposals in mind. If the site slopes, then the drive itself becomes even more prominent, and must be given even more careful attention.

New homes on sloping sites often involve special foundations. In simple terms, your designer has to make sure that the whole building will not slide down the slope. The cost of such foundations is one reason why many sloping sites have not been developed in the past, and are only now coming onto the market now that property values make the cost acceptable

One advantage of a sloping site is that it invariably comes with an interesting view. If you are deciding whether or not to buy a plot on a hillside, remember that the view that you have from the ground level is not going to be the same view that you will enjoy through the windows of the finished house. If the outlook is very important to you, do not hesitate to take a couple of step ladders to the site and make some sort of platform that will enable you to stand at a level from which you can see the view as it would be from the windows of a finished home. You may look rather ridiculous at the time, but if the view is a key factor in making your decisions, make sure that you see what you will really be getting!

Access

Whether your site is as level as a bowling green or best tackled wearing climbing boots, the arrangements for the access are a very major design consideration.

The whole question of access is dealt with at length in the chapter on finding somewhere to build, and you may find that physical or legal constraints on the access will determine both the level and the position of the new home. The smaller the site, the more critical this can be. If you turn back to page 50 you will see a typical access layout required by a Road Authority, and while this may be easily arranged on a large level site with a building line set back some distance from the road, it could present major problems with a small sloping plot. Every such situation is different. If you fear any sort of problem with the access it will be a major factor in the design of your new home.

Drainage

You will have a similar situation if you have found a marvellous plot on which to build, but it has not got any mains drainage. The person who is selling it tells you there will be no trouble, as everyone in the area has septic tanks. This may be so, but it will be a major consideration in your development of the land, and it is important to understand clearly what is going to be involved. First of all, the good news is that all the ways of dealing with the problem are now much cheaper as a percentage of the total cost of building a home than they were a few years ago, and the current solutions are also much more effective. This is because of the use of plastics and fibreglass for the underground tanks that are involved, which have replaced the old brick tanks. All the tanks arrive as prefabricated units which are simply dropped into place in a suitable hole and connected up. On the other hand, concern with pollution by the National Rivers Authority means that everything to do with the discharge of sewage is much more carefully controlled, and there maybe a considerable amount of paperwork involved. Let us examine this first.

It may seem logical to look up the address of the National Rivers Authority in your local telephone directory and to contact them for advice. Beware. They have a number of important roles to play in our society, but keeping down the cost of your new home is definitely not one of them. The time to get in touch with them is when you have taken the best advice and are making a submission for the solution to your problem that suits you, hoping that they will find it acceptable.

If there is an existing drainage arrangement on site, perhaps an old septic tank, and it has been discharging the treated effluent for at least twenty years and is still working, it may be that

Building on a slope

You have four options

Build up from the slope

You can build up from the natural ground level, and import soil to conceal the foundation walls below floor level . . .

Build into the slope

. . . or you can excavate a level plinth for the home, which will usually give you a retaining wall to look at out of the rear windows, and also involve you in finding somewhere to put the excavated soil . . .

Cut and fill

. . . or you can 'cut and fill', using soil excavated at the back to build up a platform at the front. This is the usual approach . . .

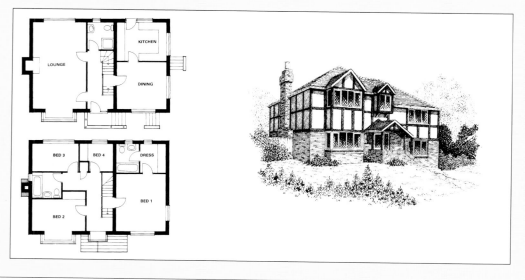

Multi-level

. . . or you can build a split level home. This house is built sideways across a slope, and has four levels. Inside the front door you turn left into the living room, or you go down three steps to the dining room and kitchen. If you go up the stairs you pass the master bedroom suite after nine steps and find the other bedrooms after four steps more. This makes an interesting home, but may make it difficult to sell.

This home built on a slope has four levels — the garage level, the ground floor levels, then the master bedroom suite over the garage, and then the remaining bedrooms over the drawing room and kitchen.

This home built across a slope has a level plinth giving exposed under-building on the right. This is fine with stone but rarely looks right in brickwork.

you have an implied right to the discharge, and can simply put in a new septic tank to replace the old one as routine maintenance of the system. This is a very uncertain area in the law, and the best way to deal with it is simply to make your planning and building regulation applications showing the drainage going down a pipe which is marked on the drawings "discharge to existing septic tank".

Now at this point you may stop to consider that you are not moving into the area in order to initiate a cholera outbreak, and that you do not intend to do anything anti social. Is this concern with the cheapest way of dealing with the problem entirely ethical? Are you taking the right approach to all of this? Well, the Rivers Authority has plenty of quite separate powers to deal with polluters: you are simply trying to keep your costs down by not spending money on a system which is more elaborate than necessary.

But what do you do if there is no existing tank? Basically there are four possible ways in which you can deal with the situation.

If there is a main drain in the area which is too far away to be reached by ordinary drains, or which is above the level of the new house, you can install a pumped system, and this is by far the best thing to do if it can be arranged. It involves a small holding tank which is fitted with twin pumps and macerators which break up the raw sewage and pump it along a small bore pipe to the main drain which you cannot reach with ordinary drains. These packaged units are hidden below ground and can pump to a drain which can be half a mile away, or perhaps 50 or 60 feet uphill. There are twin pumps and switching gear so that one pump takes over from the other if there should be a problem. All that is visible of such a system is a small slab of concrete with an access hatch at ground level.

Although they can move sewage for a long distance, these sewage package pumping systems are usually used where the mains drain is only a few metres uphill of the site, in which case the cost of the equipment will be quite modest. The expense of installing it is likely to be little more than the cost of an equivalent run of ordinary drains. If you are pumping a long distance, then the biggest problem will be arranging an easement for your pipe if it has to go through other people's land. Fortunately only a continuous length of small bore piping is involved, and it can be laid for you by an agricultural drainage contractor using mole drain equipment. Installing a package deal pump is invariably the best option if it is at all possible, and it is well worth paying appropriately for any easements required. Apart from anything else, a pumped drainage system is going to be more attractive than any of the alternatives to anyone who is considering buying your house in the future.

If you cannot pump your sewage, then you will have to consider a cesspool, a septic tank, or a mini treatment system. The Rivers Authority will take the final decision on which of them you install and they will consider the nature of the subsoil and whether or not you are building in the catchment area of a water supply. If your site is on ground which drains well, and is not in a water catchment area, then they will probably be happy with a septic tank. If the conditions are not favourable they may ask you to install a septic tank with an additional filter, or a mini treatment plant. If the conditions are less favourable still, they will require you to have a cesspool. A cesspool is bad news, but it will enable you to build on the plot which you want.

Cesspools are simply great big tanks which hold your sewage until a vehicle comes to pump it out and take it away. The normal size of tank for a single dwelling is 18,000 litres, and it has to be emptied once a month, by either the local authority or a local contractor. The cost of each visit by the emptying vehicle can be anything from £40 to £100, and occasionally more. A modern cesspool will cost about £2,000 and the cost of the excavation and installation may be significant. A common problem is that a cesspool is often necessary because the subsoil is rock, making a septic tank unacceptable, and this will also make excavation work difficult. You should only arrange to install a cesspool when you are completely satisfied that there is no other possible way of dealing with your drainage.

A septic tank is a tiny sewage works which

requires no external power, involves no pumps, and houses millions of friendly bacteria which break down the sewage into a sterile effluent which is discharged into the ground. Septic tanks are quite small, requiring a hole about 8 feet across and 10 feet deep, and they can cost under £1,000. They are easily installed. They need to be pumped out by a small sludge tanker once a year, which will cost between £40 and £100. All that you see of them above ground is the manhole which gives access to the interior and you will very rarely have to open it.

If the ground conditions are not ideal, then the River Authority may require that your septic tank effluent passes into a filter of some sort before it is discharged into a system of herring-bone land drains. The authority will be very specific about the design and size of this herring-bone drainage arrangement, and if you do not have room for it on your own land you may find that you will have to negotiate an easement to install it on neighbouring land.

A mini treatment plant works like a septic tank but requires electricity to power pumps which help the bacteria to do their job. These plants are at least twice as expensive as a septic tank system, and they may have to have the sludge pumped out more often. If the subsurface drainage is so poor that the River Board demands a mini treatment plant, they will probably require a very complicated system of herringbone land drains to distribute the sterile effluent.

How will you know which of these options you are going to have to use? You will get a good idea if you find out what sort of system is being installed for other new houses in the area. Bear in mind that the Rivers Authority is a new creation, and that applications to build new houses only four or five years ago may have been dealt with less strictly than they are now. If at all possible, find someone who has just obtained the relevant consents in the area and pick their brains.

The River Authority will determine the drainage characteristics of your site by considering the results of percolation tests, and it is usual for applicants to make the test themselves and to offer their results to the authority. Once you get to this stage, go along to the library and look up section H2 of the current building regulations, and also ask for the 1983 edition of British Standard 6297. Alternatively, your architect will deal with all of this for you, but if so, do make sure that he is aware of all the new up-to-date equipment which is available. At some stage you should certainly make use of the advisory services offered by the leading manufacturers. A start can be made by sending for their literature, and all the photographs of sewage equipment in this chapter are from Conder or Clearwater whose addresses will be found at the back of this book. Bear in mind that the people who actually market the equipment see more problem drainage sites than anyone else.

One or two points about siting systems of this sort. First of all, access to them is very important. It is necessary for small tanker vehicles to be able to reach septic tanks and mini treatment tanks, and for very large tanker vehicles to be able to get to cesspools. This may seem to be an argument for putting your tank near to your gate, with manholes hidden behind the appropriate shrubbery. It is not as simple as this. The first Universal Law of Sewage Sludge Removal is that the tanker vehicle will always arrive just when the lady of the house has her guests arriving for a coffee morning. Bear in mind that this performance is going to be repeated at regular intervals forever, and you will see why siting the tank should be given very careful consideration. The system suppliers will be able to advise you on this, and in many ways it is the most important decision which has to be made in the whole business.

Possibly the most convenient position in which to put the tank so that it will suit the visiting tankers is one which precludes natural drainage to the herringbone dispersal system. In this case, it is worth considering whether you can put in an additional pumping arrangement to move the sterile effluent to where the land drains can be more easily arranged. Again, the tank manufacturers will advise you and the expense of the extra pumps is certainly worthwhile. Another point is the importance of making sure that the hatches on top of the tanks are absolutely secure. There should be no possibility of them being a hazard to small children, or providing an opportunity for horseplay by those who are older but who have a warped sense of humour.

Do not be tempted to cut corners when installing any of this sort of equipment, and particularly do not skimp on or omit the concrete which should surround the tank. If you do, and the water table rises after a sudden downfall, the tank will pop out of the ground like a giant and very evil toadstool. This is all good fun if the tank is level, as all that will happen is that the contents will flow back into your bathroom. However, if the tank is installed on a steep slope it will have the potential to roll downhill like a runaway juggernaut. It sounds funny, but if you are responsible for several thousand litres of raw sewage approaching your neighbours property in a huge plastic bottle at 90 miles an hour, you may find your popularity will suffer. The writer once saw the results of such an incident when a runaway septic tank from a farm on a Welsh hillside descended on the village below.

To continue with this fascinating subject, always give your new septic tank a good start. There is a great deal of folklore concerned with septic tanks, most of it because some strains of bugs are better at their job than others. The man in the pub, who is always such a splendid source of misleading advice for those building on their own, will have very strong views on this. In some areas it is held that a dead sheep is second to none as a way of getting a septic tank working, and possibly this was true in the days of the old brick tanks. These days it is not to be recommended, but there is a lot to be said for collecting a bucket full of the contents of another local septic tank and transporting it very carefully to your own septic tank. The key word here is 'carefully'. Even better, borrow a car for the job from the man in the pub!

Finally, some good news. When you install a septic tank, you can apply for your water rates to be reduced, and the saving can be considerably more than the annual cost of having your own system serviced.

A typical cesspool, the solution of last resort to a drainage problem.

Septic tanks are a much better proposition.

Conducting a percolation test — digging a standard size hole, filling it with water, and timing how long it takes to soak away.

Special foundations

Special foundations for problem sites need to be approved by the Local Authority, agreed by the NHBC or by Zurich CustomBuild (if they are involved), be appropriate to your technical resources as a selfbuilder, and be cost effective. Make sure that the last two considerations are not forgotten by your architect or designer, and ask him to explain all your options. They may include some of these approaches.

Reinforced strips
A simple solution to minor problems. If you are pouring your concrete yourself take expert advice on how to keep the mesh in position.

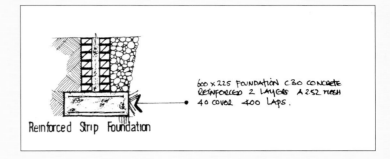

600 x 225 FOUNDATION C30 CONCRETE REINFORCED 2 LAYERS A 252 MESH 40 COVER 400 LAPS.

Reinforced Strip Foundation

Trench fill
Expensive in concrete but minimal labour costs. Popular with selfbuilders.

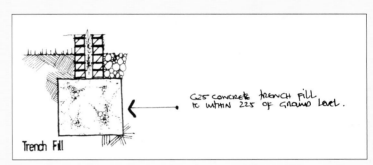

C25 CONCRETE TRENCH FILL TO WITHIN 225 OF GROUND LEVEL.

Trench Fill

Edge beam raft
Commonly used but a job for an experienced tradesman.

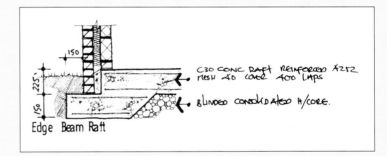

C30 CONC RAFT REINFORCED A252 MESH 40 COVER 400 LAPS

BLINDED CONSOLIDATED H/CORE.

Edge Beam Raft

Piled foundation
A specialist solution to difficult problems. Consultants and piling contractors will add significantly to your costs.

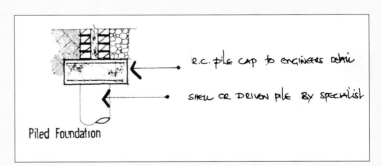

R.C. PILE CAP TO ENGINEERS DETAIL

SHELL OR DRIVEN PILE BY SPECIALIST

Piled Foundation

Subsidence raft
In coal mining areas the Local Authority will insist that you build on a raft like this, and if you still have subsidence problems you will have an automatic claim for compensation.

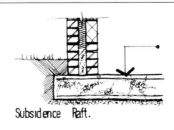

225 CONC RAFT C30 CONC. ON 25m BLINDING SAND WITH POLYTHENE SHEET SLIP MEMBRANE

Subsidence Raft.

Sketches by Robin Ashley RIBA, who sees more problem sites than most and whose work has been featured in previous editions of this book.

Foundations

Next, sites with special foundation situations. Presumably you knew about the problem before you decided to buy the plot, and you now have to ensure that your special foundations do not cost any more than is necessary. You need an expert. Where do you find him? Free advice is available from piling contractors and others who offer specialised services linked to one particular form of civil engineering, and they advertise in the Yellow Pages. However, it is wise to get a completely independent view, and this can be obtained from a soils engineer. Your architect or designer will advise you on your choice, but to enable you to surprise him with your grasp of this subject it is worth remembering that there are several ways of dealing with these problems some of which are shown on page 87.

These four approaches are just a start: there are more, and if the ground is clay and susceptible to shrinkage movement, or is on a geological fault, you may find yourself surrounding your foundations with compression strips which will deform if the ground moves sideways, or having a raft in two layers with a dry lubricant in the sandwich which will allow one layer to move over the other. None of this is as difficult as it sounds.

If you find that you are involved in a site which has real or potential foundation problems, remember that you are unlikely to be alone. Others in the area will almost certainly have met and dealt with similar problems, and invariably the local solution to a difficulty is the best one. A good place to start your enquiries is at the building inspector's office at the Town Hall. Unless you are submitting a formal planning application, the advice that you receive will be strictly informal, but you can usually expect to be pointed in the right direction.

Radon

A special foundation situation in some parts of the country results from the presence of a naturally occurring radioactive gas called radon which seeps from the ground. Radon is present everywhere in the atmosphere, and accounts for 50% of natural background radiation. (Less than 1% of background radiation comes from

Sellafield or Chernobyl or any other human activity.) Modern houses with good draught proofing can build up concentrations of radon which seep up through the foundations, and in some parts of the country this is now recognised as a health hazard, and makes a significant contribution to the statistics for deaths from lung cancer in those areas.

As a result there are special design requirements for houses built in areas where there is a high level of radon seepage, and Building Inspectors will advise on this as a matter of course. In England these are shown on the map, in Scotland these are mainly in Kilcardine, Deeside, Sutherland and Caithness, in Northern Ireland they are in the south east from Portaferry to Newry, while Wales is a natural nuclear free zone!

The precautions required involve making foundation slabs gas tight, and in some areas of high risk also providing ways for the radon to be discharged into the atmosphere. This is not complex, difficult or expensive but it has to be done. The government Radiological Protection Board has a range of free leaflets and booklets about this, and even offers test kits to indicate radon levels. You can contact them on 01235 831600 or fax 01235 833891.

Radon has been part of everyday living for the human race since the dawn of time, and until very recently our draughty houses meant only cave dwellers were at any risk from it. Modern houses built in certain areas without radon precautions would present a risk, but as the current Building Regulations completely remove the risk, which is very small anyway, all this is really of interest only to those with enquiring minds. If you go to live in Devon there really is very little risk of ending up glowing in the dark.

As long as you build on the right foundations there is no need to bother about radon at all, but if you want to know more there is a great deal of information readily available.

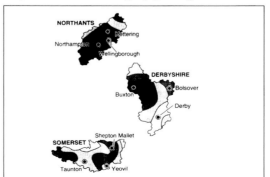

Radon hazard areas. There are others where the underlying rock is granite, but your local authority will inform you of the special foundations that are appropriate if you intend to build in a radon area.

A building site with a lake and mature trees: what more can an individual builder want? The answer is a very clear understanding of what is involved in building here! The earlier that you appraise a special foundation situation, the more cheaply you will be able to deal with it.

Trees

Trees have been discussed at length in chapter 4, and if relevant they should be a major consideration in establishing your design criteria, and if they are subject to a tree preservation order they will become part of your negotiations with the planning authority. As described on page 53, you will probably retain a tree surgeon or other specialist to handle this for you, but you will want to be involved in the negotiations.

If you are to plant new trees which will provide a back drop to the house, or just be a key feature in the garden, the choice of species is as important as your choice of bricks or tiles for the house itself, and should be made with just as much care. Trees have different shapes and grow to different heights, taking different lengths of time over growing to an appreciable size. Some are in leaf for all the year, some for seven months and some for only five months. Some prevent grass or shrubs growing underneath while others tolerate this. Make your choice an informed choice, take informed advice from a genuine expert and read books like *The Tree and Shrub Expert* by Dr. D G Hessayon in the PBI garden series or John Brookes's classic *Garden Design* published by Dorling Kindersley.

A further matter to consider if you are relating existing trees to the design of a new home, is that they change their character through the year. Dapple sunlight filtering through new leaves is very attractive in May; in December the same species may make a damp day look positively gloomy, with damp dripping from the branches onto the sodden leaves below. As with everything, the choice is yours.

River frontages

Plots with river frontages are always in demand, offering the prospect of interesting gardens and possibly fishing rights or even a boat house and river picnics. If you have a chance to build a riverside home you will have already considered flood levels and any special foundation requirements, but it is also important to know about the rights and obligations of a riparian landowner, who is someone who owns a river bank. You may be obliged to leave an unobstructed route along the bank for River Board plant and vehicles, and the board may have the right to dump mud dredged from the river on your land. You certainly cannot assume that you will he able to build a boat house or tidy up the banks without permission.

8 Designing a new home - to suit your lifestyle

Those building for themselves are able to arrange for a new home with a room layout that is exactly to their requirements. This is particularly attractive to those who have special requirements. If you are a vicar, or an insurance consultant working from home, or if you have an elderly parent who will live with you, then a home that is tailored to your own individual needs will make your life much easier. It will not be necessary to provide all of the special features at once, but it is a very good idea to give consideration at the planning stage to any future use which you may make of your home. What is involved in this will be understood by considering incorporating a granny flat in a new house or bungalow.

Granny flats

There are many advantages to having a granny flat to provide self-contained accommodation for an elderly parent or parents. The accommodation can be specifically tailored to their needs, and the younger generation are close at hand to give support when it is required, leaving granny a considerable measure of independence. It can also be a way in which money can be passed down through the generations in a tax effective way, enabling the senior generation to help with the cost of the new house by using the proceeds of selling their existing home.

As with everything else to do with individual homes, the key to success is making the right long term plans. One nettle that has to be grasped is that granny will probably not be occupying the granny flat for ever and so it is appropriate to recognise this and plan for when she will not be needing it any more. I hope that my direct approach to this matter will not offend readers.

If you are thinking of building a granny flat there are three angles which require separate and careful consideration. These are how you arrange the finances of the whole business to the best advantage, how you make sure that there will be no planning consent problems, and the design features that will both suit granny's needs and add value to the property.

Let us start with the finances. It is possible for granny to sell her existing home and to put the proceeds towards the cost of accommodation specially tailored to her needs in her children's home in a way that will largely avoid inheritance tax. This is a highly technical matter, and the advice of an accountant or solicitor is essential. Make sure that the person who you retain to advise you has up-to-date experience in this, and if he or she has not, ask them to put you in touch with others who have. You will almost certainly find that the criteria is that the granny flat should meet the special needs of the person who is paying for it, that they should have exclusive use of that part of the property, but that it does not constitute a separate self-contained dwelling. The issue of whether or not the flat has its own external door, and whether there is a linking door with the main property become determinant factors in this matter. You will probably be advised that both are essential.

If Inheritance Tax planning is a major element in the whole business, it may be that your solicitor or accountant will seek clearance on your behalf from the tax authorities so that you know exactly where you stand. This may be important. However, as the Inheritance Tax threshold is now comfortably into six figures, this is not a universal situation.

If granny is selling her existing house in order to provide some or all of the cost of the granny flat, and is perhaps also contributing towards the cost of the rest of the house, you will have to consider where she is going to live until the new accommodation is completed. This can be another major element in planning your finances.

Once you have dealt with these financial matters you should give careful consideration to the possibility of obtaining planning consent for a house with a granny flat. Unfortunately you cannot take this planning consent for granted, as it may be considered that the house plus the flat makes the building that you propose too large for the plot, or inappropriately large for the area, or that it will lend itself to being divided into two separate properties at some time in the future. All of these are valid reasons for the local authority to issue a planning refusal, and although you can challenge such a decision at appeal, this process will take at least six months and possibly longer.

A typical granny flat with a separate entrance, room for a mini kitchen in the lobby leading to the main house, and a full size bathroom. Another window in the end wall would have given a different outlook, which would have been a good idea.

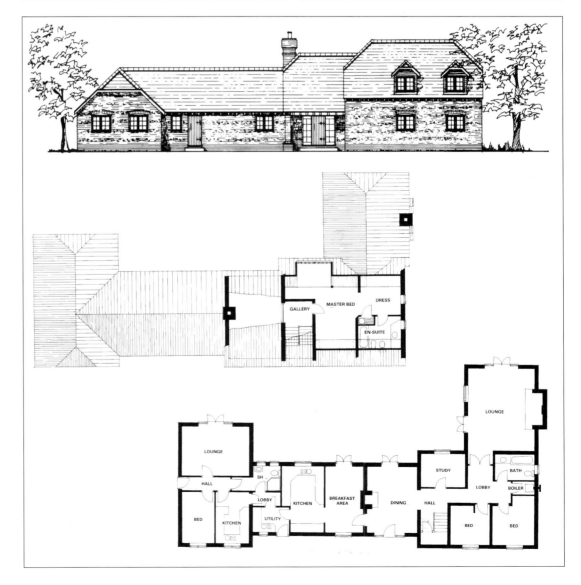

This large country home built in Dorset by Shire Homes for one of their clients has a large self-contained granny flat. As discussed in the main text, careful consideration must be given to handling a planning application for a design like this.

For this reason it is important not to raise granny's expectations of being able to come to live with you until you are certain that it is a real possibility.

A different approach may be necessary if the site is in an area where the local authority are cautious about granting permission for a building with self-contained accommodation which could lead to what is termed 'multi-user occupation', but are unlikely to object to a large dwelling of 2,000 square feet or so. In this case you may decide to apply for consent for a house with rooms appropriate to a granny flat, but which are not designated as such. When it is completed and occupied, an additional external door and some rearrangement of windows might quite legally enable you to provide the granny flat within the building. Similarly, this ability to make alterations will enable a granny flat to be turned into other accommodation when it is no longer required.

Your tactics in this should be discussed very carefully with your designer. He may have direct experience in this matter, and be able to advise you of the probable reaction of the planning authority. If he does not have relevant experience, then you should find someone who can give you informed advice. You may feel that the planners should give special consideration to your planning application for a granny flat because of the obvious social benefit and advantages to the whole community of your ability to care for an elderly citizen in this way. Do not take this for granted. It can help if your family is well established in the area, but it is unlikely to cut much ice with the authorities if you are moving into the district from elsewhere. However, this is certainly something which you should emphasize if you are lobbying the local councillor. In some areas there is a policy to encourage local tourism, and if you can present your granny flat as being potential bed and breakfast accommodation, this can be a

A design for a bungalow with a granny flat, taken from this book's companion volume *Home Plans*.

plus factor.

Another consideration in this is whether or not you are living in a conservation area. If you are, then any alteration to the building after granny is no longer using her flat will have to be the subject of a separate planning application, and this applies to alterations like providing or bricking up doors and windows or the provision or removal of a porch. Such alterations may be essential to your long term plans. If the new home is not in a conservation area, then it is a good deal easier to make any external alterations.

When designing granny flats we must first discuss what granny is likely to want. This will depend entirely on her. Will she value her independence and privacy above all else? If so, she will probably want her own front door, and she may ask that the communicating door between her flat and your house should lead from a hall and not from the living room or kitchen. How important is it going to be that the granny flat should have its own cooking facilities? Need these be any more than one of the mini kitchen units which are now widely advertised, combining a small cooker, small fridge and sink in one 60 inch kitchen unit? The right decision in these matters can make enormous difference to the whole arrangement.

There are other decisions to be made as well, many of them concerned with the possibility of granny remaining in the flat when she is very elderly and rather more frail than she is at present. Doors that are three feet wide make life in a wheelchair a great deal easier, and bathroom appliances that suit the elderly are now readily available and so it might be a good idea to install them at this stage. Electric sockets and light switches are more easily used by the elderly if they are set three feet above the floor, and these and many other matters are dealt with at length in various books on designing for the elderly. The building bookshop attached to the Building Centre at Store Street in London always has a good selection of these, and should be contacted for details of current titles.

If special features of this sort are to be built into the granny flat, it may be appropriate to give consideration to replacing them at an appropri-

ate time in the future. For instance, electric socket boxes can be provided at both a high level and in the skirting, and one or the other can be blanked off as appropriate. The tiling around a special bath can be arranged so that the bath can be changed without too much difficulty. Most valuable of all, lintels can be built into walls so that windows or doors can be inserted beneath them at a later date, and you should enure that partition walls in a granny flat can be readily moved if some rearrangement is ever going to be appropriate.

One feature advocated by architects who specialise in old peoples' housing is the provision of extra wide window sills, or window sills that continue over shelving beneath them. These are particularly valued by those who cultivate pot plants, and this is a major interest for many elderly persons. Another feature often suggested for a room in which an elderly person is likely to spend a lot of time is making it L shaped or possibly five sided. This facilitates a number of windows which give different outlooks, and also permits interesting internal arrangements of the furniture.

Finally, beware assuming that it will be easy to change a garage into a granny flat, or vice versa. Building Regulations require that integral garages should have their floors set down six inches below the level of the floors in the house, and it is also usual for there to be a slope on the garage floor near to the door to carry rainwater away. This means that any rearrangement will involve changing the level of the floor.

Home offices

Working from home is now very trendy, and has many advantages if your home has somewhere for the work to be done. Building your own house or bungalow provides the opportunity to have an office which can be tailored to your own requirements, but there are some basic considerations to be taken into account before considering which side of the room you will put the filing cabinets.

To start with you probably want your new home to be the best possible investment. If you ever sell it, you will want the best possible price. If an integral office will not affect the sale potential, or even enhance it, then it is a wise investment. If you build a home office that is likely to diminish the resale prospects of the home, you must think very carefully, and you may decide to consider working from a summer house in the garden, insulating it and installing a storage heater. This may even be considered capital expenditure within your business. However, most people want their office in the house itself, where the capital cost is most unlikely to be allowed as a business expense.

You are probably reading books on working from home, and most of them emphasise that you should be able to claim a proportion of the running costs of the home as a business cost, and this is certainly possible in many cases. Beware! You may be able to persuade the tax man that 15% of your property is really an office, and he may permit you to charge up 15% of the running costs of the house to your business. This may save a useful sum, but if you ever want to sell the property the Revenue will want Capital Gains Tax on 15% of the proceeds. This would be subject to indexing and the exemptions allowances, but it

could easily outweigh many years of tax allowances.

For this, and other reasons, any proposal to use the word 'office' on design drawings that go to the planning office should be considered very critically. It can lead to business rates, problems with old covenants that restrict business use, planning considerations and even concern about work place regulations. The word 'study' is fine: ladies and gentlemen have studies while only sordid business people have offices! Unless there is a positive reason why you really must use the word 'office' it is much more sensible to work from your study/family room instead: the same rooms with different names!

If you intend to have the public visiting your premises it is more difficult to maintain that you are not involved in commercial activities. Certainly you will destroy any credibility if you put up a sign saying *Pottery* or *Computer Bureau*. Otherwise it is a matter of whether the scale of your activities and the comings and goings which are part of it all, are inappropriate to a residence, or if you have close neighbours, to a residential

D & M Kegworth design home altered to show a prestige office suite for a self employed consultant. Note the generous storage space which is something that is often overlooked when planning a home office.

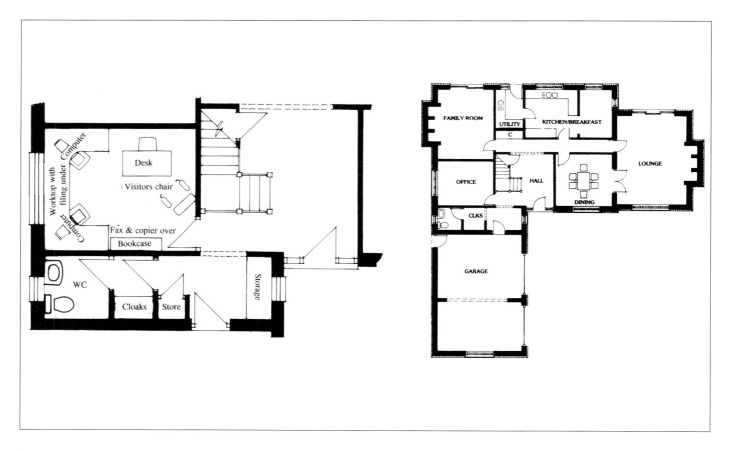

Modern telecommunications and the growth of opportunities for self-employment, make an office in a new home increasingly popular. A proportion of the running costs of the home can be legitimate business expenses, and meeting your clients in a setting like this will impress them far more than an office suite in a tower block.

A four bedroomed house in Norfolk from which an engineering contractor in the off-shore oil industry conducts his business.

A modern farmhouse by D & M. A dwelling on an agricultural holding from which a farmer manages his operations can be financed through the farm accounts.

area.

If you feel it appropriate to obtain business use planning consent for the work-related part of your home it will all be a lot more straight forward, but it will make little difference to the difficulty in being able to pay for it from the business. If, for instance, you run a boarding kennels which require your 24 hour presence you might think you could charge it to your company. Sadly this is not so, although a second home or bungalow on the site for an employee will probably be allowable.

A further consideration is whether or not there are any covenants on your land which prohibit you using the property built on it 'for trade or business'. Here again the key issue is whether you erect any signs or if you have customers whose comings and goings are noticeable. Parking a trade van may also cause problems. If the person who sold the land and has the 'benefit' of the covenant lives locally, and is personally concerned that you stick to it, you may have problems. If they live at a distance they are unlikely to take action unless your neighbours complain to them. The most difficult situation is where a field was split up into plots with covenants which are 'mutually enforceable', so that any of your neighbours can take action to enforce the covenants. This will cost them quite a lot, so that most attempts of this sort do not get any further than threatening letters from solicitors.

Bed and breakfast accommodation is another case in point, and there are many anomalies. If you buy an existing house and decide to alter or convert it to provide overnight accommodation for tourists you may qualify for a grant, but if you build a new house for the same purpose you will be unlucky. Incidentally remember that more than six B & B *guests* attract Uniform Business Rates and that more than six B & B *rooms* require expensive fire precautions. In some counties you might get a grant for the fire precautions only, but do not count on it. A useful guide to all this is called 'Starting a B & B Business' and is available from Barclays Bank.

Arrangements for grants and loans for new businesses change all the time, and many useful schemes that operated a few years ago have now been discontinued so do not rely on word of mouth for advice on this, or read old leaflets. Make sure that you get up-to-date information. General advice on the situation in rural areas can be obtained from the Rural Development Commission, while in urban centres the Development Officer at the Town Hall, or a Civic Business Opportunities Bureau should he contacted. Other information may be obtainable from a relevant trade or professional organisation. Generally speaking the current emphasis is on schemes which provide employment, and help for one-family businesses is curtailed. However , if you can get help for starting up your actual business from the government it may free up more of your own capital for the new house, so examine the situation very carefully.

Whatever you call your work space you should give a great deal of thought to how it fits into your overall design. The first consideration is whether you have business visitors who you will need to impress. These might be clients discussing a consultancy contract with you. You probably feel that you must project the right image, and

your office suite will have to be adjacent to the front door so that your visitors will not be aware of the squalor of your domestic arrangements. There should be an adjacent loo, and the view from the office window should be as attractive as possible. Storage space will be important, with samples and old records properly out of sight instead of being piled in a corner.

On the other hand, if your freelance graphic design business will never involve visitors, you can locate your activities in a room beyond the kitchen, where you can make all the mess you want. Or, if you edit a very intellectual magazine from home, you may want an office on the first floor where you can hope to enjoy peace and quiet. You know what you want your office for: locate it in a part of the house that suits this.

Next, make sure that the office space which you are creating is going to be adequate for the hoped for expansion of your business. If you intend to employ a couple of people within a year or two, but will still be working from home, you will need working space for them, an appropriate cloakroom, and somewhere for them to hang wet raincoats or umbrellas. Do not assume that they will want to use the family facilities: this rarely works out.

In the same way, services should be planned for future levels of use, so you may decide to install four telephone lines although your short term requirement will be for one only. Next year you will want a dedicated fax line, the year after a dedicated computer link onto the super highway. Beyond that, who knows — but provide the infrastructure for it now. Remember lots of electric outlets, lots of directional lights for the ceiling, and heating controls that are independent of those in the house.

The furniture layout is very important in an office, starting with making sure there will not be a reflection problem on computer screens. This is a major consideration. Built-in working surfaces along walls are space savers, while free standing desks are not. Two or three drawer filing cabinets can have fax machines or copiers on their tops, while a four drawer cabinet does not permit this. A computer printer on a low level trolley can be pushed away when not in use and does not take up any desk space. Computer screens on an arm projecting from a wall have the same advantage. If visitors are going to be important to your business success you should make sure that a chair or chairs for guests are an integral part of your layout: seating a client on a chair carried through from the kitchen will not enhance your image! Remember that a well designed office layout will improve efficiency and even if the notional improvement is only 5%, the value of this over a few years will be very significant.

Extendability

Sometimes another special requirement for a new house is that it should be possible to extend or alter it at a later date. If this is your intention, the matter should be given as much careful consideration as every other aspect of the design. You may wish to be able to put rooms in the roof at some time in the future, or to build an extra room or even a new wing. If so, then there are some basic decisions to be made at the time that you submit your planning application. Are you going to apply for planning permission for the ultimate version of your home, and build only the first phase, or will you only seek permission for the home that you propose to build immediately? The answer will depend on many factors, including whether or not you are building in a conservation area, how contentious your application will be anyway, and the nature and size of the extension which you hope to build at a later date. Remember that most extensions of up to 15% of the ground floor area of an existing dwelling do not usually require planning consent at all, although there are exceptions. A simplified note of the current position regarding this is set out below, and you may have to think long and hard about

---

Extensions

Many people build a new home which they hope to extend at a later date, and sometimes they do not include the extension in the plans which they submit for planning consent for the original house or bungalow. One reason for this is because they fear that consent for the house will not be granted with the extension, and instead they hope at a later date to take advantage of the planning regulations concerning permitted development which authorise some extensions to existing buildings without consent.

Extensions, and the planning rules for them, are beyond the scope of this book, but in general planning permission for an extension is only required if one of the following apply:

* the extension will be nearer to a highway than the original dwelling unless it will still be more than 20 metres from the highway. There are special rules for porches.

* the extension will cover more than half the land around the original dwelling.

* the extension is more than 15% of the volume of the original dwelling, or 70 cubic metres, whichever is the greater, unless it is in a Conservation Area, a National Park, an area of Outstanding Natural Beauty or the Broads in which case the limits are 10% or 50 cubic metres.

* there have been a series of extensions over the years which aggregate more than 115 cubic metres to the volume of the dwelling as it was on July 1st 1948 (which is when the planning acts came into force) or when it was built if later. Volumes are determined by external measurements.

* the extension is higher than the highest part of the roof of the original dwelling.

* the extension is more than 4 metres high and within 2 metres of your boundary unless it is a roof extension.

* the extension is a porch which has a ground area measured externally of more than 3 square metres, or is more than 3 metres high or is less than 2 metres from a highway.

Extensions which are within the above limits are usually permitted development. However this is by no means a complete statement of the law, and if you hope to extend a new house in the future by taking advantage of the permitted development rules it is important to take professional advice or ask your planning office to confirm that consent will not be required.

Building an extendable home. Two stages in building a three bedroomed house with an integral garage. Note that the foundations of stage two are used as a parking area for the stage one house.

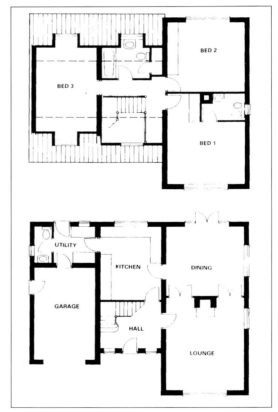

the best way to go about making sure that you can extend as you want, when you want.

Invariably a modest expenditure in making provision for a future extension will save a great deal of money later on. If the shell of the house is to be extended, then putting in the foundations of the extension at the same time as the main foundations, and getting them approved by the Building Inspector, will make everything much easier in the future. These foundations can be buried and grassed over as part of your lawn, or used as a patio, and if you install any drains for the extension at the same time, it will avoid having to excavate in the future. In the same way lintels can be built into walls under future openings, perhaps with vertical joints in the masonry below them so that doorways and windows can be installed with the minimum of fuss and mess.

Central heating installations can be arranged so that new rooms can be connected up when appropriate, and provision made for further electrical wiring in the distribution board.

The same considerations apply if you think you may want a conservatory in the future. Almost certainly it will be cheaper to use one of the specialist conservatory companies for this, in which case the consumer guide leaflets about the services available from the Glass and Glazing Federation will be relevant. The addresses are at the end of this book.

If your intention is to make use of the roof space at some time in the future, you will have to make sure that it is a purlin roof and either install stairs to the roof space or make sure that your room layout will allow for them to be installed when required. If your rooms in the roof are to have dormer windows, it will be much cheaper and easier to build them when the roof is constructed rather than to arrange for them at a later date, while if you want Velux windows set in the roof space you have a choice between just framing out for them when you build the roof, or installing them there and then and having done with it. Piping, water tanks, wiring, aerials and everything else that is normally hidden away in the roof space without much thought should be positioned with care, and special consideration should be given to insulation. Finally, if you propose to use a spiral staircase or other space saving way of getting up into the rooms in the roof, remember that you may need to be able to move beds or other large items of furniture up there. If this will present problems, one solution is to have a gable window that is secured in a sub frame with screws which will enable you take it out in order to move in anything which is too bulky to go up the stairs.

Designing for the disabled

Designing a new home for someone with a physical disability is an interesting challenge, and many people who are disabled following an accident move into a purpose built bungalow. They make excellent selfbuilders, possibly because the lengthy delays in receiving compensation payments give them plenty of time to become formidable experts on the subject, and they have an enormous enthusiasm for the job. The author has mixed memories of a selfbuilder client who used a wheelchair on the ground but who could go up ladders with ease, and who insisted on discussing progress on his bungalow with visitors while sitting astride his roof ridge. There are various bodies concerned with designing for the disabled, and they can be contacted through The Centre on Environment for the Handicapped, 35 Gt Smith St, London SW1P 3BJ. Details of the various relevant publications, addresses of suppliers, etc., can be obtained from them.

Porches

If your new home is in the country, a modest but equally useful special design feature that can be built onto almost any home is a large enclosed back porch to suit your particular style of rural living. A porch which is 4 metres x 2 metres is unlikely to seriously distort your budget, and it can provide somewhere for muddy boots, muddy dogs, children's toys, the chest freezer or even a back door lavatory. This can make the country housewife's life a whole lot easier.

Security features

All insurance companies offer discounts to those who have approved security features, usually high quality door and window locks and a burglar alarm installed by a NACOSS registered contractor.

A feature which can be built into a new home, is a secure cupboard which offers protection for valuables, shotguns and the like. If this is given consideration at the design stage, it is relatively easy for an ingenious carpenter to install a hidden door to a hidden cupboard which will escape the notice of a burglar! Those whose work involves keeping large sums of money in the house from time to time can arrange for an underfloor safe to be set in the foundation concrete, to be reached by turning back a rug or carpet in one of the ground floor rooms. This may be a condition of their special insurance.

Larders and cold rooms

Walk-in larders are very popular with housewives,

A new home in Lincolnshire built in traditional materials in a traditional style.

but rarely found in speculatively built homes. A 6ft x 6ft built-in larder with 18" shelves all round from floor to ceiling provides more usable storage space than a lorry load of kitchen units. Ideally it should be on an outside wall so that fly-proof ventilation can be arranged, or it is worth considering installing a small cooling unit which will keep the larder at the same temperature as a fridge. This will require extra insulation in the floor, walls and ceiling together with an insulated door, but this is easily arranged. Suitable cooling units are available from Interlevin on 01332 850090.

Services boxes

We seem to have a blind spot about the appearance of services boxes in this country, and one often sees an attractive new home disfigured by electricity and gas meter boxes prominently displayed on the front elevation. I have never understood why the planners do not take some sort of action that will lead to the electricity and gas authority's hideous plastic boxes not being so prominently on view.

The rules depend on the authority concerned, but generally the position is that the standard connection charge involves the services box being sited as close to the road as possible, and it is taken for granted that this is where everyone will find it most convenient and that this is the best place for it. Often those who are providing the supply will insist that this is where the selfbuilder 'has' to have his meter box. Needless to say you can challenge this, although you may be asked to pay a few pounds more for the extra length of cable. It may involve referring the matter to a senior manager, or even threatening to involve electricity and gas consumers' councils. This is very well worthwhile if it means that the boxes are installed where they are unobtrusive, or where they are easily shielded by appropriate planting.

Basements

Basements and cellars have not been usual in homes built in Britain for the last eighty years, although before the first world war they were a feature of all but the smallest homes, and are popular in much of the USA and Europe. Cellars have obvious advantages on restricted sites, but although one or two homes with cellars have been featured in case histories in earlier editions of this book, they are very unusual. No one can explain why. If you want a basement or cellar you will have to look very hard to find an architect and tradesmen with much relevant experience.

Two timber frame homes by Hedlunds in very different styles.

Any consideration of a basement starts with examination of four factors:

* The subsoil type

* The level of the water table

* The amount of natural ground water and sub surface water movement

* Any history of flooding in the area.

If the subsoil does not drain freely the excavation in which you build a basement will act as a sump, hold water around the basement walls, and subject to the tanking to high hydrostatic pressures. For this reason land drains are normally laid around the foundations, and are often installed to divert sub surface water movement right away from the building.

The way in which a basement is kept dry is described as the tanking, and there are many walling systems to provide this. Choosing the right one is something where the advice of an experienced professional with local knowledge is absolutely essential. Ask them for the arrangement of membranes and impervious coatings to be as comprehensive as possible, adopting a belt and braces approach. Remember any damp penetration will be extremely difficult to rectify, and will be detected in a survey if you ever sell the property. This would lead to the house being reported as having a damp problem, which discourages potential purchasers and their building societies!

A basement is not naturally ventilated and the tanked walls cannot breath: as a result it must have adequate artificial ventilation. If any activity is to take place in the basement which generates water vapour, as with a sauna or laundry room, special arrangements must be made to extract the water vapour.

It is most important that any fixtures and

An individual home in East Anglia, with traditional ornamental plasterwork or pargetting.

fittings in the basement should not be made into the tanking walls or the floor. Services should be concealed in hollow skirting systems, the staircase should be free standing and not secured to the wall, electrical outlets should be on the surface and fittings should be moisture resistant, and all storage cupboards should be well ventilated. If the heating boiler is to be situated in a cellar it is important that consideration is given to its specific ventilation requirements, and now that audible gas detectors are available cheaply they should be installed to detect any gas leak in the house which might lead to a build up of gas in the cellar.

Given the right attention to all of this, a basement can be as important an element in family living arrangements as in the USA or Europe.

Heating and energy conservation

In this consideration of design features to suit your own pattern of living we have ignored any mention of energy conservation, insulation or heating arrangements. This is such an important topic that it is dealt with on its own in chapter 26.

Caution over unusual features

Finally, a word of caution about arranging for design features which are in any way unusual. The Building Societies tell us that the average house is sold every seven years. Whatever your present intentions, it is likely that your new home will be put on the market sooner or later. Ensure that the features which you arrange to suit your own hobbies or lifestyle are likely to be admired by others, as this will make the home much easier to sell. For instance, if you want a small photographic dark room, do not omit a window. It will be a nuisance to have to make a screen to black it out, but with a window an estate agent can describe your dark room as either a box room or a fifth bedroom. All of this chapter has been about building what you want, but remember that your house must also be what a prospective purchaser will want some day.

Scandinavian houses are designed for cold winters and have exceptionally high levels of insulation. This Swedish kit house from Hedlunds is available in the UK.

9 Designing a new home — the external works

The appearance of your new home will be determined by the design of the building itself, the materials from which it is constructed, the way in which it is positioned in relation to the boundaries and neighbouring buildings, and the garden in which it stands. All too often any detailed consideration of the garden and the external works like the drive, a patio or the landscaping, are left until the house is finished in the belief that first things must come first, and the house itself is what really matters. Unfortunately this approach can lead to external works being carried out after the new home is occupied, when they could have built much more easily and cheaply at an early stage. Sadly it is also true that those building for themselves eventually find a limit to their enthusiasm for small scale civil engineering, and after they move in they usually want to sit back and relax rather than embark on a major landscaping project. This adds up to a very strong argument for deciding to undertake at least the basics of the external works at the same time as the house or bungalow itself is built. The rest of this chapter is concerned with the opportunities to arrange these works to the best effect, so that they enhance the appearance and the value of the property while reducing the maintenance commitment.

Much of what follows will also be found in this book's companion volume *Plans for Dream Homes*, which is specifically concerned with designs for new homes. I hope it bears repetition.

Consideration of external works and landscaping has to start by examining your living pattern. How much time are you going to want to give to looking after your garden? At one end of the scale you may think the less bother the garden is, the better it will suit you. Fine - carefully chosen shrubs, the right grass, no borders, no edges, paths set flush in the grass, and you will have a minimum maintenance garden which can look very attractive indeed. At the other extreme, you may intend that the garden will be a very important part of your life. If so, it is even more important that you analyse what you are going to be able to do, and plan accordingly. You should consider not only the gardening that you can currently undertake, but

also the work load which you are likely to be able to enjoy in the years ahead, bearing in mind that keen gardeners who are unable to keep their gardens as they wish have many regrets. If you are fortunate enough to be able to pay someone to look after your garden for you, careful consideration of the labour requirement is just as important You will want a perfect garden for your money, and an acre can be laid out either to need only half a day per week to maintain, or to occupy someone nearly full time.

The next thing to consider is your timetable. This is important, and is often overlooked. Do you look forward to watching the garden slowly taking shape, and hope to see your dreams come true in three or four years, or do you want visitors to be admiring it in six months' time? Remember a wisteria will take seven years before it makes a show but a container grown cherry will be in flower next spring. This can be an important issue if your lifestyle is such that you might move at any time, and need your home always to be a good resale prospect. An established garden may or may not add to the value of a property but it certainly always makes it more saleable.

This leads to the next consideration, which is that of cost. It used to be said that anyone building a house should be prepared to spend at least 10% of its value on its surroundings. The cost of building is now so high that this would require a large plot on which to spend the sum involved, but the principle that the garden deserves its place in the budget is important. This may be a cash budget if you intend to have a landscaping contractor do the work, or a budget for your own labour plus some cash if you intend to do all the work yourself. In this latter case, beware of overestimating your own capacity or the time that you will have available, particularly if you are going to do much earth moving by hand.

Finally, having made a realistic appraisal of the amount of time you wish to give to the garden, of how soon you want it to look established, and of the budget, you can think about what you hope it will look like. This involves a plan. Even if your garden is unimportant to you, it is worthwhile making sure that it is going to be as little trouble as possible, will show off your new home to the best advantage and will enhance its value.

If you are building on an eighth of an acre in a built-up area, your choices are limited, and you will almost certainly have a marvellous opportunity to look at all of your options by simply going for a walk to see what your neighbours have done with their own front and back gardens. Look over their hedges in winter as well as in summer, take local advice on what takes time and trouble, and what is easily looked after, sort out your own master plan, and away you go. Your garden can complement your house very well, but its size will impose a limit to what you can do.

With anything over a quarter of an acre the challenge is much more complicated. At this size your garden can be given a distinct character, and the house and garden can relate to each other in a very special way. If you have this opportunity, then you need to be a very experienced gardener to be confident that you can design the right garden. Even if you are, it is worth spending a relatively small sum in buying ideas from someone else as well, if only so that you can be absolutely sure that yours are best. Unfortunately, the right garden designers may be hard to find.

Local landscaping contractors, nurserymen and garden centres usually offer to design gardens, but this is inevitably linked to their own services. If you are wanting to place a single order at a lump sum price for all of the work, materials and the plants involved, then contacting one of them is an obvious way to make a start. Their proposals should come with an attractive layout plan, a list of plants, details of how the ground will be prepared and of any top soil to be provided, guarantees to replace trees and shrubs which do not take, and arrangements to maintain

Planning regulations for ancillary buildings

You will need planning permission for structures which were not shown on the planning application if your proposed structure would:

* be nearer to any highway than the nearest part of the original house (unless there would be at least 20 metres between the building and the highway.

* cover more than half of the area of land around the original house.

* be used other than for domestic purposes.

You will also need to apply for planning permission if:

* you want to put up a building or structure which is more than 3 metres high, or more than 4 metres high if it has a ridged roof.

* your house is in a Conservation Area, a National Park, an Area of Outstanding Natural Beauty or the Broads, and the proposed structure would be more than 10 cubic metres in volume.

* you want to install a storage tank for domestic

heating oil with a capacity of more than 3,500 litres or more than 3 metres above ground level.

* you want to install a tank to store liquefied petroleum gas (LPG).

When putting up fences, walls and gates, you will need planning permission if:

* your house is a listed building.

* the fence, wall or gate would be next to a road and over 1 metre high; or over 2 metres high elsewhere.

You do not need planning permission for hedges or tree screens unless there is a condition attached to the planning permission for your property which requires you to obtain the council's consent (for example on an 'open plan' estate, or where it might block a sight line).

lawns until they are established. Everything should be as detailed as possible.

However, landscaping is one area where the cost of a package service is likely to be more than the cost of the component parts, and where many people prefer to make their own arrangements for all of the various elements involved. If this is what you want, you can ask your local landscaper if he will provide you with a design for a fee, with the option to use his services or not as you wish. Keep in mind that all landscaping contractors make their livings from supplying plants and making gardens, and not from simply designing them. There are pressures on them to design what they find most profitable to handle, and your only safeguard in this is their local reputation, and the advice which you will seek from others. Landscapers should invite you to go to look at other gardens that they have made: if so, do not visit the one 'show garden' that they suggest to you, but ask them for the names of other private clients and ring them for a reference.

An alternative is to use the services of a garden designer. This makes the whole business more complicated but need not cost very much, and you will know that the advice that you are getting is free of commercial pressures. Plans and a list of plants from a garden designer can be used to obtain competing quotations from landscaping contractors, or can be your blue print for managing the job yourself. The problem is finding a designer.

Personal recommendations are invaluable, but most people will have to resort to the Yellow Pages or similar directories. The best approach is made by letter, explaining that you would like your garden designed on a plot of xxxx square yards (or x acres) and that you will be interested in what the designer can do for you and the cost involved. After that, leave them to sell their services to you and evaluate the proposals very carefully. Again, ask to be put in touch with previous clients. Finally, in explaining your requirements to a designer make sure he or she understands your own situation with regard to maintenance, timetable and budget.

Some gardening magazines offer a garden design service and these are usually so cheap that it is fun to sign up for one or two of them anyway. The proposals are invariably for an enthusiasts garden with something of everything: lots of quarts in a pint pot. These will give you plenty of ideas, and if you have explained your maintenance/timetable/budget requirements firmly, you may even find that they have come up with the answer!

All of this needs to be done at an early stage, just as soon as you have settled on the house or bungalow plans themselves. Certainly you want to know exactly the shape and contours of your garden while the builder still has earth moving plant on site, so that you can ask him to dump soil where it will be required and to leave the site with ground at levels that will suit the landscaping work. If you have a large site it is a very good idea to fence off the builder's theatre of operations and to start making the garden on the rest of the land while the house is being built. In this way you will almost certainly be able to move one season ahead. Having one part of the garden finished will encourage you to make a start on the area left by the builder as soon as you move in.

To be realistic, if you are managing the building of a new home yourself, it is unlikely that you will have the time or energy to spare to give any of this the attention it deserves, but even so it is important that you do have a landscaping scheme and that you try to give it consideration in spite of the overriding priority of making sure the roof is tiled on schedule. If you have placed a contract with a builder, then involvement with planning the garden will serve as an outlet for all your frustrations while the building work seems to proceed so slowly. You will have time for visits to garden exhibitions and garden centres, a great deal of reading, and you can even venture to plant trees amid the builders' confusion if you protect them properly. However much time you have for this, or however little, the important thing is to have a plan at an early stage, and to ensure that it is suited to the long term relationship between you and the garden which is going to be outside your window for as long as you live in the house.

Given the determination to make an effective plan for the new garden, and preferably before the builder moves his earth-moving equipment away from the site, what are the features to be considered? The landscaping itself is a matter for your expert, but all that he does will be framed by the way in which you have dealt with your boundaries, and will have to take into account all of your external works. Among the more important of these are the following.

The drive

Unless you have a very big garden indeed, the drive is the largest single feature in your garden, and once it is built, it will be very difficult to move it. It is likely that your planning consent will require that the gates, if any, are set back a fixed distance from the road, that the slope for the first few yards should not exceed a certain gradient, and that there is provision for a car to be able to turn on the drive without backing out into the road. You will also be concerned that there is a large enough parking area for visitors and delivery vehicles. On most plots there will only be one practicable layout to meet these requirements, but if you do have alternative alignments for the drive, consider the feel that they will give to your garden as well as thinking about them simply as a way of getting the car to the garage. If you have a very large garden you may wish to consider a sharp bend in the drive to slow down vehicles which are using it, but make sure that the council refuse vehicle and large delivery vans can negotiate it too.

There are a number of ways in which drives are surfaced and there is no doubt that paver bricks are now at the very top of the market. These are also known as pavoir bricks - exactly the same thing. They should be laid by specialist contractors, and a great deal of care should be given to choosing colours and the pattern in which they are to be laid. An effect that is superficially similar can be obtained by paving a drive with large concrete blocks that have a moulded surface to look like paver bricks. These offer cost savings, and are widely advertised. It is important to insist on an opportunity to see them laid on an actual drive rather than making your choice after seeing them in a display area. Warranties should be sought from both the manufacturers and from the paving contractors for all paved drives, and almost certainly the guarantees will be void unless the edges are properly retained. If your drive will only be used by cars this can be done with 8" x 2" path edging set in concrete, but road curbing looks far better and is essential if your drive will be used by heavy delivery vehicles.

A tarmac surface is usually cheaper, but everything depends on the area involved. The problem with tarmac is finding a reputable contractor to have it laid for you, as the 'Boys from the Black Stuff' are to be avoided. Again, seek personal recommendations or contact the surfacing contractors association which is the BACMI at 156, Buckingham Palace Road, SW1W 9TR, telephone 0171-730-8194. They will send you a list of members.

The cheapest of your options is to have a gravel drive, but this can involve a significant amount of maintenance, raking the surface and killing weeds. If the cost of a prestige drive is beyond your budget, then have the curbing installed to retain chippings and let your car act as a roller for a year or two until it suits you to have it surfaced.

All drives are no better than their foundations

Drives are very important. The gentle curve of this one is just right: consider how different it would look if it was straight. This is a selfbuilt house in Cornwall.

and the drainage arrangements made for them. The specification should be drawn up by your architect, and should be appropriate to the subsoil. It should certainly not be left to a tarmac contractor!

The approach to the front door

Most people expect their front door to be the focal point of their new home, and the way in which it is linked with the drive should be as carefully designed as the doorway itself. Almost certainly it deserves something more than the skimpy door step and the narrow paved path which is the usual approach, although it is important to avoid a fussy appearance.

If you look at the front door paths to your friends' homes you will probably consider that they should be twice as wide as they are, and that the door step should be much larger. Anything beyond this is a matter of taste, but whatever you decide for your new home, it will cost very little and contribute significantly to the appearance of the property. Ask your architect or designer for their suggestions.

Trees

If you are lucky enough to have mature trees on your site, you are probably determined to retain them and to make them a key element in your landscaping. There are lots of factors involved in this. Tree preservation orders have been mentioned on a earlier page, and it is also likely that steps will have to be taken to preserve the house from being damaged by the trees. Tree roots affect foundations in a number of ways, and precautions against damage can involve deepening the foundations or digging a deep trench between the building and the trees and filling it with concrete. Conversely, the building work may have an effect on the trees by altering the subsurface drainage as well as by interfering with the roots, and it may cause them to die. If this happens, then the voids left by the decaying roots will affect the foundations of the house. If a tree is a key element in your landscaping, or is subject to a tree preservation order, it is a good idea to get it surveyed by a qualified tree surgeon. He should be able to advise on the size it will

reach, its probable life and any work necessary to keep it healthy and safe.

If you are planting trees that will grow to any size it is important to take advice on the appropriate clearance from your buildings to avoid problems in the future. This applies only to trees of indigenous species such as beech, lime and particularly poplar, and not to fruit trees, flowering cherries or the smaller weeping willows.

Hedges

If you have an existing field hedge on a boundary to your land, you are very lucky, but it may need a great deal of work to put it in good order. First of all, check with your solicitor whether it belongs to you or your neighbour. If it is your neighbour's, all

A new gateway in an old wall. It looks right because the pillars are substantial. They took a lot of very expensive bricks, but the gateway will still be there when the cost has been forgotten.

Water, whether to look at or to swim in, can be a key feature in the garden, but involves a significant maintenance commitment.

much as an appropriate one will enhance it.

The material to be used has to be right. If you use the wrong bricks or stone your wall will look wrong and will get worse as it weathers, while the right materials will look even better as time goes by. Walls should give a feeling that they are part of the landscaping: they will not do this if they are inappropriate to the surroundings. In particularly, this means that concrete blocks of all sorts, particularly the ornamental pierced walling blocks, should only be used in the sort of urban or seaside situation which suits them, and never as part of a rural scene.

It is equally important that walls should also look massive and permanent, built in a local style, with an appropriate coping. They need a proper foundation, and there are tables which determine the minimum thickness for different heights of a wall if it is to be stable: 215 mm solid walls can be built up to a height of 1.35 M without piers, and 1.8 M with piers at 3 M intervals. This is largely irrelevant: to look right any wall has to be thicker than this, with 325 mm (13in) as a minimum. To build to this standard, with the right coping, is expensive. If it is outside your budget, then plant a beech hedge instead!

If you have bought the site with existing old walls, or best of all have found a site in a walled garden, you are very lucky. It will give your home a very special character. Unfortunately the walls may also have a tendency to start to fall down the moment that you assume responsibility for them, especially if the ground is higher on one side than on the other. The worst potential problem is in the foundations. If a wall leans at all you should consider finding old bricks or stone to match the original material, and have a buttress or buttresses built to support the wall before it becomes unstable. This is cheaper than rebuilding, and if the buttress is really massive and in character, and quickly clothed with climbing shrubs, no-one will know that it is not an original.

The stone coping or coping brickwork of an old wall is another weak point: fortunately it is easily replaced and it is important to get this done before the rain gets into the masonry. Re-pointing is less essential than dealing with any water seeping in at the top, but if it needs to be

you can do with it without his agreement is to keep it trimmed from your side 'in accordance with good practice'. If it is not regularly trimmed on the other side, it is worth considering whether you want to try to take on this work, because a hedge which is regularly cut on one side only will soon become misshapen and unattractive. Again this is a matter on which to get expert advice, particularly if your hedge is of mixed species.

A new hedge always present a difficult choice of exactly what to plant. Fast growing hedges like privet and leylandii require frequent trimming, and are very greedy, starving neighbouring plants of nutrients. More attractive hedges, like beech and yew, which require little maintenance, take some years to become established. If you can take the long term view and plant a slow growing hedge you will probably never regret it.

Walls

Walls are a very special feature in a garden, and building a new one announces that you are seriously into landscaping. Walls also involve you in serious expenditure. If you want to build a garden wall it is important that you make no attempt to cut corners, as a cheap looking wall will damage the appearance of the garden as done, make sure that the work is carried out in exactly the same style as the original, with mortar made from the same sand so that the colour matches. If the original wall was built with lime mortar, use lime mortar for the repairs.

Fences

The only type of fence which is in any way suited to today's housing styles is a timber post and rail fence. This is attractive, will last for a very long time, and can be fitted with wire mesh at a low level to keep your dog at home. All that can be said about close boarded or chain-link fences is that the sooner they are hidden by creepers, climbers and appropriately planted shrubs the better. If they are unavoidable, and they sometimes are, consider whether or not you can arrange for garden features to give them additional support. A brick built cold frame or compost bay halfway along a fence, firmly fastened to it, will help it to stand up to the wind long after it would otherwise need replacing.

Patios

A patio or terrace is a feature that needs to be as large as possible. If big enough it will enhance the appearance of the property and make it look larger and more impressive. Unless it is large it will do nothing for it at all. A paved area should be at least a third of the size of any adjacent building. It is also important to consider the use to which the patio will be put. If it is principally a feature to be seen and admired, but not regularly walked on, then gaps between paving slabs can be filled with alpine plants, and dwarf walls can be of a height to suit the outlook, with everything arranged for visual effect. For most of the year your view of this area will be from inside the house through a window, and this must be kept in mind. However, if your patio is to be used for sitting out, or for parties with lots of guests, it is important that there are no gaps in the surface to trap high heels, and it will help informal gatherings if any walls are at a convenient height for sitting on them. As with walls, materials for patios and terraces should complement the materials used for the main building, and cheap or inappropriate materials should be avoided at all costs.

Garden paths

The right garden path gives a period feel to a garden, enables you and your guests to walk around it when the grass is wet, and somehow help your garden to look larger than it really is. On the other hand, all paths need maintenance to some extent or other, and if they are not made on properly laid foundations they will become uneven and dangerous. Some stone flags become very slippery in wet weather, particularly if they are in the shade, and if elderly people are likely to walk around your garden paths at all times of the year it is important to remember this.

The first consideration is whether or not you are going to build your path with a properly compacted foundation, possibly with a concrete base. If you do this, and have proper path edging, there should be little risk of it becoming uneven and hazardous. However, this involves significant costs, and you may prefer simply to level flags that become uneven or replace gravel that has disappeared into the ground. However, a proper sub-base is essential if you are using paver bricks. In this case careful consideration of the finished level of the path is needed, and if it is set in a lawn, of whether or not you are going to mow over the top of the finished surface.

Pools

Water in a garden involves a maintenance commitment and if you intend to swim in it you can expect to spend as much time looking after the pool as you are likely to spend actually in the water. If you are an enthusiast you already know this, and will enjoy it. On the other hand, a pool as a status symbol requires careful consideration of who is going to look after it, and of its effect on the value of the property. Except in a few areas, where high value properties are expected to have pools, they can make a house difficult to sell.

If this does not discourage you, then make your plans for your pool at an early stage. If it is to be built at the same time as your house, it is up to the pool contractor to arrange an effective liaison with your builder. If the pool is to be built later, then it is important to ensure that an excavator will be able to get round to the site of the pool in due course. You will probably want to retain the excavated soil in your garden, piling it up as a bank to be planted as a shrubbery, possibly to give some privacy to those using the pool. If so you should allow for this in your landscaping plan. It is probably a good idea to make all of the arrangements for the pipe runs for any future swimming pool while the house is being built, particularly if the pool is to be heated using your domestic boiler.

Water storage

Now that water shortages and hosepipe bans seem destined to be a normal part of our summers, building a new home, with excavators on the site, provides an opportunity for underground water storage arrangements to be made very cheaply. Tanks and pumps for this are already on the market, although they are very expensive. However, most people building for themselves are versatile enough to find a suitable tank and equip it with a pump for next to nothing. 2,000 square feet of roof receiving 18in of rain a year will discharge 30,000 gallons, and if you can catch and store only a tiny fraction of this it may enable your borders to be the envy of your neighbours at the end of a long dry summer.

Summer houses

Summer houses, perhaps built as an artist's studio or a writer's study, and other permanent buildings have their own rules under the planning acts. The general requirement is that you can do what you like within certain limits provided that the building is constructed 'for a purpose incidental to the enjoyment of the dwelling house.' This does not include any sort of business use, or any building which is used as residential accommodation, such as guest accommodation. It does permit quite elaborate structures, such as a changing room for a swimming pool incorporating a sauna and a gym. The limiting requirements are that:

* No part of the dwelling may be constructed within twenty metres of any highway or nearer to the highway than any part of the main house.

* The building is not more than 4m high if it has a ridged roof or more than 3m high with any other sort of roof.

* Total area of ground covered by the new building is not more than half the total area within the curtilage of the house.

In addition to this, an extension to an existing property can be constructed up to 15% of the size of the original property, subject to various provisions of the same sort as those that have been already listed. The permitted development clauses in the planning acts are usually invoked when a conservatory is constructed. However, none of this applies to a conservation area, or in an area of outstanding natural beauty. If you have plans to build ancillary buildings in your garden, then your best way of checking what is permitted is to wait until you have already got your planning consent for the main house, and then to check your proposals at the planning office. It is also necessary to make sure that there are no covenants in your title deeds which restrict buildings of this sort.

Brief details of the planning regulations for this can be found on page 105.

Electricity

Electrical sockets in the garden can be very convenient for electrical garden tools, garden lighting and barbecue accessories, and if they are arranged when the contract is placed for wiring the house, and you use casual labour or a machine that is otherwise standing idle to dig the trenches, the cost will be very reasonable. Special circuit breakers and weatherproof sockets will make everything safe. Garden lighting, and any lighting used with ornamental pools usually operate at a safe low voltage, and the transformers and cable for this are supplied by the manufacturers concerned. Security flood lighting requires mains voltage. All of this requires discussion with an electrician, and it is important to remember that the work involved is little trouble at the time the house is being built, but will be a nuisance to carry out afterwards.

10 The planning scene

A person intending to build a house must understand the various approvals necessary to construct a dwelling. Dealing with these approvals, without which nothing can be legally built and which are essential for outside finance, is a job for a professional. This is normally an architect, either one who is in private practice or one who is working for a firm which provides a package service, but it can also be a design firm which is not a registered architects practice, or sometimes an individual who has relevant experience and submits applications as a part time job. Getting the application right is important, and an understanding of what is involved will help in choosing the right man to handle the matter.

Having said this, there are many people who build for themselves who successfully handle their own planning affairs, much in the same way that there are those who handle their own conveyancing without using a solicitor. If this is what you want to do, go ahead. However, it should be remembered that many planning applications for homes on individual plots are contentious, and that experience is invaluable in dealing with planning problems. The cost of using a professional to make a planning application is a negligible proportion of the cost of a new home, and any saving made by adopting a DIY approach is not really worth the risk of your application, and of negotiations stemming from your application, not being handled professionally.

The Planning Acts are concerned with whether or not a dwelling can be built at all in a particular locality, and with its appearance and the way in which it will relate to its surroundings. This control is exercised by the Local Authority, to which an application has to be submitted to erect any new dwelling. Members of the public are entitled to appeal to the Minister for the Environment against any decision of the local authority in a planning matter. In theory all planning applications are considered by a committee of councillors who are advised by the council's professional planning officers, but in practice they often just rubber stamp the recommendations of the planners for run of the mill applications. The planners make their recommendations in accordance with set criteria after going through set procedures. The way to obtain a planning consent quickly and easily is to ensure that it meets all the established criteria for an approval. This fact is often forgotten. A planning officer does have the discretion to make recommendations which are at variance with planning policy, but this is unusual. The golden rule is to avoid applications which may be contentious, and to present anything unusual in a non-contentious way.

It is always the land which acquires planning consent, not the person who applies for it. When land with planning consent is sold, the consent is available to the new owner. You do not even have to own the land to apply for planning consent on it, and it is quite usual for a prospective purchaser to make an application on a site which he or she is buying before obtaining the title to it, although there is a legal obligation to tell the owner about the application. There are various types of consent, and it is important to appreciate the differences between them, which are quite specific.

Outline planning consent is authority for a dwelling of some sort to be built, and it is always subject to a condition that full details of the proposed development will have to be approved before any work can start.

Approval of reserved matters gives permission to build a specific house or bungalow on a site which already has outline consent. An application for approval of reserved matters has to give details of the design, materials, access etc, etc, and it must never be assumed that because there is outline consent on a building plot that the approval of reserved matters is just a formality.

Full planning permission is granted when an application covers all the matters regulated by the planning authority in one application. It can be considered as a combined outline consent and approval of reserved matters. There may be conditions attaching to full consent, such as that subsequent approval has to be given for the actual type of brick to be used, or regarding details of an access, but these are invariably minor issues which can be dealt with directly with the planning officer and do not have to go to the planning committee.

The words *permission* and *consent* are interchangeable in all that you read here (and in anything else) about planning matters.

Conditional consent is given when the planning authority approves a particular development subject to it being for a specific use. A typical situation is where consent is given for a farm cottage which is conditional on the dwelling being occupied by a person who works on the farm. The actual wording used is often critical, and in a case like this would typically read "Occupation shall be limited to a person solely or mainly employed in agriculture, or last so employed, or the widow of such a person.".

Conditional consents have a fascination for prospective self builders who often have ideas of establishing a connection with agriculture to take advantage of them, perhaps by raising a few calves in an outbuilting or having ponies in a paddock. This course of action is not recommended to any but genuine sons of the soil,

and anyway long term finance for a dwelling with such a consent is usually only obtainable from agricultural credit sources. Also the fascination may lead to building a house and being in contravention of the consent, and local authorities have excellent track records in winning confrontations in such situations.

Planning applications

A Planning Application has to be acted on by the planning authorities within a set period, and once a decision has been made it is on record. However, an application can be withdrawn by the applicant before a decision is made and this is useful in some circumstances.

Planning refusals

A Refusal of a planning application will be taken into consideration by the planning authority whenever any further application is made, but does not preclude a further application being made to do the same thing in a different way. However, a refusal of an application for approval of reserved matters can mean that any re-submission must be a full planning application, so it is usual to withdraw such an application

The various leaflets which you can obtain at your local planning office describe the planning law very well, but are written from the official viewpoint. They do not explain the best way to avoid a refusal.

before a decision is made if bad vibrations are detected in the planning office. More of this later.

Conservation areas

These are areas of particular architectural importance, where all proposals are subject to special scrutiny and where any new building is expected to blend in perfectly with its surroundings. Conservation areas are usually to be found in town centres, in particularly attractive rural villages and in beauty spots. Building in a conservation area can cost a great deal over the odds because the planners invariably require expensive material to be used, but the resulting property is likely to be worth correspondingly more. Building plots in conservation areas rarely come on the market, and the prices they command are high.

National Parks

Building in a National Park, or in a site of Special Scientific Interest, or in other areas which are designated under conservation legislation is subject to even stricter control than building in a conservation area. Professional advice from someone who knows the appropriate local planning scene is essential.

Green Belt, white land and residential zoning

These are phrases which are used in different ways and are usually not very specific. They relate to existing development plans for towns or villages, and are used as a basis for dealing with planning applications. "Green Belt" refers to land where the policy is that no development will be permitted, and "White Land" is land not zoned for residential development, but which has not been specifically reserved as green belt land. This seems straight forward, but not all areas of the country have development plans, and these plans are constantly being revised. Furthermore they are regarded by different authorities from anything from a vague statement of principle to Holy Writ. Terms such as these are really only specific to a particular development plan, and the standing of such plans can vary.

Making a planning application

The majority of people who buy a building plot ensure that it already has outline planning consent. There must be very special circumstances for anyone to get involved with buying land that does not have such consent. However, other individual builders already have land, usually because they have owned it for a long time or have inherited it, and want to get permission to build on it.

When one looks at the different types of consents, it seems logical to apply first for outline consent to establish whether a house or bungalow can be built at all, and then, only when this is granted, to move forward to preparing detailed plans and submitting an application for the approval of reserved matters. This course of action is frequently advocated by estate agents, solicitors, and accountants as it facilitates planning a client's affairs without committing him to detail. This approach can be disadvantageous.

Architects often prefer to make a full application. This saves time, and can demonstrate that an application is being made by the person who actually intends to build on the land and live in the house, and is not simply made to establish an enhanced value for the plot. Architects also suspect that planning officers may impose expensive conditions in an outline consent, and might have accepted a cheaper alternative if it had been presented to them as part of a full application. For instance, an outline consent for an infill site in a rural village might be qualified: "the building shall be of two stories, shall conform to the building line of adjacent properties, and shall match them in style and character".

It may have been the applicants intention to build a bungalow hidden away at the back of the site, and this possibility may have been overlooked by the planning officer. Any subsequent application for the bungalow will be judged on its merit, but the planning officer is unlikely to see anything which does not conform to his original recommendations as an improvement.

Whether or not to make an outline or full application can be a difficult decision and can depend on the council involved. Some rural councils have so few applications that the planning committee is able to see the drawings for most new dwellings, and a well presented set of drawings, with a perspective sketch, will create a favourable impression at first sight. In busier areas where the drawings are likely to be seen only by the planning officer, the presentation is relatively less important. Be guided by your architect, or whoever is to submit your plans.

If you are dealing with a situation where a plot already has outline consent and you want to know what you can build on it, you must make a start by obtaining the actually outline planning consent document, or a photostat of it. Under no circumstances should you rely on the account of it that is contained in the estate agents particulars. When you have got it you can read for yourself exactly what the conditions are, and in particular you can look at the expiry date. Outline planning consents are usually valid for five years, and approval of reserved matters are usual conditional on work being started within two years, but nothing can be taken for granted, and you should make sure that you check the actual documents. Failure to do this is a common source of disappointment for those hoping to build for themselves.

The public have free access to planning officers, who invariably have set times when they are available to discuss the relevance of planning policy to specific sites in their area. Their treatment of visitors is more friendly and open than is usual in local government. However, in dealing with them you must recognise two things. Firstly, they cannot discuss planning policy, but must restrict themselves to advising how planning policy will affect your proposals, without committing themselves or the council in any way. A planning officer will advise that an application "could normally expect to be approved" or that he "does not anticipate being able to recommend approval". He is not allowed to be more specific, and it is pointless, and rude, to try to tie him down. What he will do is explain planning policy in such a way that there are plenty of lines to read between. What he will not do is debate the policy itself, or give any sort of promise regarding an application. The only exception to this is when he deals with a minor reserved matter in a full

consent, such as a requirement that the type of brick should be approved. Approval of these matters is usually delegated to the planning officer by the council, and here he will be specific and will often negotiate and compromise.

The second point to consider before entering the planning office is that if you ask for advice, you will get it. Planners have an interesting job, and, almost without exception, they are interested in it. They are anxious to influence development by advice as well as by control, and there is usually a notice at the reception desk saying they are ready to advise on proposed developments. This means what it says. Unfortunately this advice is usually a council of perfection, and it may not suit you to take it. For instance, suppose you are buying an attractive site with outline consent. As a first stage in establishing a design it may seem sensible to call at the planning office to ask for the free advice on offer. You will be well received, will be impressed by the trouble taken to explain how the site "needs a house of sensitive and imaginative design to do justice to its key position". You may be shown drawings, a sketch may be drawn for you, and it is not unknown for the Planning Officer to drive to the site with you. This is splendid, until you realise the ideal house being described suits neither your lifestyle nor your pocket. You will wonder what the reaction is going to be to your application for the quite different house which you want, and which is the one which you are able to afford. The simple answer is that your application will be dealt with on its merits, and that the planning officer has an obligation to approve what he thinks acceptable, and he should not insist on what he thinks is best. However, presumably he will be disappointed and this disappointment must colour his thinking.

From this it will be seen that a preliminary discussion with the planning officer is not to be taken lightly. As a general rule, the best person to deal with the planning office is the professional whom you employ to submit your application.

When your planning application is made you will have to pay a standard fee to the local authority and this is payable whether it refers to an outline application, a full application, or an application for approval of reserved matters. There is no refund if the application is refused. The level of fees is reviewed at frequent intervals, and always adjusted upwards!

Your application will be formally receipted by the local authority, who are legally obliged to advise you of a date by which they are obliged to let you have a decision, which is usually eight weeks after their receipt of your application. The receipt will quote a reference number, and after a week or so it is perfectly in order to ring the planning officer to ask if he is happy with the application, or if he would like to meet you to discuss any aspects of it which require further clarification. Your first approach of this sort can be the start of a huge file of correspondence, often involving a letter from the Council asking you to agree to an extension of the decision date to enable the negotiations to continue at leisure.

The planning officer is very happy with these delays, but you will find them most frustrating. Some idea of what sort of matters with which you will be involved is shown in the extract from an actual letter received as the result of a straight forward application made to a West Country Authority which is reproduced opposite.

Dealing with questions of this sort from a local authority, within an established budget, confident of winning points as well as having to concede others, is a job for a professional.

What else can you do to ensure that your planning application is granted besides retaining the best person to act for you, and discussing with him how it is submitted in the way most likely to succeed? One important thing is to ensure that any queries or requests are dealt with in a very business like way. Letters should be replied to promptly. If a sample brick is required,

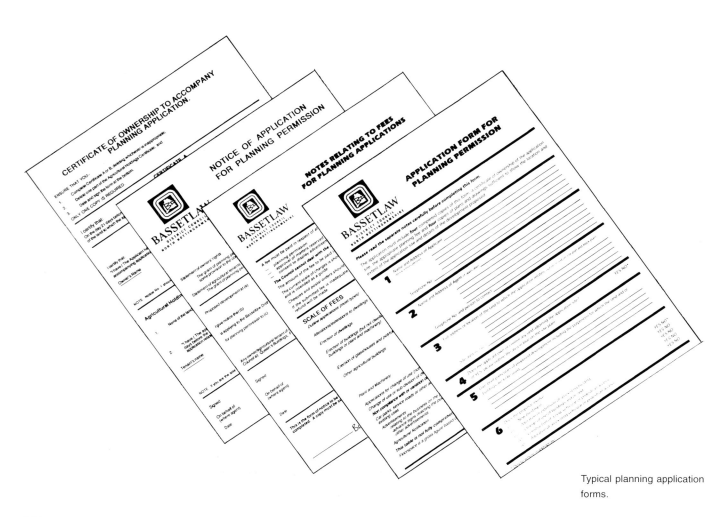

Typical planning application forms.

Negotiating on a planning application

The following is an extract from an actual letter received as a result of a straightforward application made to a West Country local authority:

"I have a number of observations to make regarding this application, and you may wish to comment on them before I make any recommendation to the planning committee.

1. While I appreciate that the dwelling has been sited to obtain a view from the ground floor windows over the trees to the south east, it is now well behind the building line and this does little for the street scene. I suggest that moving the building forward by four metres would effect a satisfactory compromise between user and environmental considerations.

2. The proposal to fell the existing horse chestnut tree in the proposed driveway, and to replace it with a new tree two metres to the south, is only acceptable if the proposals specifically relate to a well established indigenous tree of an approved species.

3. The proposed visibility splay to the access is not in accordance with the requirement of the county highways department for an access into a road with the width of less than 6 metres. Presumably details of this requirement are available to you, but if not they can be obtained from County Hall.

4. The proposed roof pitch of 35 degrees is not acceptable and I am not prepared to recommend approval of any roof in this village with a pitch of less than 40 degrees. 45 degrees would be preferable. This also applies to garages, even if situated well behind the building line.

5. While the proposed fenestration is in character with adjacent buildings, the pair of french windows is completely out of character. Cannot they be moved into the rear elevation? If access to the side garden from the lounge is considered essential, a single glazed door would be preferable.

6. Are the soldier course of bricks above windows as shown on the drawing to be finished as segmental arches as in the chapel on the opposite side of the street? This is not clear from your drawings,, but would be welcome.

7. To the best of my knowledge the bricks proposed have not been manufactured since 1985. Is your client proposing to use second hand bricks of this type? Alternatively a full specification and samples of the bricks proposed will be required.

As the proposals stand I do not feel that I can recommend the committee to approve your application. However, if you can be available to discuss this I am sure that we can achieve a satisfactory compromise, and look forward to hearing from you with a view to making an appointment.

Yours faithfully,

Planning Officer."

Dealing with questions of this sort from a local authority, within an established budget, confident of winning points as well as having to concede others, is a job for a professional.

send it in a plastic bag with a proper label with your name and application reference number on it, and do not simply hand in a brick at the reception desk and say it is for Mr. Bloggs who will know what it is all about. Make everything easy for the planning officer, so that it is easy for him to say 'yes'. Next, keep copies of all letters and make a record of meetings, writing to confirm any agreement reached. Such a letter might read:

"I am writing to confirm that, as discussed at our meeting on January 23rd, we are agreed that the building line should be set at 6.8 metres behind the front boundary, etc., etc."

If the planning officer is promoted to a different desk on January 25th, his successor will normally feel bound by your letter.

It is usually important to suggest that the planning officer may wish to visit the site and to offer to meet him there. This offer should be in writing. He may or may not meet you as you suggest, but if he does not, and you decide to appeal against a refusal, your letter is very useful in suggesting that he did not give your application proper consideration!

Next, consider whether you should enlist the help of your local councillor, or of any other councillor who is on the planning committee. This is not always simple. First of all, no councillor is likely to want to get involved unless you live in his ward, or he is a member of a relevant committee, so do not waste your time with a councillor whose name was not on your ballot paper and who serves on the finance committee. You can get details of which councillors do what at the Town Hall, including their addresses and telephone numbers. How you approach them and what you say is up to you. However, before you reach for the phone remember that they would prefer to see you at a ward surgery rather than receive a phone call while they are watching 'Coronation Street'. Secondly, think about whether or not you may stir up a hornets' nest by drawing a councillor's attention to your application. If other people are going to be affected, or think they are going to be affected, he may ask them their views, and they may have more

influence with him or her than you have.

This leads to consideration of whether or not you are going to seek support for the application from neighbours. This can often be very useful — see the story of Colin Spooner's home on page 179 — but it can backfire badly. If you are going to approach others for letters of support, then have a nice sketch of the new home to show them, illustrating it in a way that will help them to appreciate that it will be small, unobtrusive, and will enhance the appearance, authority and value of their own properties! Remember no one wants a new house next door that will be more impressive than their own home.

Finally, if you are not satisfied with a decision made about a planning application, you have a right of appeal, but it is often better to make the same application again in a way which you think is more likely to be approved. If this seems a possibility, visit the planning officer and ask his advice on this in a straightforward way. If you do not get anywhere, only then should you consider an appeal.

Caravans and the planners

Caravans in general are unpopular with the planners, and even more unpopular with those who live next door to them. If you are putting a caravan on your site, either to use as a site hut or to live in while you are building, it is useful to know where you stand, and what your rights are.

Basically, providing that the caravan is there to facilitate 'building or engineering works' you do not need any consent for it even if you decide to live in it. The key here is the phrase *building works*, and you should avoid putting the caravan on the plot before the work starts, as if you do you probably would have to apply for planning permission. Assume that the neighbours will be very apprehensive about it when they see it, and find an opportunity to reassure them that it will be taken away just as soon as the house is finished. However, they may still go to the planning officer to complain, and it is not unusual for the planning officer to write to you to ask that it should be removed or a planning application made for it to remain.

If this happens write back pointing out that

General Development Order 1988 Part 5 exempts caravans 'employed in building works' from control, and that the application of this to selfbuilders was discussed in the Journal of Planning and Environmental Law 1982, pages 726 to 729, of which you are sure he is aware. Further refer him to Adams v. Shady 1985. If he perseveres you can string out the correspondence knowing that you will have completed the house and sold the caravan long before any action can be taken against you.

Other advice about caravans on a selfbuild site will be found on page 174.

Appeals

Before considering an appeal, you should seek an interview with the planning officer after you receive the refusal notice. At this meeting it is most important that you are not aggressive, sarcastic, or waste his time by explaining how unfair you consider the planning laws to be. You should be seen to be a most reasonable person, and take an early opportunity to say that you know he cannot commit the council in any way in his discussions with you. This will make him much more likely to be helpful, either by suggesting a way in which you can frame a further application which he may be able to support, or by giving you a clear idea of exactly what the council thinks about the issue, which is going to be useful when you are considering appeal tactics.

If the planning officer will not discuss the matter in a helpful way, do not assume that he is being deliberately unfriendly. It may be that your category of application is a very hot political issue locally, and that he feels obliged to deal with you in a very formal way. If this is the case, you can at least ask him to advise you about the planning history of the area and ask for a copy of the recommendation which he made to the council in respect of your application. Of course, you should have been aware of the planning history before you made your application, but you may find to your surprise that there have been a string of previous refusals on applications made by other people, and that this was a major factor in the councils decision.

It may be that the council considered that the design, siting, materials or some other feature of the proposed dwelling was inappropriate. If this is the case, it is even much more important that you establish an effective relationship with the planning officer to see if you can reach agreement on features which he will find acceptable.

All of this adds up to the advice that you should try every way of getting approval by making further applications. A planning appeal should be your last resort, both because it effectively closes the book if it is unsuccessful, and also because it will take between five and nine months to get a decision. This time lag is usually critical to those building on their own. However, if you do decide to appeal, you will be pleased to know that the odds on winning may be better than you think.

About 40% of all planning appeals are allowed, which means that they are won by the person making the appeal. If your own case is a strong one, your chances are considerably better, because many appeals are made by those who are trustees or executors and who want to establish the planning situation on a piece of land beyond all doubt for legal reasons. Others are made by those who fail to realise that their original application did not have any chance at all. Take them out of the statistics, and you probably have a 50-50 chance. In some rural areas planning applications for individual houses are such a contentious issue that councils have got into the habit of issuing refusals knowing that there will be appeals which will probably be allowed, thus avoiding trouble with local conservation groups etc. This is no way to run a planning system, but such is life!

If you are going to appeal, the first thing to do is to write to the appropriate office of the planning inspectorate asking for their guide to planning appeals, and the appropriate forms on which to make an appeal. A leaflet giving details of the address to which to write will have arrived with your notice of refusal. You have six months from the date of the refusal notice for your appeal to reach the inspectors office, but you probably want to get things moving as soon as possible. The guide to planning appeals is 30 pages long, written in plain language, and very user-friendly. You will discover from it that there are two sorts of appeals, written appeals and public enquiries. The booklet explains the differences between them and gives excellent general advice.

At this point you have to decide whether you are going to handle the appeal yourself, or whether you are going to retain someone to deal with it for you. Many people do handle their own appeals, and the procedures are not difficult. The inspectors are not influenced one way or the other by finding they are dealing with the appellant rather than with a professional. However, the professional may be much better than you at marshalling and presenting the facts, and it is the facts on which the appeal will be determined. You should be very sure of yourself before you decide to handle your own planning appeal.

If you do not do so you will have to find someone to deal with it for you, and it is important that he or she has a great deal of relevant experience of handling appeals against refusals to allow individual houses to be built on individual sites in your local area, preferably with a track record of winning! It may be difficult to find the right man: your solicitor or architect should be able to advise you, or if you are on good terms with the planning officer he may be able to point you in the right direction. If you then take your papers, or better still a summary of your papers, to the person concerned they should be able to quote you a fixed fee for which they will handle everything for you. Beware anyone who says they will require a fee just to read your papers and appraise your situation. He or she is probably far too high powered for your job, and is probably more used to conducting planning appeals for supermarkets. You can simply ask a planning consultant "are you the right person to deal with this for me? If not, can you suggest someone else who I should go and see?".

Any full description of how to conduct a written planning appeal is beyond the scope of this book, but the key points are as follows.

An appeal is against the reasons for refusal, and starts off with your written submission explaining why these reasons are inappropriate.

This is on an appeal form which will be sent to you from the inspectors office. You do not have to set out the reasons why your application should have been granted. You must deal only with the actual reasons for refusal listed on the refusal certificate, and explain that they are unreasonable. It will require a considerable mental discipline to restrict your submission to this simple formula, but anything else which you write is irrelevant. If a professional is putting this document together for you, you should ask him to let you have a look at it before he sends it off. If you are dealing with the appeal yourself, then somehow you should try to look at papers relating to other appeals. The English used is not important, nor is the quality of the typing or handwriting, but it is essential to restrict yourself to dealing with the reasons for refusal. Avoid any extravagant language. Do not describe the councils decision on your application as a diabolical liberty: the correct phrase is that it "failed to take all the circumstances into account"!

A copy of your opening broadside is sent to the local planning authority, which then has four weeks in which to produce its own written reply. If it does not reply within four weeks, then the appeal carries on without the council having a say. It is therefore worthwhile finding out when the area planning officer who dealt with your application is going on holiday, and make sure that this coincides with your appeal. Everything will then be dealt with by his overworked colleagues who are standing in for him. You will receive a copy of the council's written statement, and you have two weeks in which you can submit your comments on the council's comments. It is unwise to arrange to be on holiday during this period.

While all this is going on you have to consider mobilising public support for your appeal. This requires even more thought than drumming up support from neighbours for a planning application, and if the appeal site is in a conservation area or is otherwise of general interest, you may find that someone who is opposed to your plans will involve the conservation societies and others. Discuss all this with

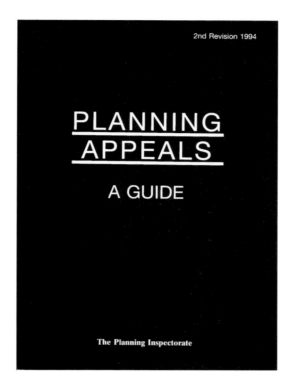

whoever is acting for you, or if you are on your own, proceed with caution.

Following this exchange of statements there is a long pause, and then after some months you will receive a letter from the planning inspector concerned saying that he proposes to visit the site on such and such a date. The site visit is important. The local authority will be represented, and of course you will go along yourself, accompanied by anyone representing you. At the visit the inspector will not allow either party to make any further submissions, nor will he discuss their written submissions with them. His purpose in visiting the site is simply to see the situation on the ground, and you and your agent, and the planning officer, are simply there to answer his questions. These will take the form of "I take it these are the three trees to which you refer" or "thank you for putting in these pegs showing the proposed site of the house. I presume both of you are agreed that they are in the correct place?"

It is most unwise to try to influence the planning inspector at the site visit, although the temptation to do a little stage management is often irresistible. If proposals involve turning a piece of derelict land into a beautiful garden with a handsome house set in the middle of it, some appellants decide to make sure that there is even more offensive rubbish than usual on the site at the time of the visit. You may time your planning appeal so that the visit takes place in summer when trees will be in leaf and conceal problem aspects of your proposals. However, bear in mind that the inspector has seen all of these subterfuges many times before.

Finally, about a month after the site visit you will receive the inspector's findings, and these are final. If you have won, they act as your planning consent. If you have lost you must realise that you have come to the end of the road. Only if there is a major change in the local planning situation is any further planning application likely to be successful. The whole of this process will probably take between five and nine months. You can do nothing to hurry it up. There is provision in the legislation for the inspector to give an Advanced Notice of Decision before he issues his official finding, but this is very, very unusual. In over twenty years of dealing with appeals for selfbuilders the writer can only remember one such occasion. A couple wanted a house on their successful small holding. The council said that the holding could not be economically viable, refused the house but gave temporary consent for a caravan for twelve months, indicating that consent for a house would follow if the venture had not failed. Four years later the small holding was prospering, but four further applications for the house had all been rejected on the grounds that the holding was not economically viable. There was an appeal. At the site visit the inspector was totally noncommittal. When he had seen everything he turned to the lady who had coped with living in a caravan for four years with small children and said that he had seen all that he wanted, and that he would issue his decision in a month's time. He paused, and for the first time he smiled. "It will seem a long month", he said "I suggest you spend the time choosing curtains".

But it isn't always like that.

11 Having your plans drawn

You have found a site, you know roughly what you can build on it, and now you need some plans. Hopefully you have realised that the design of your new home will be largely determined by the characteristics of the site where you intend to build it, by the policy of the planners who will approve your proposals, and by your own concern to strike the right balance between getting the features that you want regardless of expense, and your overall concern with building costs. You have been buying books of plans, collecting catalogues from the package companies, and have probably been looking at every other new house that you see with a critical eye. What you have to do next is to put your ideas in order, and to draw up the list of design features that you want.

This is the essential first step. Until you have got hold of this list, and have discussed it with your nearest and dearest and made sure that a veto is not going to be applied at a later stage, you should avoid considering any particular plan from a book or catalogue as the answer to your requirements. Above all you should avoid sketching out possible layouts on a piece of graph paper. We are all experts at communicating with words, while very few of us are experts at graphic design. It is important to make your list first, differentiating between things that are essential to you, things that you hope can be arranged, and things that would be very nice if they are in any way possible. When you have drawn up this list, by all means have a go at designing a house yourself. However, you should not show your amateur effort to the professional who you are going to retain to handle the actual drawings, as you are paying him good money to exercise his own design skills. Give him your list of features, show him the site, and wait to see what he suggests. You can compare it with your own sketch later, and possibly you will think that yours is far superior. At that stage you can show him what you have drawn yourself, and invite him to criticise it. Whatever the outcome of the debate, you will be discussing design concepts that are specifically related to your own requirements, which have been clearly defined.

The next step is to obtain a design study. This is a simple drawing which shows a proposed design in sufficient detail for it to be the basis of full design drawings. Typical design studies are shown on pages 124 to 127. A design study is meant to be discussed, altered and changed until it shows the home that the client wants. Then, hopefully there will be no more changes. It has to show the building on its site, so that distances from boundaries, access, shape of garden and many other matters can be considered, and it may incorporate a perspective sketch. The designs in the books of plans published by the design firms are design studies without site plans.

How do you choose the man or woman who is going to provide you with this study? They will be either an architect, or a designer who is not an architect, or someone from a company that is offering you a package that includes architectural services. Obviously you are concerned to make the right choice.

Architects design less than one in ten of all new individual homes. This is a pity, because they are certainly the best qualified people for the job. There are various reasons why they deal with such a small proportion of this work, and these include the fact that the cost of running their offices is such that they may not find it profitable to handle commissions for individual houses, or more simply that they price themselves out of the mass market. Also, in the opinion of many, they are so lacking in communication skills that they fail to establish a rapport with prospective clients. This is a pity.

An exception to this are the architects who promote their services as Associated Self Build Architects, who specialise in selfbuild clients. They are all in separate private practices, but their association produces fact sheets, runs advice centres at exhibitions and organises local seminars. They offer free initial consultations and you can contact them on Freephone 0800 387310.

It is not usual for architects to suggest to clients that a design study should be the first stage in a commission. This is because an architect does not want the client to ask himself "do I want to use this architect's services if this is how he interprets my requirements?" He would far rather keep the arrangement moving gently along, until it reaches a compromise between his professional advice and the client's ideas, and he is established firmly as the client's professional agent. Maintaining this role is difficult because the client, wanting a modest house within a fixed budget, feels that he wants a clear idea of what he is going to get before he commits himself. Also, at some point in the design process, costs have to be quantified, and the decision made that the design proposed is practicable within the budget. An agreed fee for a design study enables this to be done, and avoids the risk of a design developing to full working drawings for a building which the client cannot afford. This may seem unlikely, but it often happens and is only rarely the fault of the architect. It usually stems from the client not telling the architect exactly what he can afford, and from both parties avoiding the ungentlemanly topic of money in their discussion. If money is a key factor in the whole business it should be treated as such, and at an early stage the client should insist on costs being discussed. It is also essential to discuss fees generally and to ask if the architect will provide a design study at a fixed price.

Established practices will not like to be asked how much design a fiver will buy, but an enquiry explaining that one is seeking preliminary sketch designs for a new home to enable budget costs to be considered, and enquiring what the cost of such a design study would be, indicates that you are likely to be a realistic client. The reply may be that the practice prefers to be retained on a firm basis, in which case it is probably so successful that it does not touch anything in your price range at all. The reply is more likely to suggest a meeting at which a figure of between £50 and £200 to cover the study will be proposed. It may equally lead to a phone call from a newly qualified architect offering to come and talk about what he can do, and to show you preliminary sketches absolutely free. Whatever happens, the prospective client should realise that it is wrong to commission a design study, decline to go further with the person who drew it, and then have the same design drawn out in detail by someone else.

With the package companies, or those selling sets of plans, things are more cut and dried. They all publish details of their standard designs, and set out clearly the commercial basis of their arrangements with their clients. If designs can be modified, or drawn to a client's own requirements, this is detailed at length. Most have experienced field staff who call on enquirers without obligation, and who will arrange site surveys and other work for a fixed fee. Above all, they offer prospective clients the opportunity to see show houses or other completed buildings for which they have been responsible. They do not accept all the responsibilities borne by architects, and they sell a standardised service which may or may not suit an individual's requirements. For the average selfbuilder they have one advantage which may not at first be obvious; as commercial operations, with no professional trappings, it is easy to reject their services after a preliminary meeting. The service they offer varies in competence, but with well established companies this only varies from adequate to excellent, as the need to get drawings approved under the building regulations filters out the substandard. Even so, it is worth asking if drawings are prepared under the supervision of qualified staff architects. A staff architect is important, as a consultant architect to a package deal company may not handle the affairs of individual clients. If possible you want the signature of a qualified man on the bottom of your drawing.

It is easy to get the idea from publications that the great majority of individual homes are built either to architects' designs, or come from the package companies. If you go to the Town Hall and ask to see the planning register you will quickly discover that the overwhelming majority of applications are made by designers who are not architects, many of them working on a part-time basis in the evenings when they have finished their other jobs. Many of them are council Building Inspectors and the like who will only design for the next door council's area, not the one that they work in. Some of them are listed in Yellow Pages, but all of them get most of their business by personal recommendation. If they have been providing this service for any

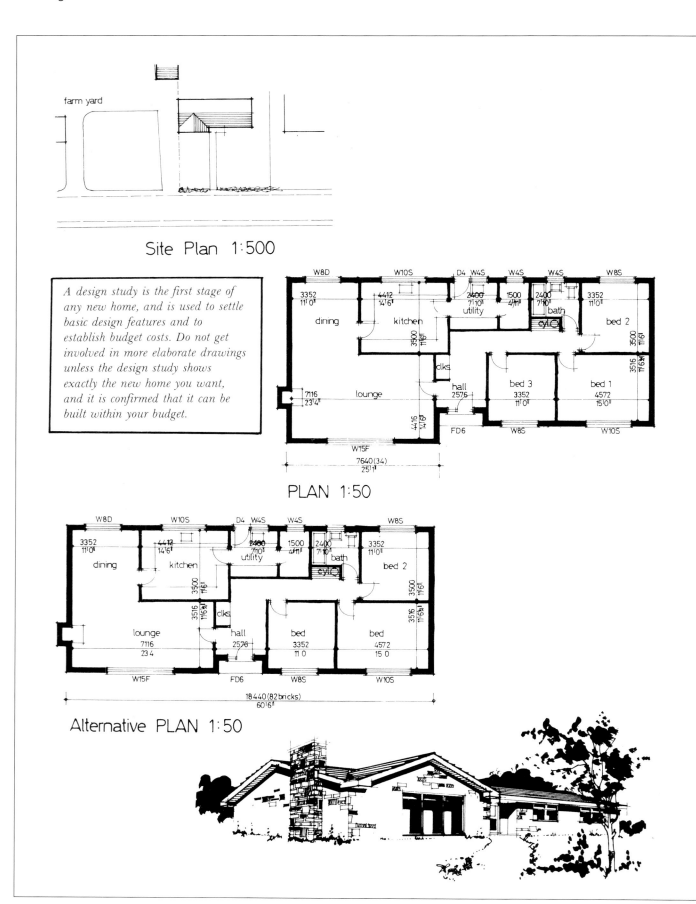

Site Plan 1:500

A design study is the first stage of any new home, and is used to settle basic design features and to establish budget costs. Do not get involved in more elaborate drawings unless the design study shows exactly the new home you want, and it is confirmed that it can be built within your budget.

PLAN 1:50

Alternative PLAN 1:50

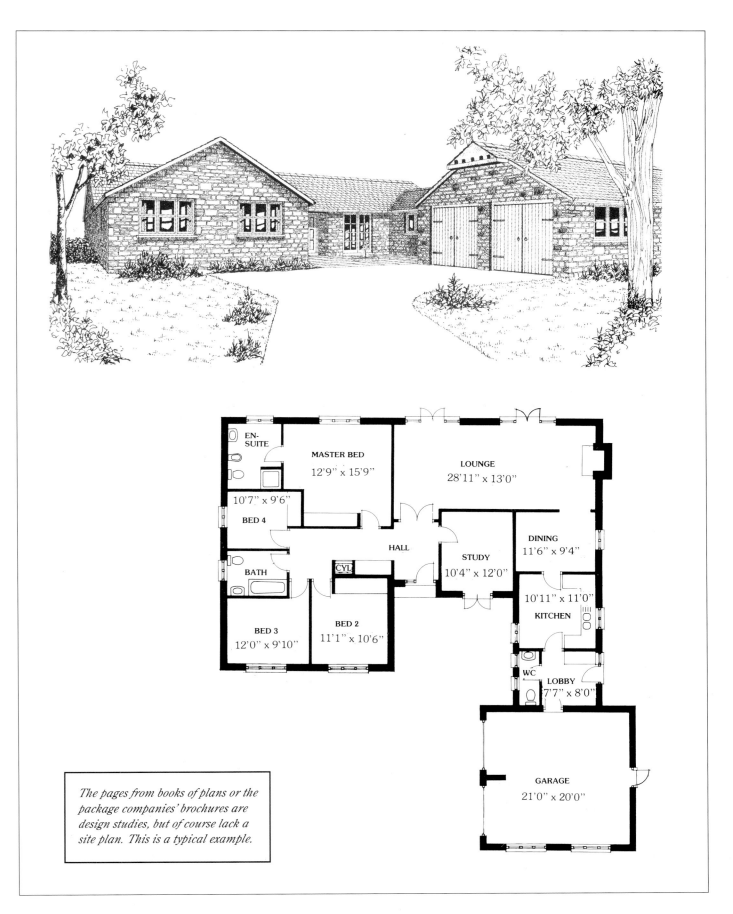

ENSUITE

MASTER BED
12'9" x 15'9"

LOUNGE
28'11" x 13'0"

10'7" x 9'6"

BED 4

HALL

STUDY
10'4" x 12'0"

DINING
11'6" x 9'4"

BATH

CYL

10'11" x 11'0"

KITCHEN

BED 3
12'0" x 9'10"

BED 2
11'1" x 10'6"

WC

LOBBY
7'7" x 8'0"

GARAGE
21'0" x 20'0"

The pages from books of plans or the package companies' brochures are design studies, but of course lack a site plan. This is a typical example.

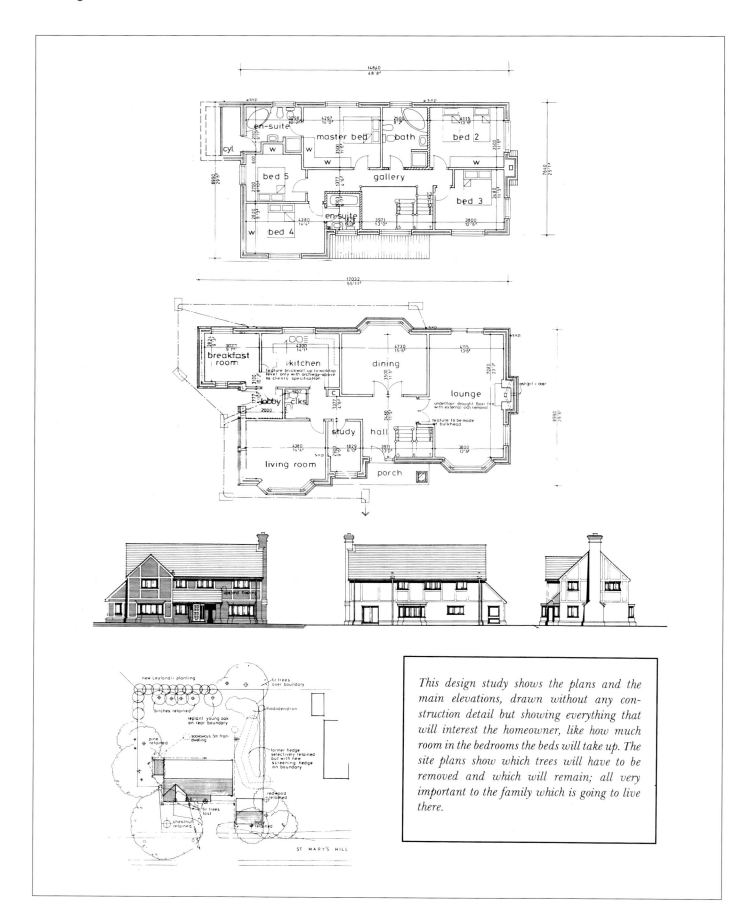

This design study shows the plans and the main elevations, drawn without any construction detail but showing everything that will interest the homeowner, like how much room in the bedrooms the beds will take up. The site plans show which trees will have to be removed and which will remain; all very important to the family which is going to live there.

PROPOSED DETACHED HOUSE for Mr A.WILLIAMS

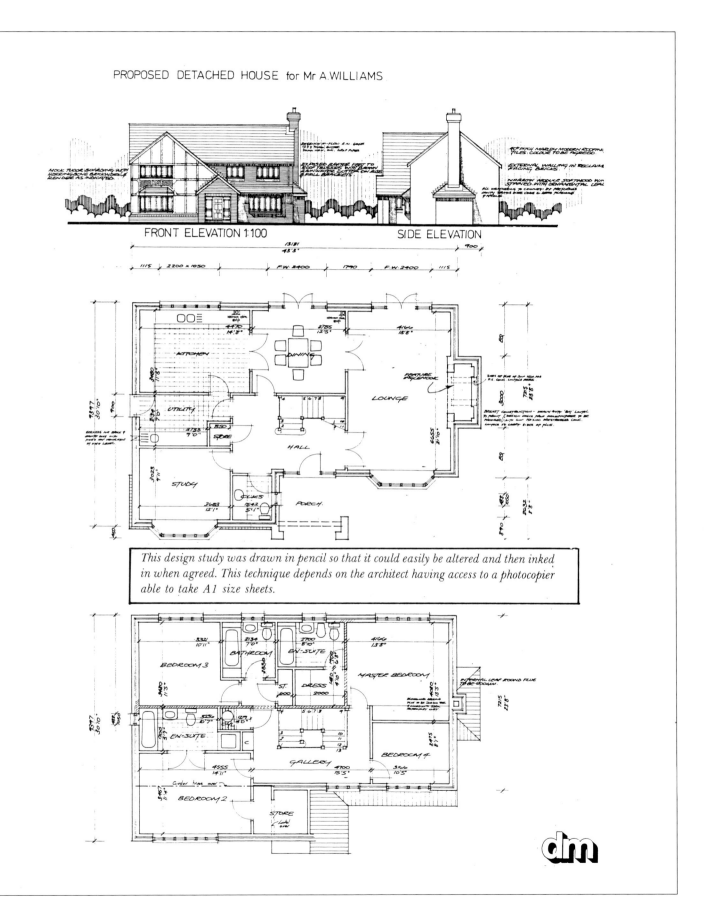

FRONT ELEVATION 1:100   SIDE ELEVATION

This design study was drawn in pencil so that it could easily be altered and then inked in when agreed. This technique depends on the architect having access to a photocopier able to take A1 size sheets.

length of time they are pretty good at it, although of course they do not have the professional indemnity insurances of the architects and the companies. What they are very good at is gaining the confidence of clients, who happily put their trust in them however well this may be deserved.

Most selfbuilders make their choice of someone to handle their design, drawings and planning and building regulations application from the personal recommendation of a friend, or after deliberately seeking out other selfbuilders and asking them for advice. The package companies understand this very clearly, and will encourage you to meet their satisfied clients. Whoever you choose there are some golden rules that you must follow.

First of all, you need a letter setting out a firm basis on which your designer or architect is going to work for you. This should set out what he hopes to do for you stage by stage, with a clear statement of the costs involved. Either he writes it to you, or you write to him confirming verbal arrangements. Above all, it must set out what your financial liability is going to be if at some stage you decide that his design is not right for you, and you wish to go elsewhere. The best way of dealing with this is to insist that you simply pay for the design study as a first stage. This study should be detailed enough for you to be able to understand the designers proposals, and if appropriate, to get the opinion of the planning authority regarding them. As already discussed, this is not the way that most designers like to work. They would far rather let the thing develop, stage by stage, with them firmly in the driving seat. As it is your money that is going to be spent you may think that you wish to be in the driving seat yourself. The choice is yours.

Once you have a design study which meets your requirements, then you have to discuss with the designer your tactics for obtaining planning and building regulation approval. This is a subject to which considerable thought should be given. You may live in an area where planning consent for any new property that is in a local style can be taken for granted providing that you already have outline planning permission. On the other hand, your site may be in an area where

various authorities are concerned and take months to come to decisions, usually after a number of re-submissions. Presumably you have chosen your designer because he has local experience and knowledge of the local planning scene. Make sure you take his advice concerning the best way of handling your application.

For instance, there is the whole question of whether the planning officer should be shown your design study in order to give his opinion on the general design concept, or whether in all the circumstances it is better to send in fully worked up drawings accompanied by a letter that points out very firmly how they meet all the local design criteria, implying that you will fight any suggestions for any changes. Decide which route you want to take with care.

Before your architect or designer does any work beyond the design study, you should establish a contract with him or her for whatever you decide you want them to do for you. With an architect this is easy. The architect either gives you three booklets produced by the RIBA for architects' clients, including one called 'Architects Appointment', or else you send to the RIBA Publications Office for them. The address is at the back of this book. The architects' appointment booklet describes all the different stages in an architect's involvement with a building project, and you can discuss with the architect which of these you want handled for you. Most individual selfbuilders retain their architect in one of the following ways. These are:

*   To establish a design, prepare working drawings, obtain planning and building regulation approvals, and provide the client with ten sets of approved drawings.

*   Or to handle design, drawings, planning and building regulation approvals, and find a builder for the client and help the client to place a suitable contract with the builder.

*   Or to handle design, drawings, planning and building regulation approvals, and to find a builder and supervise the work of the builder in all respects.

*   All or any of the above plus the issue of

A house in the Georgian style by Potton. If you can give your architect or designer a clear idea of your stylistic preferences it will help them to do their best for you.

architects progress certificates at stages required by a bank or building society.

Whatever you decide you want of your architect, you should discuss in detail with him exactly what is involved, and the fees to be paid. The Architects Appointment booklet lists standard fees which are a fixed percentage of the value of the building work, but they are not mandatory and it is much better to agree a fixed fee. Establish that this fee will include all travelling and other expenses: the RIBA suggested terms allow for such expenses, but it is far better to have a lump sum arrangement. All of this should be confirmed in writing. If you are using a designer who is not an architect it is even more important that you establish exactly what he is going to do for you, and the fixed fee arrangements, as there are no standard conditions of engagement as with architects.

The package companies will present you with very clear and unambiguous quotations which will carry their conditions of business. You are most unlikely to be able to negotiate on these: if you are in doubt about them, consult your solicitor. Selfbuild consultants who provide you with an architectural service will probably give you a loosely worded quotation with a number of standard conditions which will be worded to limit their liability. If the consultant is using an outside architect, who is not on his staff, it is far better to

try to insist that you deal with the architect direct as only in this way will you have the benefits of the architect's professional indemnity insurances. However, you may discover that a consultant's 'architectural services' are provided by a designer who is not an architect. If he does the right job this is immaterial, but make sure that the contract suits you.

Finally you will want whoever is drawing your plans and dealing with your planning and building regulation applications to get on with the job to your programme and not to put your affairs on the back burner while he gets on with something else. For this reason a note of your required timetable should be part of the contract. You also want to be advised of how things are progressing. The easiest way of dealing with this when given a standard contract to sign is to write in beside your signature 'on the basis that final drawings will be available for approval 1/10/93, submissions made by 14/10/93 and that we are advised routinely of progress with the submissions'. This will help to keep you in the driving seat!

Other professionals

Besides an architect or designer you may require the services of other professionals. The essential solicitor has been discussed on page 56. You may also need planning consultants, civil engineers, soils engineers, quantity surveyors and others, although it would be very unusual for a self-builder to become involved with all of them. If you need them, how do you find them and how do you ensure that they do the right job for you?

As with so many others who you need to find to be able to build for yourself, the best way to start is with a personal recommendation. If this is not possible you can follow up advertisements, or take names from the Yellow Pages, or ask the professional bodies like the Institute of Town Planners to send you list of their members who work in your area. You might get help from the planning office or the building inspector: They will not recommend anyone, but if you take a photostat of the relevant pages from the classified phone directory they can usually be persuaded to mark the people who regularly work for self builders.

Next, write to the people or firms you wish to approach, and make sure that your letter gives them enough information for them to decide if *they* want to work for you. A letter to an engineer might read:

I have a building plot at Dogswood with planning approval for a house which I wish to build before Easter. The ground is suspect, and I wish to arrange for appropriate investigations and foundation designs. If this small contract is appropriate to your practice I would like to hear from you as soon as possible, with details of the service you can offer.

Now sit back and wait to see what happens. If you get a prompt reply, perhaps with a follow up phone call, you can know at least that the engineer replies to letters. If you do not get a reply within two or three days you can draw your own conclusion!. The right letter will lead to a meeting and, if appropriate, you can suggest that this is at the site, which will tell you if the individual concerned is interested enough in working for you to leave his office.

At the meeting you should explain very clearly what you want to achieve and then expect to be told in detail how the expert can help you, how long it will take him, the fees that he will charge, and any warranty or insurance cover that comes with his services. He should take great care to explain everything to you in detail, and apart from asking questions there is no point in parading any little knowledge that you may have yourself. Your job is to evaluate how well you think he will suit you, and how well he explains everything is a good test of this. Do not take up too much of his time, thank him for his suggestions, and say you will consider all that he has explained and will write to him soon. Then consider carefully if he seems able to do the job you want, to your deadline, whether you can afford him, whether his qualifications and insurance cover are acceptable to your building society if this is relevant, and simply if you like him. Liking him is probably just as important as all the other issues. If everything fits, write and ask him to let you have a detailed quotation for

the work as you intend to retain him, or else write promptly to thank him for his interest but say that you are making other arrangements.

When the quotation comes, think carefully whether it covers all the important aspects, and preferably show it to someone with experience in these matters. If it refers to standard conditions of engagement appropriate to the profession, a copy should be attached: if it is not there, ask for it. There will probably be things that you want to elaborate, and if you accept the quotation you can either send a letter saying that you accept subject to XXX, or you can simply scribble your acceptance and special requirements on the quotation and return it, keeping a photostat for your records. Everything must be specific: the time for the work to be done clearly stated, the cost quoted as a lump sum or on some other basis, and the arrangements for any other payments, such as travelling expenses, set out in detail.

These are the sort of arrangements which you may make with a soils engineer, but the same routines hold for the others. They should all work to an agreed brief at an agreed fee.

It is important that professionals should report back to you at specific intervals or stages. This will be important to you: planning a selfbuild project can be worrying, and you will want to know how things are getting on. This does not mean you should ring your architect every evening for a chat, but it is certainly appropriate for him to let you know how your planning application is progressing on a stage by stage basis. Contact with him will keep you in the picture, and will also reassure him that you are still alive and thus likely to pay his bill!

All of this makes the initial choice of the right experts to act for you very important indeed. Choose with care, and be reassured by remembering that the longer the fellow you have chosen has been in business the more likely he has given a good service to thousands of happy clients!

12 The Building Regulations

Approval under the Building Regulations is concerned with the design and siting of a dwelling from every angle except that of whether it should be built at all and its appearance, both of which are planning issues. A mass of health and public safety legislation is now administered by the local authorities through their building inspectors, and for many years it has been a requirement that drawings for proposed buildings should be submitted to local authorities for examination and confirmation that they conform to the building regulations. This confirmation is known as Building Regulation Approval. The building inspectors also check that all dwellings under construction conform to the regulations, and it is mandatory to advise them when building work is under way.

Since 1985 architects with appropriate professional indemnity insurances, and other bodies such as the NHBC, have had the authority to inspect and certify the construction of new buildings as an alternative to the local authorities. The various bodies concerned have only recently started to handle this, and it is likely to be some years before selfbuilders have any easily available alternative to the Local Authority and Building Inspector route to Building Regulation Approval. What follows is concerned with this local authority approval of plans, and the Building Inspector's approval of work on site.

Fees are payable with building regulation applications and it is usual for an application for building regulation approval to be made at the same time as the planning application. Approval of a Building Regulation application is a matter of fact - whether the drawings of a building conform to the regulations or not. Virtually every detail of the building, of its siting, and of its services is subject to the regulations and it is not enough simply to advise the council that everything will conform to the rules. All the salient points must be detailed individually.

Having checked these the authorities will invariably write asking for further details, and they may suggest a meeting. Finally, they will issue an approval form, which states that the plans inspected conform to the regulations, together with a set of postcards to be sent off to the council at various stages as construction progresses. On receipt of these cards, and at any other time he thinks fit, the Building Inspector will call at the site to inspect the work. His concern will be that all construction complies with Building Regulations, and not simply to enforce only the details shown on the approved drawings.

An application for Building Regulation Approval starts with your architect or designer completing a very simple form and sending it to the council offices accompanied by two sets of drawings and the fee. The application is acknowledged and the council then has 5 weeks in which to issue either an approval or a refusal notice. Unlike a planning application there is no provision in the law for an extension of this time, so that if there are any queries which cannot be resolved within the 5 week period, your application will be rejected. You then immediately re-submit it so that the debate can continue. Unlike a planning refusal this is not a matter of record, and there is no disadvantage to you in this. No further fee is payable.

The timing of an application is important. It is obviously preferable to have your planning approval dealt with and agreed before submitting the plans for Building Regulation approval, but most applicants are short of time and send in both applications at once. Subsequent alterations to plans to suit the planners may mean a re-submission at the building control office, or, more rarely, vice versa. You would think that the two offices would work together in this, but this is rarely the case. At the same time as all of this is being dealt with, you are possibly seeking approval of your proposals from others, possibly a river board or someone who has a covenant on your land which requires them to approve the design of anything built on it. Obtaining agreement from everybody has been likened to driving a flock of sheep through a gate: you will not get everything dealt with in an orderly fashion, but you will just have to push everything forward as best you can and round up anyone who wants to go their own way!

Once your Building Regulation application is registered a single building inspector is appointed

Negotiations on a Building Regulations application: what to expect

The letter below is a typical communication from a local authority following a Building Regulation application. It is in no way unusual, and irrespective of the amount of detail shown on the drawings submitted, queries of this nature can be expected from most local authorities.

"I refer to the plans referenced above which have been deposited for consideration under the Building Regulations. Your attention is drawn to the following items regarding which I require clarification to enable further consideration to be given to this matter.

1   Further details required of the porch roof construction. Denote siting and method of supporting cold water storage cistern, with calculations to confirm that roof truss design is adequate. Detail wind bracing. Denote waste sizes to sanitary appliances.

2   Confirm arrangements made for cavity to be closed at cills and reveals.

3   Confirm flashings as shown to roof to provide 150 mm upstand.

4   Buttress walls to be taken down to similar foundations as external walls. Indicate strutting proposed at mid span to first floor joists over lounge area. The first floor gable return appears to be undersize, and if not in accordance with that deemed to satisfy provisions, calculations will have to be provided. Detailed proposals for lateral restraint to gables at first floor ceiling and roof level. Detail size of trimmer below bulk head studding over stairwell.

5   Detail section of stairs, including head room to stairs measured vertically above pitch line.

6   Confirm that proposals for balustrades conform with Building Regs.

7   Confirm relationship between ventilation openings and floor area to bedrooms 1 and 4.

8   Full details of the prefabricated septic tank are required, and we will require a standard percolation test to be carried out in accordance with section 3/5/6.2 of Code of Practice 309 1972.

I regret that unless I receive an amended application taking all the above matters into account I will have no option but to reject the plans in fourteen days from the date of this letter."

to deal with it, and it is not usually possible to play one inspector off against another. Before the man looking at your drawings can grant approval he has to be satisfied that the building will conform to the regulations. Your architect or designer will have made this clear by the drawing itself and in the notes written on the drawing. It is not sufficient for these notes to say 'all construction to conform to the building regulations', and many of the regulations have to have their own notes. Which regulations? This depends on the particular authority and the particular hazards of building in a particular area. If you live in a particular part of Hampshire where the longhorn beetle lives, your plans must carry a note about how your roof timbers will be treated to make them unpalatable to it. If you live in South Yorkshire this note is unnecessary, but you will certainly be asked for details of your special anti-subsidence foundations.

However thoroughly your architect or de-signer has compiled the notes on your drawings, the building inspectors can be relied on to raise queries. The letter on page 133 is a typical communication from a local authority following a Building Regulation application. It is in no way unusual, and queries of this nature can be expected from most local authorities.

It will be appreciated that dealing with this sort of query, within the Building Inspector's deadline, which he cannot alter, requires profes-sional advice in almost every case. An architect would consider all the queries raised above to be perfectly straightforward and would expect to deal with them as a matter of routine. Inciden-tally, the seemingly haphazard order of the matters raised in the letter above is the order in which they are dealt with in the regulations themselves.

Many of these matters are covered by stand-ard approved ways of designing a building, and reference to this is all that is required. Timber frame buildings from an established manufac-turer often have a blanket 'type approval', and it is usual for these manufacturers to be closely involved in applications to erect their buildings. However, if your proposal for something unusual is not covered by a standard method of construc-

tion, you will have to provide full structural calculations or other technical details. In this case find out first if a manufacturer involved will provide the details or calculations for nothing. If this is impossible, get an estimate of the cost involved, and then think very hard about a more acceptable alternative.

A special situation will arise if you are building on a site where the load bearing charac-teristics of the subsoil are suspect. In this case the building inspector will want you to submit details of a site investigation and may ask for a qualified civil engineer's proposals for meeting any hazards disclosed by the investigation. This sounds very frightening and can it be extremely expensive, but it is usually dealt with quite easily by someone with the right experience. A site investigation can be anything from your architect and the building inspector peering down four test holes dug eight feet deep on the site by a JCB hired for an hour, through to having bore holes sunk and the cores professionally analysed. Specially designed foundations can be a standard raft with type approval, through to something that has been designed for you by a consultants computer. This is all a matter of local knowledge, negotiation with the local authority, and being aware of the alternatives. This is another reason why you have to have the right man to handle your building regulation approval, and it is also the reason why the need to consider possible foundation problems is emphasised in the chapter on assessing a potential plot. Anyway, remember that if other people have built in the same area they have probably met and overcome your problems, and they may be very pleased to tell you how they did so.

If your proposals involve any unusual drain-age arrangements the local River Boards will be involved. They will require a percolation test, described on page 85, and may tell you that you must provide special soakaways or herringbone land drains which will affect the positioning of the dwelling on the site — which is why any possible drainage problems have to be assessed before you buy the site.

When you submit a building regulation application you will probably also send your drawings to Zurich Custom Build or the NHBC for the building to be registered for a warranty. If there is a special foundation situation or other unusual design features, they will want to see the same paper work as the building inspectors. If these matters involve special meetings it may be worth while trying to persuade the Custom Build or NHBC engineer to attend them and deal with everything at the same time. At this point it must be emphasised that the building regulations derive from health and safety legislation, and are concerned that you will neither build an unsafe structure or start a typhoid epidemic. They are not concerned with standards of finish, which are the province of the NHBC and the Custom Build regulations. Neither do the building regulations have much to say about the plumbing, electrical or gas installations in a new house. These all have their own regulations which are administered by the appropriate services authority. There are many popular guides to these regulations in the bookshops, but they are differently interpreted by different authorities. In particular, the current change from hot water systems that involve an open header tank in the roof to closed systems is being dealt with differently in different parts of the country. Your sub-contractors will talk to you about all of this, and it is essential that you do not make a final payment to them until their work has been approved by the water, electricity and gas boards and the appropriate service connected. If you are in any doubt about any of this it is worth while calling at the local office of the authority concerned. They do not provide an advisory service but if you make your approach in the right way you will probably find that one of their inspectors will be very happy to take you under his wing.

Once you have obtained your building regulation approval your liaison with the Building Inspector continues until the dwelling is occupied. At this stage it must be emphasised that for the selfbuilder his Building Inspector must be a friend. He is responsible for ensuring that construction is as required by the book, and he has no authority to deviate from it. He knows the book backwards, which gives him an advantage over anybody who has not made a special study of the subject. He spends his life dealing with those who are trying to ignore or bend the rules, and dealing with problem buildings. In spite of all this, it is likely that he will have remained a sympathetic, friendly and helpful character, and he can become a guide, mentor and friend to the selfbuilder. Officially they will advise on any detail of the construction, and unofficially they will often recommend sub-contractors, suggest suppliers, and generally make their unrivalled experience available. No one else in the building industry sees all the work on all the sites in an area.

Two aspects of routine inspections which have to be taken very seriously are drain tests, and inspection of the work below ground level before it is covered up. There are various ways in which your drains may be tested, usually involving pressure hoses and pressure gauges. Pipe runs have to be exposed for this, and you must not cover them up even if the inspector's car breaks down and he is unable to keep his appointment for the test, for which he provides the equipment. He will also want to inspect your foundation bottoms before your concrete is poured, and he will be quite ruthless about requiring you to expose work for his inspection which has been deliberately or inadvertently covered up before he has approved it.

13 Timber frame construction

At an early stage in their project planning everyone building a new home has to decide whether they are going to build traditionally, or with a timber frame. In considering this they will quickly become aware of a rather unusual situation. The timber frame option is very heavily promoted, and the advertising, exhibition displays and literature from the companies concerned is very informative and will be most useful. On the other hand, apart from two companies that specialise in a package service for traditional construction, there is no comparable promotion of the old fashioned way of doing things. One could be forgiven for coming to the conclusion that building with a timber frame is rapidly taking over the housing market, and that traditional construction is on the way out.

By 'traditional' the writer means homes with two leaf masonry walls. However, timber frame companies have sometimes used this expression because they claim "traditional" medieval buildings had timber frames! Because of this, construction with two leaf masonry walls is often referred to as 'brick and block' construction. One word being better than three, we shall use 'traditional' construction.

Most prospective selfbuilders are well aware that they see many more traditional homes under construction than timber framed homes, which in fact have about 10% of the whole of the new housing market. If they seek advice from friends about this, they are likely to be told that one or other is the 'only' way to build and that one system or the other has enormous advantages or disadvantages. This is simply not true.

First of all, neither system of construction is either better or worse than the other. The finished homes both look exactly the same and are just as good an investment. The costs are generally comparable. The difference is simply the way in which they are constructed.

Traditional construction is based on load bearing walls which have two skins of masonry with a cavity between them. This cavity may be either wholly or completely filled with insulating material, and the two skins are tied together with metal ties. The external skin is built in brick, stone or concrete blockwork for rendering to give

the building its character, and the internal skin is invariably built with lightweight insulating blocks. In modern construction it is this internal skin which supports the weight of the first floor and of the roof, although the external skin does play some part in ensuring the stability of the whole structure. The partition walls between the rooms of a traditionally built house are built in lightweight blocks where they are required to contribute to the structural stability of the building. Partition walls which are not 'structural' walls are built either in the same lightweight blocks or in timber partitioning, as determined by whoever draws up the specification for the building. In general terms, the more expensive the home the more likely it is that all the partition walls will be in masonry.

Timber frame houses and bungalows have a wooden frame to support the roof and to take the weight of the first floor of a house. This structural frame gives the whole building its strength and it has to be designed and built to very high standards. For this reason the components for the frame come from a manufacturer who has copyright on the details of the engineering calculations of his frame, which are required by local authorities before they issue Building Regulation approval.

The majority of manufacturers have systems based on walling panels which are fastened together to provide a rigid structure which is enormously strong in spite of being constructed of relatively light timber. The panels are factory made and delivered to the site ready to be assembled. Some systems use panels which are only a few feet wide, while other manufacturers supply large panels right up to the limit of the largest unit that can be loaded onto a lorry. Prestoplan, Medina-Gimson and Scania-Hus homes are built in this way.

Another system of timber frame construction uses massive timber uprights supporting beams which make a skeleton frame to take the weight of the superstructure. These are called Aisle Framed buildings. The walling panels are made in a similar way to the panels for the modern stressed skin systems but they are not necessarily load-bearing. The well known Potton homes

The illustrations below demonstrate the difference between traditional and timber frame construction. Remember that it is the inner skin of a building which supports the roof and provides most of its structural strength.

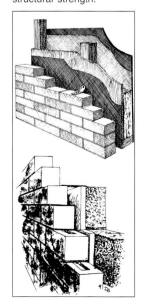

136

Timber frame house by
Prestoplan.

are engineered in this way.

Neither of the systems of construction should be confused with the Tudor style timber frame houses with an external skeleton of heavy timbers to provide frames which are built using fifteenth century building techniques. The Border Oak homes are built in this way.

All of these ways of building require the involvement of a company which will provide the timber frame, and this means that they also have to design the structure. It is not practicable for most selfbuilders to make their own timber frames, however well they think they could do this, as the local authority will require the design calculations to be submitted with the building

regulation application, and it is not generally practicable for anyone who is not a qualified timber engineer to provide the design calculations himself.

Timber frame buildings are usually finished by cladding the frame with masonry, which can be brick, stone or lightweight blocks for a rendered finish as determined by the style of the building. The inside of the walls are lined with plasterboard, and the internal partition walls are built with more plasterboard on a timber studding. These internal surfaces are finished with either a traditional wet plaster, which will take a month or so to dry out, or are given their final surface using a 'dry lining' technique which will not involve any delay before you can decorate.

Both traditional and timber frame homes usually use factory made prefabricated roofs based on complete roof trusses. These are made from relatively lightweight timber so that they can be hauled up onto the roof where they are fitted together in just one or two days. If there are to be rooms in the roof then special trussed rafters called attic trusses may be used, although some traditional homes still use purlin construction where the roof is actually built in situ. This gives more room in the roof than attic trusses, but requires skilled carpenters and takes longer to build.

There are various compromises between the two systems of construction. Traditionally built homes can be dry lined and their first floor walls between bedrooms are often timber framed. Sometimes a whole timber frame first floor is built above a traditionally built ground floor. As far as the ground floors themselves are concerned, a choice between solid concrete floors, beam and block floors and suspended timber floors is usually made to suit ground conditions irrespective of whether the building is to have a timber frame or to be built traditionally, while first floors are now almost universally constructed in timber.

Virtually none of this affects the way in which the new home is going to look, or how pleasant a building it is to live in. The timber frame house will last just as long as a traditional home, will be no better and no worse a fire risk, and will be just as easily mortgaged. Decisions by selfbuilders to adopt one system of construction or the other are usually based on which system of construction gives them the easiest way of getting exactly what they want. Paramount in this is the role of the package companies, which are described in a later chapter. If you want to build a timber framed home you are going to have to use a package company unless you have an architect who will provide the plans and help you place a contract with a timber frame manufacturer which may, or may not, also operate as a package company. If you decide to build traditionally you have a choice between using a package company that deals with traditional homes or going it

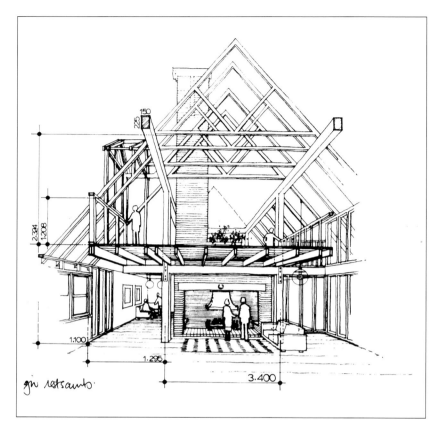

A section of a Potton house showing the aisle frame construction which is completely different from most other timber frame systems.

alone. Most individual builders go it alone.

One advantage of building traditionally is that there is a far larger pool of sub-contract labour who are accustomed to traditional construction. A selfbuilder is most unwise to engage workmen to erect the shell of a timber frame house unless they have the relevant experience, and, unless you are using the construction crews available through the manufacturers, in most parts of the country it is not easy to get competitive prices from a selection of companies with appropriate experience. In the same way there are plenty of plasterers able to work with traditional materials to a high standard, but far fewer are skilled in dry lining techniques. As a result, when selfbuilders are discussing their ambitions with people in the building industry whom they may know socially, perhaps in the pub or at the cricket club, they tend to be influenced towards traditional construction. With only 10% of the market in most of the UK, building with a timber frame is unusual, and there are many influences towards conformity. It is interesting that in much of Scotland, where timber frame is becoming the standard way in which a selfbuilt home is constructed, the same urge to conformity works the other way round.

A further consideration in the traditional versus timber frame argument is that timber frame homes are generally less easily altered than traditionally built homes once they are completed. Many selfbuilders have long term plans for extensions or for a granny flat, and masonry walls are much more amenable than timber frame walls to having doorways cut in them, or being tied into new extensions and annexes. However, if you are building with a timber frame you can easily ask for provision to be made for alterations at a later stage.

Which of these different systems of construction is best for the selfbuilder? As already explained, both traditional and timber frame homes meet very strict building regulations in every way. They last just as long as each other, are no better and no worse a fire risk, and attract the same insurance premiums. Above all, they look exactly the same, and are equally good investments. Values of houses sold through estate agents depend on their situation, and their size, appearance and condition, and few purchasers pay much attention to how they are built.

As far as building costs are concerned, there are no advantages either way, although it is probably true to say that both the very cheapest and the most expensive new homes tend to be traditionally constructed, while timber frame homes sit happily in the middle of the cost bracket.

People considering a new home at this time when we are all energy conscious are naturally concerned about insulation. Mandatory insulation levels required by the Building Regulations are now very high, so high that any improvement on them is really a matter for the energy conservation enthusiast rather than the ordinary selfbuilder. Any required level of insulation can be built into a traditional home, although of course it has to be paid for. The timber frame manufacturers provide a level of insulation as standard in their homes which is often better than the building regulations requirements, and they emphasize this in their marketing.

Traditional construction has an advantage with the other sort of insulation, sound insulation, particularly if all the internal walls are in blockwork. It is up to you to decide how important sound insulation is to you. Extra acoustic insulation can be fitted in timber frame homes, but, as with the extra thermal insulation in a traditional house, it will add to your costs.

Speed of construction is another issue which features prominently in the debate, and a timber

Timber frame house by Potton.

Is this new individual home timber framed or traditionally built? Choose your way to build because it suits you. Both types of construction are equally good value.

frame home usually saves a couple of months in the construction time. The average family building a new home spends about six months putting the whole project together from the time that they first decide to go ahead, and the building work typically takes from four months to eight months. The homes that are built in four months are more likely to be timber frame. Speed of construction is very important to the commercial builder as it enables him to turn over his money more quickly. Whether or not saving a couple of months in construction time is important to you depends on your circumstances.

This leads to consideration of designs. Some timber frame companies will only provide kits for houses from their own range of designs, or straightforward modifications of their standard designs, although a few will prepare drawings to your own requirements. Companies which supply kits for traditional construction will normally draw a design specially for you, although they hold a wide selection of standard designs to whet your appetite.

For most selfbuilders the most important factor in making the decision about how they will build is whether they think they will get the design that they want, with the help that they want, by using a particular company, a particular

architect, or a particular way of building. It all adds up to personal choice, and, very often, to personal relationships. A salesman from a package company, or an architect or a friend who has built for himself before will often give a strong lead to someone who is uncertain about their options, and often he or she will follow that lead. If you are undecided which system of construction you are going to use for your new home, make sure that you make your choice after having visited houses or bungalows of the type that interest you and having met the people who are going to help you make it all happen. Once you have made your decision it will be very difficult to make changes, so give the matter the consideration it deserves. And remember, all the well known ways of building will result in you ending up with a very well engineered home.

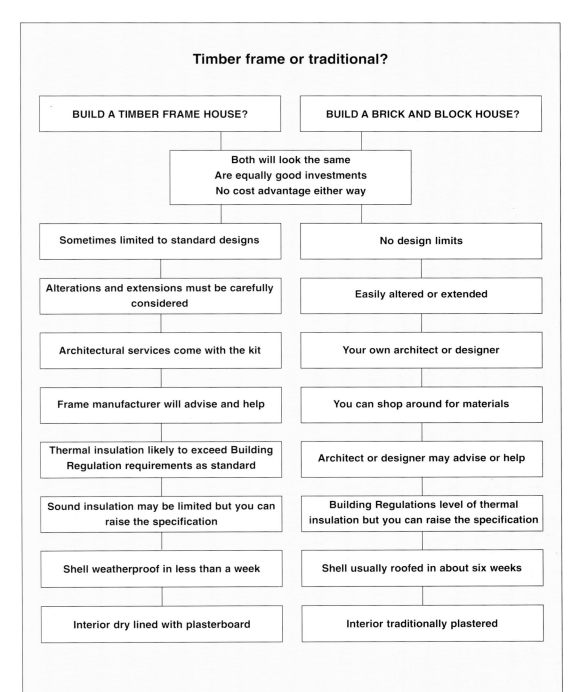

Timber frame or traditional?

BUILD A TIMBER FRAME HOUSE?	BUILD A BRICK AND BLOCK HOUSE?
Both will look the same **Are equally good investments** **No cost advantage either way**	
Sometimes limited to standard designs	No design limits
Alterations and extensions must be carefully considered	Easily altered or extended
Architectural services come with the kit	Your own architect or designer
Frame manufacturer will advise and help	You can shop around for materials
Thermal insulation likely to exceed Building Regulation requirements as standard	Architect or designer may advise or help
Sound insulation may be limited but you can raise the specification	Building Regulations level of thermal insulation but you can raise the specification
Shell weatherproof in less than a week	Shell usually roofed in about six weeks
Interior dry lined with plasterboard	Interior traditionally plastered

This chart is based on the experiences of most selfbuilders. However, the distinction between the brick and block and timber frame routes to a new home is not clearcut. Many Tudor style homes are built with traditional ground floors and timber frame upper floors. You can get an architect to design you a timber frame home and shop around for a frame supplier. You can get packages for traditionally built homes. And you can make special arrangements to have just as much thermal or sound insulation as you want in any sort of home.

14 Working drawings and specifications

Most medieval cathedrals were built with the aid of fewer drawings than are now considered essential for a municipal public convenience, and many selfbuilders follow the former example. However, formal drawings are essential to obtain planning and Building Regulation consent, and if a full set of construction drawings is not going to be prepared for use on a site, the decision to dispense with them should be a conscious one, taken with a clear understanding of the options.

First of all, the term 'drawings' refers to a full set of drawings for both the planning and Building Regulation applications and the construction drawings which will be used to obtain prices and for use on site. These are much more elaborate than the design studies already discussed.

The drawings required by a Local Authority must include:

* A location plan to a recognised scale to identify the site.

* A site plan, normally to a scale of 1/500, to show all boundaries, the position and siting of the building, drainage arrangements, access, etc.

* A building plan at a scale of not less than 1/100, together with at least two elevations of the building.

This detail can be presented on a single sheet, or in any practicable way provided that it is to an appropriate scale. At least six copies are needed, and some authorities require one printed on linen, meaning a cloth reinforced paper. If alterations or additional information are required by the authority, then a further set of prints will be needed. To facilitate this, drawings are normally prepared on tracing paper or tracing film, from which prints are made by the dyeline printing process, although some are printed by computers or special photocopier type machines. Most towns have document reproduction centres where copies of drawings can be made while one waits. The cost is quite small.

It is usual for one set of the approved drawings to be returned with the planning and Building Regulation Consent certificate, with the approval stamp endorsed on it. This approved drawing must be followed when building, and any subsequent alterations agreed with the authority should either be shown on a revised drawing sent to the Local Authority for approval, or else clearly detailed in an exchange of letters. It is important to understand that the details shown on the drawing which has been approved under the Building Regulations are only particular aspects of the work which it is considered should be specially emphasised. All relevant requirements of the Building Regulations have to be complied with, not only those shown on the approved drawings.

The construction drawings for site use may be the same as the ones submitted to the council, or they may be drawn separately. If the latter, they must contain all the details shown on the approved drawings and remember that they will probably be used by sub-contractors as well as by the builder or selfbuilder himself. There are many advantages in using fully detailed drawings. They can save a lot of argument if a problem arises from the work of any one trade. For instance, the way in which windows are set in a window opening determines the width of the window sills. There is no standard way of doing this, and if the drawings define exactly the way in which windows are to be set there will be no confusion. If they are wrongly fixed it will be clear that it is the bricklayer's responsibility to take them out and build them in again as per the drawing. If the right setting was not clearly shown in the drawing the issue would not be easily resolved. There are literally hundreds of such details in every construction job.

Construction drawings are also used by sub-contractors to design their services. An electrician will require a drawing which he will mark up with the wiring layout, and the plumbing and heating engineer will want drawings for the same purpose. Others will be required when the kitchen is being planned. Central heating drawings are often provided free of charge by the fuel advisory agencies, and the kitchen layouts can be obtained from various bodies, but they all start with a print of the actual construction drawing.

The drawings will have been drawn to either metric or imperial scale, although both dimensions may be shown. These days imperial design work is unusual, and may be identified by the scale, typically 1/4" to 1' or 1/8" to 1'. The more usual metric scales are typically 1/100 and 1/50. All setting out of construction work should be done in the units used for the design, and the converted dimensions should be used with considerable caution as they are invariably 'rounded off' and, if added together, will give rise to significant errors.

Remember that room sizes on construction drawings are masonry sizes and that the finished dimensions from plaster surface to plaster surface will be about 25 mm (1 in) smaller. Carpet sizes will be about two inches smaller still, allowing for skirting thicknesses.

A full set of drawings for a traditionally built home is reproduced on the following pages, showing miniatures of metric Al size drawings. Drawings with this amount of detail are only normally obtained from the specialist package companies, whose very full drawings are a selling point for their services. Some of the information on the drawings may be superfluous for experienced tradesmen, but it is essential for selfbuilders.

Revisions to drawings are normally made by altering the master drawing. When this is done the fact that the drawing has been altered should always be noted on it, and the date added. Prints of the unrevised drawing which are now out of date should be carefully collected and destroyed to avoid confusion.

Drawings for complex projects are normally accompanied by a specification (written by an architect), and a bill of quantities (compiled by a quantity surveyor). Between them these highly technical documents describe and define every detail of the building but they are only relevant to a building contract that is being supervised by an architect. However, a simple specification, usually referred to as a short form of specification, can be of use to an individual builder who is not using an architect and who needs to place a contract with a builder. A short form of specifica-

tion will be found on pages 152 and 153. This sample specification should be used as a guide for drawing up a specification that relates to a particular project, and not simply copied!

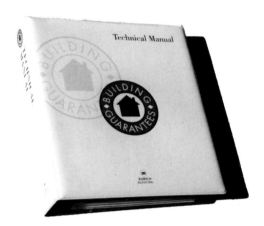

This series of drawings by Design and Materials Ltd are reproduced at less than a quarter of their real size. As such they are barely legible, but they are useful to show the sort of drawings which are required for a house built with level foundations on a level site. This drawing and the one opposite show the ground and first floor plans with the elevations. The notes are to demonstrate conformity with the Building Regulations.

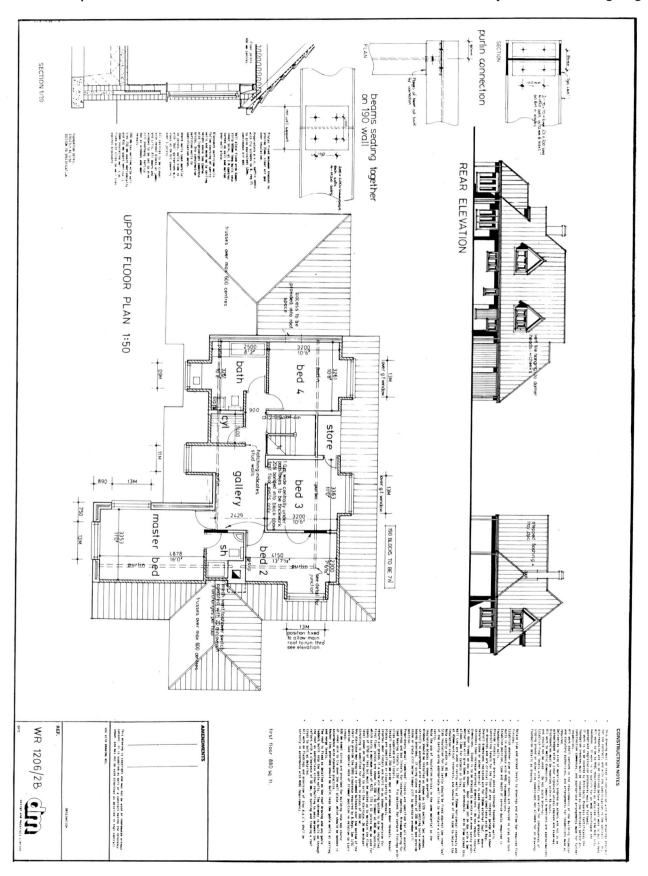

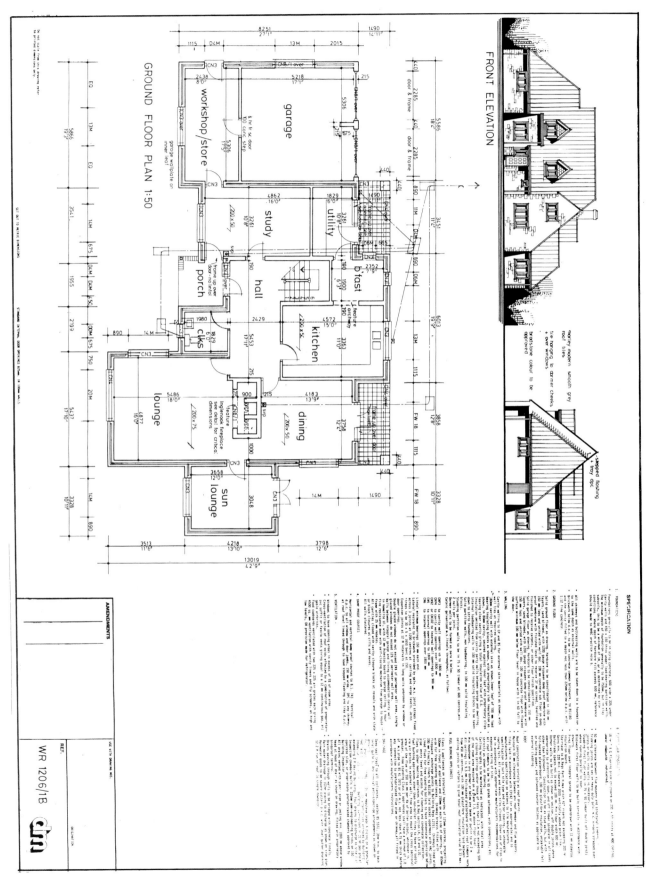

This is the site plan, and includes a location plan as required for the planning submission. It shows the position of the house on the site, and the access, drainage and landscaping details.

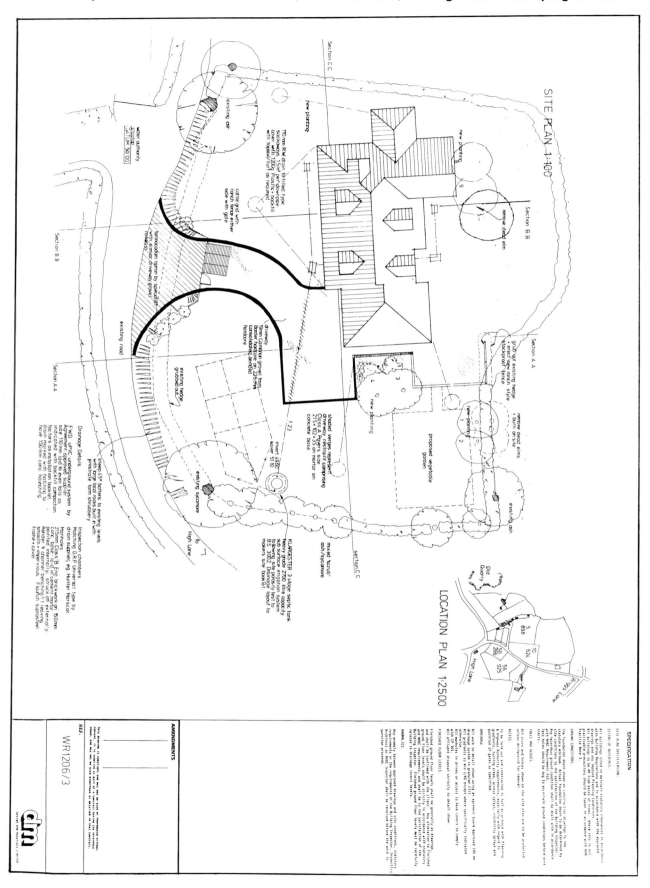

This is the foundation plan, with all the dimensions for setting out, including the key diagonal measurements. It also shows sections through the building.

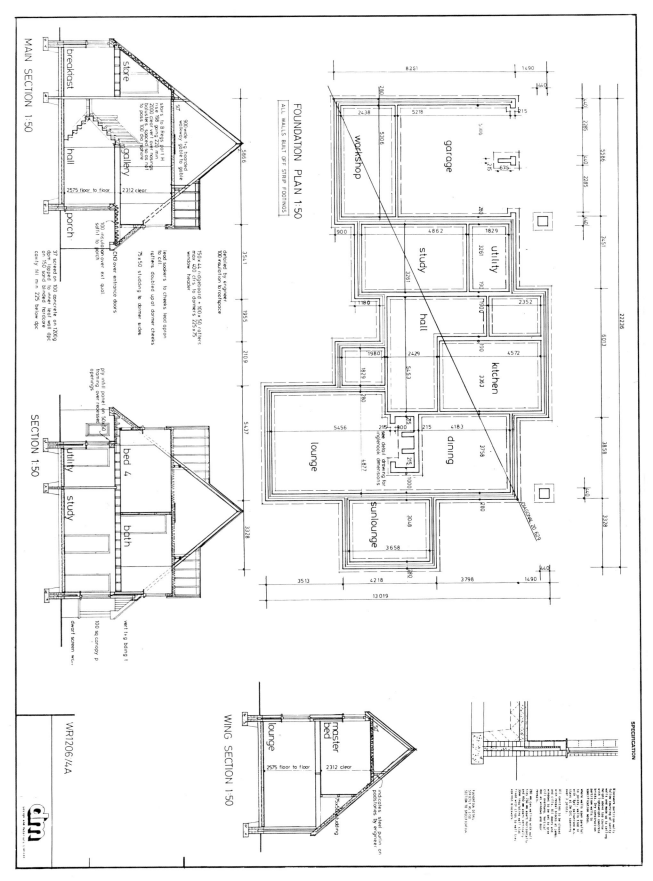

This is the drawing which gives details of the main purlin roof, with a large scale section to show the key dimensions to which the carpenter must work to ensure that the purlin roof marries in with the trussed rafter roof for the garage.

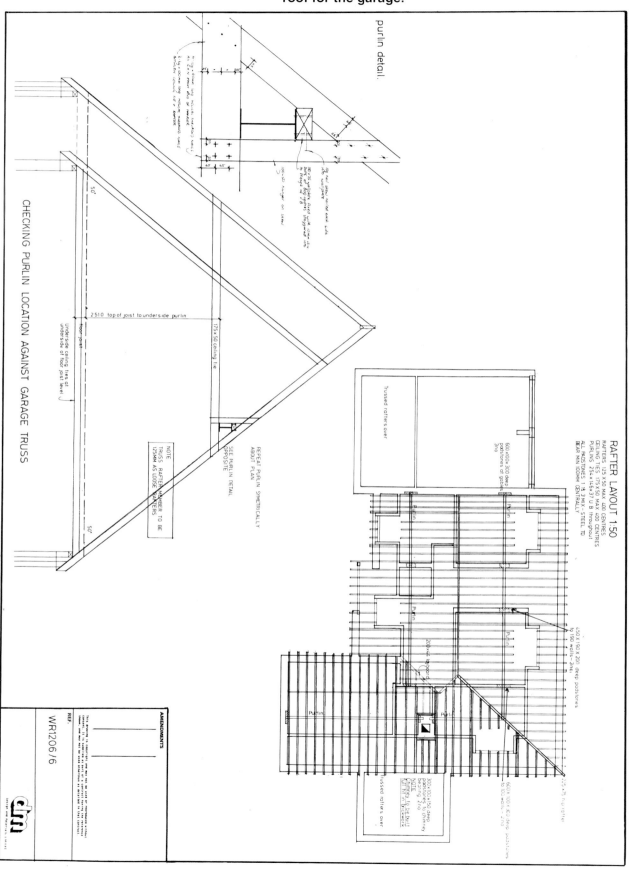

This drawing gives details of the position of the first floor joists with various construction details required to demonstrate conformity to the Building Regulations.

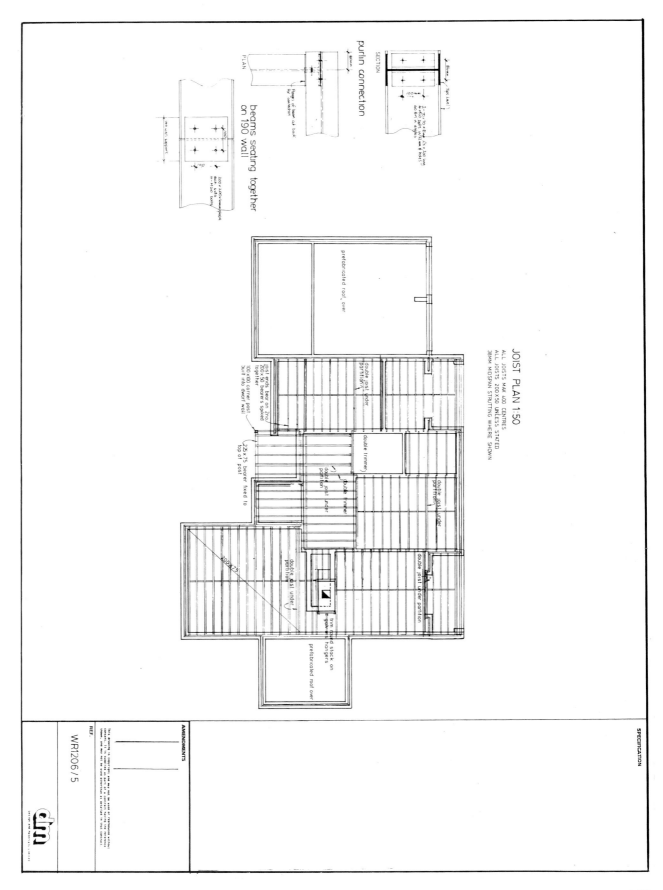

This sheet shows two sets of details which involve key dimensions in the structural shell of the building. One relates garage floor levels to the garage door frame position, and the other is concerned with the conformity of the inglenook fireplace to the Building Regulations.

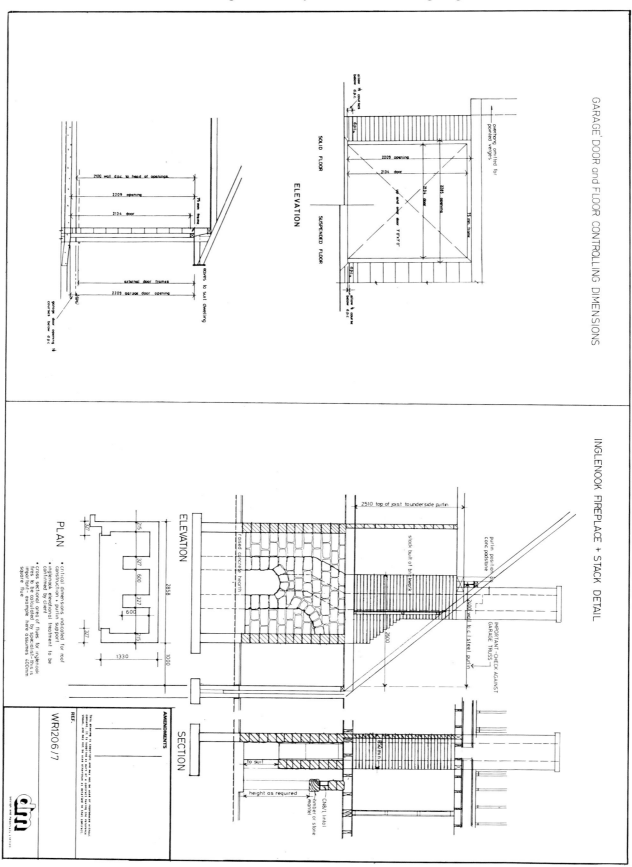

GARAGE DOOR and FLOOR CONTROLLING DIMENSIONS

INGLENOOK FIREPLACE + STACK DETAIL

REF. WR1206/7

This drawing shows how to fix the trussed rafter roof over the garage. It is a standard drawing for use with trussed rafters in different applications, and contains rather more information than is required in this case.

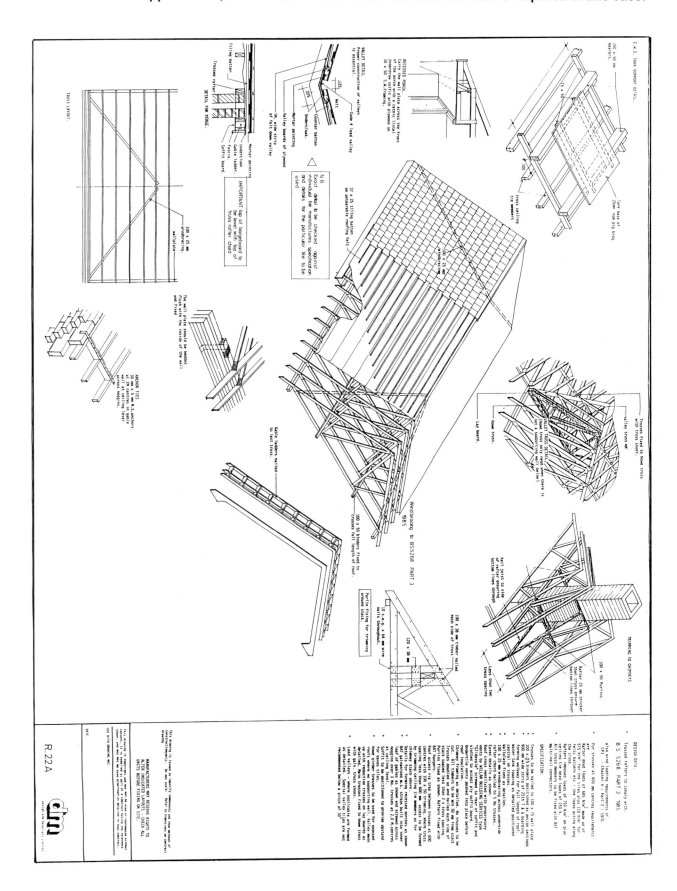

Typical short form of specification

A full architect's or quantity surveyor's specification is a long and complex document, couched in technical jargon, and defining materials, methods of construction and standards. Relatively few new homes on clients' own land are built with full specifications, and many are built without the essential definitions of key elements in the builder's contract. This short form of specification covers these definitions, but avoids technicalities by referring to well-known published standards, particularly the NHBC or Zurich Building Manuals. **This specification is an example only, and is intended as a guide to drawing up a specification for a specific contract for a specific house.**

1. This specification relates to a contract established by ... (detail form of contract, if any, or exchange of quotation and acceptance which establishes the contract) and is a schedule to that contract. The contract is between the client ... (name and address) and the builder ... (name and address) and is a simple contract between the client and the builder. Neither the designer, any supplier of materials or any sub-contractor are party to this contract.

2. This specification refers to a house to be built to the drawings attached, which have been initialled and dated by both parties, and all notes on these drawings are part of the specification.

3. The builder shall obtain a NHBC or Zurich Newbuild certificate for the property, and shall provide the client with the documentation relating to this in accordance with standard NHBC or Zurich Newbuild practice before work commences. All materials and work shall be to the requirements of the NHBC or Zurich Builders' Handbook, and shall follow the further recommendations laid down in the NHBC or Zurich site manuals and practice notes.

4. Time will be the essence of the contract and the builder is to start the works on ... (date) or as soon as practicable thereafter, and shall finish the whole of the works in the time stated in the tender.

5. Six sets of working drawings will be furnished to the builder for site use, and any further prints reasonably required shall be supplied on request.

6. The term prime cost when applied to materials or goods to be fixed by the builder shall mean the list price of such goods as published in the supplier's catalogue; any trade discounts obtained by the builder shall be an advantage enjoyed by the builder. Prime cost sums shall include suppliers' charges for delivery. All expenses in connection with the fixing of such goods shall be allowed for by the builder in the contract sum.

7. All work and materials shall be to British Standards and Codes of Practice and shall comply with Building Regulations. Proprietary materials and components shall be used or fixed in accordance with the manufacturers' recommendations.

8. The builder shall be responsible for the issue of all statutory notices and shall comply with the requirements of the Local Authority and statutory bodies. The client warrants to the builder that all necessary planning consents and appropriate building regulation approvals have been obtained, and shall be responsible to the builder for any delay or cancellation of the contract consequent on their not being such consents or approvals.

9. The builder shall be deemed to have visited the site and to have satisfied himself regarding site conditions.

10. The builder shall be responsible for all insurances against all risks on site, including public liability and fire risk, to date of handover. The builder shall make security arrangements for the proper storage of materials on the site as appropriate to the local circumstances. The builder is to avoid damage to public and private property adjacent to the site, and to make good or pay for reinstatement of any damage caused. The builder shall extend to the client the guarantees available to him on proprietary materials and fittings, and shall provide the client with documentation required to take advantage of such guarantees.

11. The builder is to cover up and protect the works from the weather, and to take all action for the protection of the works against frost in accordance with the requirements of NHBC or Zurich Newbuild.

12. Top soil shall be stripped from the site before commencing excavation of the foundations in accordance with the requirements of NHBC or Zurich Newbuild, and shall be ... (spread or left heaped). Any trees removed shall have the whole of their roots excavated, and the back fill shall be with materials appropriate to the works to be executed over the excavation.

13. The builder is to set out and level the works and will be responsible for the accuracy of the same.

14. Foundations shall be as per the drawings with footings under partition walls taken down to solid ground. Depth of the foundations shall be as per drawings, with any additional depths required by the Local Authority paid for at measured work rates.

15. Concrete for the foundations and solid floors shall be truck mixed concrete as specified. Foundation brickwork shall be in bricks or blocks to the requirements of the Local Authority. Fill shall be clean material to the requirements of NHBC or Zurich Newbuild.

16. Ducting shall be provided for service pipes and cables through the foundations, and chases shall be formed in concrete for pipework inside the building in accordance with good building practice.

17. The ground floor, if to be of solid or beam and block construction shall have a sand cement screed, finished to receive ... (tiles as defined or carpeting), or if suspended floor construction shall be as per drawing with floor boarding to be ... (define whether tongue and grooved boarding or interlocking flooring panels or other material).

18. Mat wells shall be provided at the front and back doors to be ... (define type and size).

19. The shell of the building is to be built with the materials specified on the drawings. The external walling material shall be ... (make and type) and shall be laid and finished in a manner detailed on the drawings. The windows shall be ...(make and range) and shall be finished ... (define finish). The external doors shall be ... (make and types) and shall be finished ... (define finish). Internal door frames shall be ... (material and finish). Window boards shall be ... (material and finish). Staircases shall be ... (material and style, particularly style of balustrade and rails).

20. The roof and any vertical tiling shall be ... (define tiles by manufacturer, type and colour) and the tiling work shall be carried out by a tiling contractor approved by the tile manufacturer so as to obtain the most favourable guarantee available from the tile manufacturer. The sub-contract for this shall be between the builder and the sub-contractor.

21. First floor boarding shall be ... (define whether tongue and grooved boarding or interlocking flooring board or other material).

22. Access to the roof shall be provided to NHBC or Zurich Newbuild requirements and a loft ladder shall be fitted within the contract sum.

23. All walls shall be plastered in ... (define lightweight or traditional plaster) to the plaster manufacturer's full specification, and all materials used shall be from the same manufacturer. Caving and other plaster features shall be extras to the contract, to be specifically defined in a quotation and ordered with a written order.

24. Ceilings shall be boarded to suit the ceiling finish specified, which shall be ... (define).

25. The under surface of the stairs shall be ... (define arrangements for below the stairs if a visible feature).

26. Architraves and skirtings shall be ... (define material, size and moulding shape after discussion of samples).

27. Internal doors to be ... (define doors specifically, by manufacturer and model).

28. Any sliding patio doors shall be ... (manufacturer and type).

29. All windows shall be double glazed with sealed double glazing units to be ... (manufacturer and type). All glazed doors shall be single glazed. Obscure glass and shatterproof glass shall be used where required by Building Regulations or where shown on the drawings.

30. Garage doors shall be ... (manufacturer and type).

31. The door furniture to be as the schedule attached. (The schedule should detail which internal doors are to have latches, which are to have locks, security locks to external doors, letter plates as required, plus any other fittings. Windows are supplied complete with furniture, but if security bolts or special fittings are required these should be specified.)

32. The central heating system shall be installed against a prime cost sum of £... and the proposals for this system shall be as detailed separately. The system shall be designed to meet the heating requirements of the NHBC or Zurich Newbuild, and all work shall be to the appliance manufacturer's requirements. If the heating system requires the installation of an oil tank the position and height of this shall be agreed, and the structure to support and/or conceal the tank shall be ... (define).

33. The chimney and chimney breast shall be built as per drawing, and the fireplace opening provided shall be for a ... (name appliance). This appliance and the fire surround shall be provided against a prime cost sum of £... All work to the fireplace opening and chimney shall be to the appliance manufacturer's requirements.

34. Sanitary ware and bathroom fittings shall be provided against a prime cost sum of £....

35. The cold water tank shall have capacity of ... litres and shall be situated in the roof in a position agreed, to give ease of access, on a stand to NHBC or Zurich Newbuild requirements. It shall be fitted with a lid, and frost protected as required under the Building Regulations. The hot water cylinder shall have a capacity of ... (discuss) in a position to give ease of access while providing for the maximum space for shelving alongside. The hot water cylinder shall be fitted with an immersion heater, to be ... (discuss, including whether this is a dual model to provide both full and top-up heating). The cylinder shall be lagged to NHBC or Zurich Newbuild requirements.

36. The kitchen fittings shall be provided against a prime cost sum of £... and this shall include all sinks shown on drawings. Hot and cold water and drainage connections to a washing machine/dishwasher situated ... (define) shall be provided.

37. Wardrobes, cupboard fronts and other fitted furniture shall be provided against a prime cost sum of £...

38. The electrical installation shall allow for lighting points, power points and switching arrangements to be to NHBC or Zurich Newbuild minimum requirements, and the builder shall quote the additional sum required for each extra ceiling light, extra wall light, and each extra socket outlet. Light switches and socket outlets shall be ... (manufacturer and range). Simple pendants shall be provided at all lighting points, or alternatively the client's fittings will be fixed if provided to programme. The fuse box shall be ... (define fuse board and circuit breaker system). Provision shall be made for television sockets and telephone points in ... (define rooms and position in rooms). Electricity meters shall be in a meter box fitted ... (define position: the Electricity Board may try to define where this should be). Provision shall be made for bells at the front and back door, and a simple bell shall be provided, or alternatively the client's chimes or other fittings to be installed if provided to programme.

39. All interior plaster surfaces shall be finished with ... (define emulsion paint and colour) which shall be applied in accordance with the manufacturer's recommendations for new work to give a consistent colour.

40. Wall tiling shall be quoted as a prime cost sum of £... per sq. yd. for a stated minimum area.

41. All softwood joinery shall be knotted, primed and treated with two undercoats and one gloss finishing coat of interior paint to be ... (define paint and colour).

42. All timber surfaces which are not to be painted shall be protected by using protective stain to the manufacturer's requirements, and shall not be varnished (or other requirements as considered appropriate).

43. Foul drainage shall be as detailed on the site plan, in the materials detailed on the site plan and all work shall be to the requirements of the Local Authority.

44. Rainwater goods shall be ... (manufacturer and range) and shall discharge into open gullies or via sealed drain connectors ... (as defined). Surface water drains shall discharge into soakaways or elsewhere as detailed on drawings.

45. External steps, and the porch or step at the front door, shall be finished with ... (quarry tiles or finish required). A path shall be provided around the whole of the perimeter of the building, to a width of 2 ft., to be ...(specify concrete surface or paving slabs) and shall be laid to the full requirements of NHBC or Zurich Newbuild for external works.

46. Other external works, including any work on the drive, or any fencing, shall be considered as extras to the contract, and shall be quoted for in writing, and the order for them placed in writing.

47. Any detached garage shown on the drawings is outside the contract, and any work to construct such a garage shall be an extra to the contract, to be quoted in writing and any order placed in writing.

48. Any other work required or fittings to be installed in connection with the contract shall be quoted in writing and any order placed in writing.

49. Any defect, excessive shrinkages or other faults which appear within 3 months of handover due to materials or workmanship not in accordance with the contract, or frost occurring before practical completion, shall be made good by the builder, and payment of the retention detailed in the payment arrangements at paragraph 50 shall only be made on completion of this making good.

50. Payment to be made on a progress rate of:

20% of contract price at dpc
25% of contract price at roof tiled
25% of contract price at plastered out
30% at handover.

The above all subject to a 2% retention as provided in paragraph 49. All payments to be made within 7 days of notice that payment is due.

51. The client may but not unreasonably or vexatiously by notice by registered post or recorded delivery to the builder forthwith terminate the employment of the builder if the builder shall make default in any one or more of the following respects:

(i)   If the builder without reasonable cause fails to proceed diligently and properly with the Works in accordance with this specification, or wholly suspends the carrying out of the Works before completion.

(ii)   If the builder becomes bankrupt or makes any composition or arrangement with his creditors or has a winding up passed or a Receiver or Manager of his business is appointed or possession is taken by or on behalf of any creditor of any property the subject of a Charge. Provided always that the right of determination shall be without prejudice to any other rights or remedies that the client may possess.

52. The builder may but not unreasonably or vexatiously by notice by registered post or recorded delivery to the client forthwith terminate the employment of the builder if the client shall make default in any one or more of the following respects that is to say:

(i)   If the client fails to make any interim payment due within 14 days of such payment being due.

(ii)   If the client or any person for whom he is responsible interferes or obstructs the carrying out of the Works.

(iii) If the client becomes bankrupt or makes a composition or arrangement with his creditors.

Provided always that the right of determination shall be without prejudice to any other rights or remedies which the builder may possess.

53. In the event of a dispute between the parties arising out of the contract, the parties shall agree jointly to engage an architect independent of either of them to arbitrate between them, and shall be bound by the architect's findings as to the matter in dispute and to his apportionment of his fees as an arbitrator.

15 The package companies and consultants

The architectural services provided by the package companies have already been discussed, but they are only part of a contract that you might place with such a company. What the package companies offer, and how they work, needs to be clearly understood.

First of all, they advertise their services in a very polished way. This can give the impression that using a package service is the usual thing for a selfbuilder to do. In fact, although statistics cannot be obtained, it is certain that no more than 20% of those who build an individual home use a package service, and the total may be significantly lower. However, while the package companies do not offer the usual way to build for yourself, it is certainly true that many people find that they provide an excellent way of going about things, and, significantly, they obtain a very high proportion of their business via personal recommendations from satisfied clients.

You may wonder why the package companies only have such a small share of the total market, in spite of the very real advantages that they offer. It is not because of the cost of their services, and there is keen competition between them. One is driven to the conclusion that for many selfbuilders the independence of spirit that leads them to strike out to build on their own also leads them to disdain help from others in doing so. They are determined to do entirely their own thing, and they value being completely independent and do not want to hire others to tell them how things should be done. This outlook leads inevitably to traditional construction, where the individual is free to make his own arrangements for the supply of materials and where, within limits, he can make changes as he goes along. For instance, while the walls are being built he can decide to build an arch between the dining room and the drawing room, or to combine his kitchen and utility room, or to build an alcove into the wall at the bottom of the stairs, and if he changes his mind he can change things back again. But this book is about how selfbuild works, and is not a study of the mental processes of selfbuilders, which would require a very long book indeed.

A package company is generally understood to be an organisation that provides an integrated service of design advice, the provision of a full set of drawings, handling planning and building regulation applications, and the supply of at least the structural shell materials for a house or bungalow. It is the supply of materials which characterizes them: companies offering design and management services, or financial and management services, are better described as consultants. Incidentally, all the package companies will either find a builder for their clients, or will tailor their service for a selfbuilder. Their advertisements may emphasize one or the other depending on the media in which they are advertising.

The package companies often provide their clients with informal advice on many other aspects of building a home, from recommending the right soil engineers to design special foundations, through to helping with VAT claims. However, they will take care to ensure that this sort of help is outside their contract, and that they are not legally responsible for the advice that they give. Obviously this cannot be otherwise. They may introduce you to a builder, but are not party to the arrangements that you may make with a builder.

It is unusual for a package company to offer to arrange finance for their clients, although they may introduce a broker who is able to help with this. Their view seems to be that a client who cannot sort out his own financial arrangements is unlikely to get to the point of making a start on site, and that this is a useful way of sorting out runners from non-runners.

A contract with a package company invariably covers the whole of the package, which a client pays for in stages, usually paying all or a large proportion of the money up front. Package companies do not take kindly to people using only their architectural services and then obtaining materials elsewhere, and they will promptly take legal action against those using their drawings in a way which has not been agreed. For this reason the small print of any arrangements with package companies should be studied with great care, and if you have any doubts it is worth while asking your solicitor to explain to you exactly what it means, and how you will be situated, if, for instance, you suffer a change in your circumstances after you have signed.

How reputable are these companies? Some are owned by major industrial or commercial groups, and some are relatively small private companies. As the average package probably costs around £20,000, and selfbuilders and their friends are shrewd people, package companies have to ensure that their credit references are impeccable. They invariably are. The services which they provide are complex, and it is not easy for an entrepreneur to set up a package organisation, which in itself is some guarantee. However, the best assurance which you can get will come from tracking down others who have used a particular company, looking at their finished homes and asking how everything went. You will usually get a very positive report. Some package companies have been operating for 20 years.

Most package companies specialize in timber frame construction which is discussed in detail in another chapter, and their advertising is usually concerned with extolling this way of building. This is because anyone building with a timber frame has to use a package company to provide the design calculations for the frames, so that if the company can persuade you to build in this way you are firmly tied to them. However, in choosing a package company it is the designs which are offered and the service which is provided that is more important than how the building is constructed. Anyway, there are companies which offer a package service of traditional materials. Notes on using a package company are set out on pages 158 and 159. For most people the most important person in making it all work is the company rep/area manager/site liaison person who will deal with them. These individuals are a remarkable lot, and those who have been in the job for any length of time have enormous experience of selfbuild, and indeed of the building scene generally. They often make firm friends of their customers, possibly because those who are building a new home for themselves feel the need for a confidant who knows all about it. The moral is that the oldest, most weather beaten and cynical area manager is probably the most experienced, as the furrows in his brow come from sorting out selfbuilders problems.

Payments to package companies

The kits sold by package companies are specific to a particular house or bungalow, and the elements like the roof, walling panels etc. have to be made weeks before they are delivered. For this reason the companies expect to be paid a substantial deposit when an order is placed, and usually require the balance of the contract sum in advance of the delivery of materials. These sums may be over half of the total cost of the new home, and selfbuilders are naturally concerned that this arrangement is 100% safe.

To meet this situation the Individual House Builders Association advises that customers of package companies should pay nothing up front beyond the deposit, and should lodge other monies required in a solicitors' client account for release in stages when a kit has been delivered and then when satisfactorily erected. They also stress that the customer should go for a single line of responsibility, i.e. one company to supply and fix. This advice should also be followed when ordering other products — negotiating a supply, fix and commission contract. They specifically warn selfbuilders not to permit themselves to be railroaded into authorising early release of such payments. For the firms concerned, these arrangements are a condition of IHBA membership.

Some companies operate their own client accounts into which they expect customers to pay money in advance, but where it still belongs to the client until authority is given to pay it across to the company. This has advantages to the company because they can raise money against the sums in the client accounts that they operate. If you accept this arrangement, make sure you ask your solicitor or bank to check out the status of this client account, always make out cheques to the client account and not simply to the company, and if the money is going to stay in the client account for any length of time enquire who gets the interest.

The established package companies will take

Most, but not all, package companies are concerned with timber frame construction. However, this striking house was built with a kit of traditional materials from D & M Ltd.

An individual home supplied as a package by Scandia-Hus.

This photograph was taken by D & M to help explain their service. Anti-clockwise from the clients in the foreground are the Contracts Manager who provides the link with the company, the Architect who provides the professional services and the Despatch Clerk, who make sure the right materials get to the right site at the right time.

care to explain this to you. Less well established companies may not. If in any doubt you can consult the IHBA, and seek the advice of your bank manager or your solicitor.

The I.H.B.A.

In September 1992 the major package companies formed the Individual House Builders' Association to promote the concept of self build on a generic basis, and to raise standards across the industry. The association's long term aims include providing reassurance for individuals embarking on their own house build project, as well as providing an independent source of information. For details write to the IHBA at the address given on page 294, enclosing a large stamped addressed envelope.

Consultants

The term 'consultant' is used to describe those who provide integrated services to those building on their own but who are not simply architects or designers, nor package companies which supply materials. In recent years the number of consultants has proliferated, and it now includes land finding agencies which will also arrange finance, management agencies which may or may not offer architectural services, and small building firms offering a design and build service on a national basis. Some are one man companies where the proprietor works from his home,

others present a very polished image with impressive office accommodation and a glamorous receptionist. Some have beautifully printed literature, others little more than headed paper. The important thing to remember when evaluating what they offer is that the best recommendations come from satisfied customers, and that you should seek out such customers yourself and speak to them personally. Remember show houses and those who live in show houses are always interesting but are really adjuncts to a marketing operation. The houses that you want to inspect are the ones built by clients of the consultant who are people just like you!

Well established companies should be able to offer appropriate financial references, while newer companies may prefer to offer a performance bond. The best safeguard of all is that your contract does not involve you in paying any money in excess of the value of the work done. Some consultants specialise in managing projects for clients who work overseas, and this usually involves special financial arrangements. The safest arrangements for both parties involve the whole of the project finance being on deposit with a UK solicitor who releases it in stages against appropriate certificates from an independent surveyor. Sometimes overseas banks are involved in this, but at this level both parties are invariably well able to look after themselves and do not need advice from a consumer guide like this one!

Using a package service

* Collect all literature and study it. Particularly note the companies with the standard designs that suit you, or which offer to design to your own requirements.

* Before you find a site you can visit package companies show houses if they have them, or you can go to exhibitions where they have a stand, but they will prefer not to discuss details of your requirements until you have a site, as nothing more than a preliminary chat can really be relevant.

* When you have a site, or are arranging to buy one, contact the company or companies which interest you and ask to meet their local representative. Do not be frightened to ask the person who comes to see you whether they will be the person to provide your liaison with the company if you place a contract, and ask them what their qualifications and experience are. You will depend enormously on the individual concerned, and you will have to decide whether he is the right man for you.

* If you reach a stage where you have decided on the company which you are likely to use, make sure you visit recently completed homes for which they have been responsible, and meet the householders. Make sure you go to typical new homes, not just show houses.

* If you are interested in a standard design you will be able to get a firm quotation: in this case you should study the order form or contract presented to you with the quotation with great care.

* If you want a company to draw up plans to your own requirements, you will need to be sure that the first draft of the plans is either prepared free of charge or at a fairly nominal fee, and that there is no further commitment if you do not like the plans. This first draft of the plans may be called a design study, and if you approve it you should speedily get a firm quotation and an order form or draft contract.

* Before you sign an order you must
    * Take up references if you will be paying in advance of delivery, which is very common. Your bank will do this for you if you are unsure about how to set about it.
    * Make sure you understand exactly what the company is going to do for you, and what you are contracted to pay, and when. Remember you will have to arrange for prompt payments.
    * Make sure payments in advance are made into a clients deposit account - see the main text.
    * If everything is conditional on you completing the purchase of a plot, or buying an access strip, or obtaining a statutory consent, or getting finance signed up, make sure that this is written on the order form, or write it down yourself.

* There is a legal implied assumption that the package service will have to meet the requirements of the Building Regulations. However, if the work is also required to meet the requirements of the NHBC or Custom Build or any other approving agency, or

particularly if you are obtaining architects' progress certificates, a requirement that the package service should meet the requirements of the agency or the architect should be written into the small print. If it is not there you can add it when you sign.

* Make sure that you confirm any offers of help from the company representative. Promises like 'I shall be round to see how you are getting on every fortnight until we have delivered the last of our materials' are not part of the contract, but you can send a letter saying that you are relying on this and it was a major factor in deciding to use XYZ Limited.

* The company will not be able to wave a magic wand to obtain your planning consents, etc. any more quickly than anyone else could, and there may well be problems in this. Keep a note of all phone conversations and if possible send a letter confirming what was arranged, and keep a copy. Typically this might read:

> 'Dear Jeff
> Confirming our phone conversation last night, if the planner really insists on the 6m distance between the garage and the north boundary, so be it. I am sure you did your best. But the front building line remains where it is on the revised drawing 3a. No question of negotiating on that. Keep at it.
> Regards.'

* When you get your planning consent, etc., and if you are pleased with the way this has been dealt with, send a letter to the managing director expressing your appreciation and saying how you are looking forward to everything else going as smoothly. This simple ploy guarantees special consideration, and the managing director will fall for it every time.

* The programme for delivery of materials should be established in writing, and the inevitable alterations confirmed in writing even though they will be arranged on the telephone. More importantly, if there are any minor changes to the plans which will affect the materials to be delivered, you should have a firm record of what has been agreed.

* Package companies provide excellent services to thousands of selfbuilders, but in the natural way of things there are sometimes problems or disagreements. If you meet difficulties which cannot be resolved at a local level you should take pains to deal with them promptly, but in a way that demonstrates that you are a most reasonable client. Avoid letters that are written in anger. If appropriate, arrange to call at the Head Office by appointment and explain the difficulty in a friendly but very firm way to someone at director level. No firm can afford to ignore a customer who has a five figure contract, and you will be treated accordingly.

16 A contract with a builder

Whenever you arrange for someone to do some building work for you, you make a contract with them. In it they undertake to do the job, and you undertake to pay them. You cannot escape it: even if you simply say *'get this done, Ted, and I will see you right'*, you have established a contract. However, you will want to make sure that the arrangements that you make to build a new home are a good deal more specific than that!

Now there are many textbooks on the law of contract, and they are both heavy going and also omit to mention that very few people arranging to build their own homes establish contracts in the way that the textbooks advise, or indeed in the way that their solicitors would advise. There is a great gulf between theory and practice.

A contract is a way of expressing an arrangement which both parties enter into without reservations, believing that they know exactly how everything is going to happen. When they make the contract, whether verbally or in writing, they regard it simply as a convenient way of recording what they have agreed. If all goes well, everything is fine. If there are unforeseen circumstances or problems, they turn to the contract to see where they stand in the matter. If they should fall out over this, it is the contract that determines their legal position. The contract should thus define exactly what the parties have agreed and, if there are problems, how they are to be resolved.

The best way of establishing a formal contract that deals with all of this involves solicitors, quantity surveyors, and documents that are dozens of pages long. If you ask a solicitor what is the best way to arrange a contract or contracts to build a new house, he must recommend these involved procedures. However, such contracts will scare off most small builders and using them automatically puts you in a special league. For this reason only a very few of those building for themselves use them. The choice is yours. This book cannot advise you to ignore the best legal advice, but it does describe how most people arrange these affairs. If it is a matter which worries you, then discuss the whole business with your solicitor.

What follows is about single contracts placed to arrange for the whole of the work involved in building a new home. Contracts with sub-contractors will be dealt with in the next chapter. Technically anyone placing a contract for the whole job is an individual builder, and not a self builder, as a self builder is someone who reclaims his own VAT. An individual builder has the VAT reclaimed by his building contractor, and his bill from the contractor will not include VAT.

There are two very different ways of arranging for a builder to build a house for you — using an architect to establish and supervise the contract, or arranging and supervising everything yourself. If you use an architect he will invite tenders from builders, advise you which one to accept and will draw up a suitable formal contract which he will supervise on your behalf. This is the Rolls Royce way of doing things. The architect will charge fees of up to 10% of the value of the contract, and although he or she will be concerned that you get the best value for money, he does tend to operate at the very top of the market. If you are using an architect then you must make sure that you have made proper arrangements with him, and the basis of this will be the RIBA standard arrangements described in an earlier chapter.

If you are making a contract with a builder yourself it is important that you do not simply accept any arrangements that he suggests, and that you settle things in a way that you yourself are happy about. Negotiating this in an amicable way may not be easy, but you should insist on what you want while avoiding giving the impression that you are going to be a difficult customer who should be charged extra for being a potential nuisance! If your builder is taking the initiative in this matter he is likely either to present you with one of the standard building industry forms of contract or with a detailed quotation which he will hope you will accept by sending him a written order.

Beware the pre-printed contract: these are excellent for professionals but are unintelligible to the layman and contain all sorts of clauses which you might not want if you knew of them. If such a contract is to be used then ask your solicitor to advise you of any clauses that should be struck out. A contract established by exchange of letters, or by a quotation and an order, is much better provided that you refer in it to an agreed specification and approved drawings. You must make it your job to ensure that the specification and drawings deal with everything that you want to establish, and they should be initialled by both parties so that they are firmly part of the contract.

Some of the matters to be dealt with in the specification are listed in the draft specification on pages 152 and 153. Obviously special conditions require special clauses in the specification, and the best person to advise you on this is whoever drew up your plans.

The draft specifications make extensive use of Prime Cost Sums, otherwise known as PC sums. At the stage when you are negotiating the contract you have probably not decided on the particular fixtures and fittings that you require, and so a 'Prime Cost Sum' is allowed for the items concerned. A PC sum of £2,000 for a kitchen means that the builder must allow for kitchen fittings to this value: if you want to spend less than this the contract price will be reduced: if more it will be increased. Typical features covered by PC sums are kitchens, bathrooms, central heating, fitted furniture, fireplaces, feature staircases, wall tiling and floor tiling. It is a very useful way of going about things, but make sure you know the basis on which the PC sums are calculated — list price, trade price, special offers or perhaps you are going to buy the item and supply it to the builder. In this case you must define who insures it on site.

Another important matter to deal with is the cost of any alterations to the agreed work, or any extra work. This is a potential minefield: a simple request from you that something should be fixed the other way round can involve the builder in a great deal of expensive work, and, unless it is agreed in advance, the cost can be a source of dispute. The cost of all alterations and extras should be discussed and should be confirmed in writing, and the specification should say so.

Assignment of the work is also something to be discussed. If you take on builder A because you have admired his work on another house, you want his same workmen to build your own new home, and you do not want him to assign the contract to builder B, or to use other workmen. If this is important to you it should be set out in the contract or in the specification.

The stages at which payment is made, the arrangements for payment, and retentions to be held for a maintenance period should also be clearly established and never, never should any payment be made other than in accordance with these arrangements. Beware any requests for a payment in advance to enable materials to be purchased at a particularly advantageous price, or any other good story. If your builder needs money in advance then it is 100 to 1 that he is in financial difficulties, and you are not there to bail him out.

This leads to the question of what your position is if the builder fails, or dies, or just does not get on with the work. It does happen. On his part, what does he do if you disappear? All this has to be part of the contract. The way to deal with such a situation is discussed on pages 201 and 202.

Finally, when you have the best contract which you consider to be appropriate to the way in which you want to go about things, for goodness sake stick to it. Be punctilious about making payments on time and generally fulfilling your part of the bargain, as to do otherwise may make things difficult if the worst happens and you have to establish your contract in law. This is a frightening thought: console yourself with the fact that nearly all individual builders end up by having their new home built without dispute, and retain a good relationship with those who are building for them. Having the right contract is a very good beginning to this.

17 Selfbuilding with sub-contractors

Selfbuilders come in all shapes and sizes, are of all ages, and have only one common attribute: a capacity for getting things done. One large category are farmers and smallholders, and certainly they are the most successful group. Their strength lies in their versatility, since dealing with new situations and acquiring new skills is commonplace for them. They are used to handling money, can be philosophical about the risks involved, and are familiar with dealing with both officialdom and sub-contractors. Self employed artisans and businessmen have the same attributes, the key being that they have experience of being responsible for getting things done. All selfbuilders have enormous self-confidence. Knowledge of building techniques is certainly not the essential ingredient of success, while a knack for finding the right answer to every problem definitely is.

The success of a project is usually determined by the project management before any work starts on site. By the time the technical/legal/financial infrastructure has been established you may be forgiven for thinking that now surely is the time for simple enthusiasm and hard work. After months of managing paper and people, formalities and delays you will look forward with pleasure to the actual building work - hard, unremitting toil but very worthwhile - the creative part of the whole exercise.

This is fine since you will want all the enthusiasm you can find, but it is essential that the hard work is managed as carefully as all the earlier stages. As stated so often before, by building on your own you are working outside the system, and the only way to do it successfully is to plan every move - including the donkey-work. 'Get stuck in' is a useful motto only as long as you get stuck in to a programme based on a careful analysis of the job. The phases of the construction work involving different trades have to be programmed to follow in sequence, and deliveries of materials and the provision of site services have to be arranged to suit. Nothing will destroy the enthusiasm for work on site more quickly than confusion or delays, while the best boost to morale is the steady progress of work to a realistic timetable. Management is 90% of the building operation.

As a start to this management the selfbuilder must decide how many hours he can put in, and build this commitment into his programme. If he wants to do any physical work himself he must analyse what he can do, what he can learn to do, and what he should do. These are unlikely to be the same. Anyone can learn to mix concrete. Given a set timetable and a fixed number of hours of personal involvement it will probably be more cost effective, although more expensive, to buy truck mix concrete. Anyone can dig trenches, but a JCB will dig them at a speed that will make the hire charge totally cost effective. Almost anyone can learn to bone a line along a trench and lay drain pipes, and if he applies his labour here the savings will be really significant. Amateur roof tiling is hazardous, backbreaking and unlikely to be worthwhile when a supply and fix tiling contractor costs little more than the cost of the tiles alone. On the other hand, sub-contract labour rates for bricklaying are now often over 10p per brick laid, making this the most desirable of skills if the selfbuilder can learn to lay bricks to the required standard. This is unlikely, so most people will have to pay for this expensive work. A balance has to be struck between what the selfbuilder can do, what he wants to do, and what is cost effective for him to do. This is not always easy.

All decisions about the jobs tackled, and those put out to contract, depend on individual circumstances, but there are three areas where the selfbuilder should be extremely cautious - electrical work, gas fitting and plastering. The hazards of the first two are self-evident, and in spite of the checks made by the gas and electricity boards before connecting main services, it is morally indefensible to run the risk of making installations with a built-in hazard. Plastering work is not only a particularly skilled trade, but a substandard job cannot be remedied or disguised. The author rarely sees a first-class amateur plastering job, and has met many where an uneven wall or rippled ceiling is a monument to over-enthusism.

Virtually all the other skills required in building a house can be acquired by a competent enthusiast who decides that this is what he wants to do and so sets out to learn the skills methodically. He will be exceptional, but hundreds like him build their own houses, and facilities for learning basic trades are provided at evening classes and adult education centres. Usually such a selfbuilder will develop one particular building trade, for which he has a natural flair, and he will make it his business to find out enough about the others to be able to place subcontracts and to supervise them. Besides the courses organised for those who wish to acquire building skills there are innumerable helpful books and magazines.

Subcontractors

Whether you are going to tackle a high proportion of the work yourself, or none of it, it will be a very unusual job if you are not going to employ subcontractors at some stage.

Generally they fall into two categories — 'labour only' sub- contractors who provide only their labour and tools, and who expect you to provide all materials, and 'labour and materials' sub-contractors who provide the materials to do the job for you, and will charge you a price for the job rather than for their work. Both will require you to provide plant like scaffolding, mixers, a water supply, secure storage, etc., and the distinction between the two groups is often blurred. Some jobs, like bricklaying are always labour only while some like electrical work and plumbing may be done on either basis. However, whichever way a job is tackled, some rules apply for dealing with all your subbies.

First of all, the legal situation. In the last chapter we discussed placing contracts with builders who are responsible for every aspect of the work on a new home, with clearly defined obligations all round and provision made to cover all eventualities. Detailed and legally enforceable contracts like this are rarely made with labour only sub-contractors, who often work simply on the basis of a verbal agreement. The best you can hope for is a quotation on a piece of headed paper. A quotation on a labour and materials basis may be quite detailed, but it will not deal with unforeseen contingencies in the way that a builder's contract does.

As a result you have to rely on finding the right man, coming to an amicable agreement with him, making sure that he does the right job, paying him only for work done, and terminating the arrangement promptly and without rancour if things are not working out. The key word in this is amicable. Arguments between selfbuilders and sub-contractors are rarely won by either party, as either the sub-contractor will walk off the site, or else the work will proceed in an atmosphere that does not make for a good job. It is virtually impossible to enforce an arrangement made with a sub contract workman in any legal way, and you have to handle problems on a give and take basis. Builders have experience of this: most selfbuilders have not. This is one reason why subcontractors should always be chosen on the basis of personal recommendation.

Fortunately there are ways in which you can take action to avoid misunderstandings and problems. Firstly, reinforce the arrangements made by giving the subbie a letter or a note which is either your acceptance of the written quotation if you received one, or, more usually, confirms a verbal arrangement which you have made. There is a typical note of such a verbal arrangement on page 165, and while it in no way establishes a watertight agreement it is a good basis for dealing with any disputes.

The business of payment is important. Labour only sub-contractors expect to be paid promptly, and in cash. If you do not do this you are asking for problems, and running the risk of your sub-contractors going off to other work. Although you should keep a record for your own accounts of who you paid and how much you have paid them, you have no responsibility to notify the tax authorities of the payment, although a builder is obliged to do so under what are called the '714 arrangements'. This is a complicated business, and it makes working for selfbuilders popular with sub-contractors who want to avoid paying their taxes.

Besides paying promptly, it is essential never to pay in advance. You can expect to be asked for this, either because the subcontractor is really short of money, or, more likely, because trying to get ahead on payments is a sort of game that is played on building sites. Paying sub-contractors in advance is a recipe for problems, and sometimes for disaster. So are arrangements to pay cash to a subcontractor so that he can buy materials cheaply on your behalf. Money lending is for banks, and it is not a good idea to start trying to compete with them!

Although the sub-contractors working for you will not be employees in the strict sense, you should ensure that you have employers liability insurances. The tiler who falls off your roof will decide that he had a 'Deemed Contract of Employment' with you before he hits the ground, or if he does not remember this, his solicitor will! Dealing with a resulting claim will be expensive whatever the outcome, and it is best left to an insurance company. Appropriate cover is part of standard selfbuilders' insurance policies, and is discussed in a later chapter.

Labour only sub-contractors will expect you to provide all the plant required for the job, and to have it there on time. If there is a difficulty with this, such as a mixer breaking down, they will expect you to solve the problem at once, otherwise they will want to be paid for their wasted time or will go off to another job. The applies to delays in delivering materials, or in arranging to replace materials stolen from a site. Detailed arrangements for plant and materials should be clearly understood: it is not enough to say 'I will get the scaffolding' without everyone knowing if you will have it erected by the scaffold hire company, or whether the sub-contractors are to

Building with subcontractors requires daily visits to the site and close involvement with all that is being done, even though you may not be doing any of the work yourself. This photo of Ralph Hitchcock whose new home in Surrey was featured in an earlier edition of *Building Your Own Home*.

erect it themselves. Similarly it is important to establish what unloading of materials you will expect of the men on site. In arranging all these matters it is an excellent idea to take a lead from the sub-contractor about the arrangements to be made. Unless you are very used to the standard procedures on local sites, it is better to ask the question 'what about scaffolding?' before coming up with your own ideas. However, there is one point on which you should be dogmatic.

Selfbuilders should always employ sub-contractors on fixed prices. Builders often do not, and there are complicated systems of measured work, where for example, a bricklayer is engaged at £xxx per thousand bricks laid. This may sound simple, but building industry practice is that some bricks at cills and reveals count as one and a half bricks, or even two bricks. Measured work rates for plastering are even more complicated. Get lots of copies of your drawings so that you can given them to prospective sub-contractors, and make it quite clear that you will be employing them to do the job that you want doing as shown on the drawings at a fixed price, and not in accordance with the mysterious rites of the building industry.

If things do go wrong, and you have to terminate an arrangement, try to do so with dignity. Never lose your temper, and never try to bawl out a building industry operative: his vocabulary may be far wider than yours. A useful approach is to announce that as things are not working out, you intend to ask the inspecting architect to advise you what payment is appropriate, and to arrange for this as soon as possible. Never use the word 'sack': the term is 'finishing'. It is always a difficult business, but it is far preferable to finish bad workmen rather than having people in whom you no longer have full confidence working on your new home. It is all part of the management of the building operation, which is what selfbuild is all about.

How do you find the right sub-contractors? First of all let us consider what you want of the man who you hope to find. He has to do the job to a high standard, and, because you do not have the building industry experience yourself, you want someone with the right personal qualities.

Note of an informal agreement made with a two-and-one bricklaying gang

John Smith with two others to undertake bricklaying work on bungalow as drawing W719 as self-employed sub-contractors, with all payments to Mr. Smith. This is not a contract of employment.

* All work as drawings and NHBC or Zurich Custom Build requirements and to satisfy building inspector, inspecting architect and NHBC or Zurich Custom Build Inspectors.

* Clean footings after JCB, and pour concrete for strip foundations.

* Build up brickwork to dpc max 750mm. Any extra depth at £xxx per thousand bricks laid. Fill cavity as required.

* Fill foundation with hard core, consolidate, blind, lay membrane, etc. Pour slab.

* Build shell to wall-plate. Bed wall-plate with assistance from carpenter.

* Assist carpenter to rear trusses. Build up gable walls and build in gable ladders.

* Build up partition walls to ceiling joists and take chimney through roof.

* Lay drains as drawing, with ICs and connect to existing foul drains.

* Lay surface water drains from gulleys to soakaways as drawing.

* Lay 900mm wide path on rolled hardcore all around building.

* Unload all materials and tiles. (Bricks and tiles ordered for crane unloading.)

* Employer to supply all materials, diesel mixer, roller, water supply, carpenter to assist with wallplate and gable ladders. Scaffold supplied to be Quickstage, which sub-contractors will erect and leave standing.

* Work to start (date) and continue every practicable working day. On completion of this work the site to be left clean and tidy with scaffold still standing.

* At lump sum price £xxxx.

* Payable £xxx brickwork to dpc, £xxxx slab cast, £xxxx building to square, balance when all work is completed.

Also, you want the work done at the right price.

The best way of satisfying yourself that a sub-contractor can do the right job is to go to see work which he has recently completed for others. Seek out his last employer, and ask for a personal recommendation. Many sub-contractors make a speciality of working for selfbuilders, and they will invite you to get in touch with people for whom they have worked before. As far as personal qualities are concerned, one general rule is that someone who is a householder living locally is to be preferred to a travelling workman who is living in digs. Again, meet previous employers, even if this means going to a new house and knocking on the door in the evening.

Prices that you are quoted have to be related to your budget and to the big picture. A man who impresses you as someone in whom you can really have confidence may quote more than someone else, but he may be very much the best man for the job. A golden rule is to get a clear idea of the general level of prices for a job like yours before you start discussing the matter with any potential sub-contractor. This, again, is part of learning all that you can about selfbuild before you actually get involved on the site.

Most selfbuild projects start off with ground workers or bricklayers who they have been recommended by others. It is then very common to employ other sub-contractors who are introduced by the bricklayers. Alternatively, selfbuilders sometimes use tradesmen who they know through family connections, or through their work, or as a result of some other sort of personal recommendation. Sub contractors can also be chosen from the Yellow Pages or from advertisements on notice boards in DIY superstores, but such introductions are no substitute for the personal recommendation.

Finally, most of what is written here sounds rather frightening. In practice most selfbuilders make good friends of the sub-contractors who work for them, and frequently rely on them for advice and assistance in many ways. The case histories in this book explain how many of the selfbuilders featured met their sub-contractor. Often you read how a first class bricklayer recommended a carpenter, who recommended a

plasterer, etc., etc. This is very typical. Most sub-contractors are the salt of the earth, and the wise selfbuilder will work hard at establishing a relationship that enables him to get the best from them.

The sub-contractors with which selfbuilders are likely to be concerned are as follows.

Labour and materials subcontractors

Tilers

Tilers will felt, batten and tile the roof and they usually supply all these materials. They will also do any vertical tile hanging required. You supply access to the roof (i.e. leave the scaffold up until they finish), cement and sand for pointing (but they supply colouring to match the tiles) and arrange for the plumber to work with them where any flashings or leaded valleys are involved. You also provide labour to unload the tiles and stack them near the building. All carpenter's work on the roof should be completed before they start, and valleys to be tiled with valley tiles must be boarded. Ensure that the quote is for 'tiles fixed to manufacturer's recommendations', and send for your own copy of the recommendations. Check that tiles are nailed or clipped to suit your particular exposure rating - again see the manufacturer's recommendations. Establish the guarantee offered for both the tiles and for the tiling work.

Plumber

The plumber will normally quote for all materials, including bathroom suites and also the sinks to install in your kitchen units. In his quote he is allowing for the benefit to him of the discount which he can get in buying these fittings on his trade account, so if you intend to buy them yourself you should make this quite clear. Remember - if he supplies the fittings then he is responsible for breakages until he is finished. If you supply a bathroom suite which is broken on site you are responsible for replacing the broken units. If his quote includes a prime cost sum for fittings of your choice, which he will obtain for you, this means that you will have a saving on his quotation if you spend less than this total, and an extra if you exceed it. Always ask him to define whether the p.c. sum is for the retail price of fittings, or for some other price, typically retail less 10%. Incidentally, the plumber will get about 30% discount, but he allows for this in his quotation. Always define the smaller fittings as well as the major ones, particularly taps. Establish whether tank and cylinder lagging is included in the quotation. Clear who has responsibility for arranging the water supply, and establish that the plumber has to arrange all liaison for water board inspections and testing.

A plumber does not usually give a guarantee as such, but he should undertake to return to deal with any problems which arise within a set period, typically six months. Finally, place an order on the basis that 'all work shall be to the standard set out in the NHBC or Custom Build handbook'. Reference to the handbook is an excellent way of dealing with any disputes that arise.

Heating engineer

Your heating engineer may well be your plumber. Whether he is or not, there is the same position regarding the supply of appliances, and if you intend to buy either the boiler or radiators yourself this should be made clear. Again, there are clear advantages in the sub-contractor providing everything, and being responsible for it until it is fixed.

The heating engineer must design the system to meet specified standards (which are clearly understood), or he should quote for a system designed by others, such as a fuel advisory agency. For the selfbuilder the latter is probably preferable. Ensure that the installer is responsible for commissioning the system, and if he does not have his own electrician, you should require 'proper liaison with the electrician'. This avoids the situation where the heating engineer announces that everything will start when your electrician has done his stuff, and that as far as he is concerned he has finished and wants to be paid in full. Suitable guarantees on both equipment and installation should be offered.

Electrician

An electrician will normally quote a fixed price for an installation shown on a drawing, or as detailed in a quotation, plus a fixed extra charge for each additional light or power outlet required. He will supply the switches and sockets (of a make and type which should be specified) but will quote for simple batten or pendant light fittings only. As you will wish him to get the installation tested by the Electricity Board as soon as practicable, so that the mains connection can be made, he has to provide these fittings for the Board's test. Normally you will be buying your own ornamental fittings, and if you can give them to him at the right time he will normally fix them free of charge in place of the pendants and battens. Do not ask for rebate for the savings on pendants and battens - this is balanced by the cost of involvement with your own fittings, which you will have chosen for reasons which have nothing to do with ease of fixing! The electrician will also fix TV points (but not TV aerials), telephone ducting and deal with your heating thermostat and boiler wiring. He will supply and fix any immersion heater required.

All electrical work should be offered as being to specific agreed standards, and it is important that you satisfy yourself that you are getting the best arrangement of ring mains etc. The fuse board should be accessible and have separate circuits clearly labelled. Discuss the provision of contact breakers instead of fuses: this is a very sensible thing to arrange.

Plasterer

The plasterer will quote from your drawing for all plastering work in the building, including any external rendering, and for laying floor screeds. Breaking this down into the separate jobs:

Ceilings are boarded with plasterboard ready to receive a final surface of some sort. Note that if your ceiling joists are at 450mm centres you can use 9mm boards, otherwise you must ensure that 13mm boards are specified. Ceiling boards can be foil backed to provide added insulation if this is required. The finish is usually 'skim' or 'Artex'. There are other similar materials, and they should be discussed with the plasterer. He will be able to arrange for you to see samples. If a proprietary ceiling finish is required, ensure that you specify that it shall be applied to the manufacter's instructions. If you have an integral garage the Building Regulations may require that the ceiling between it and any rooms above are double boarded. If so, it will be noted on the drawing.

Walls are either traditionally plastered, or dry lined. The former involves wet plaster being applied to the wall in two coats giving a solid feel to the resulting surface but taking up to a month to dry. The latter uses plasterboard on timber framing with a surface finish. Whoever designed the building will have specified walling materials to suit one of these plastering systems, and whichever is involved you should insist that work is done to the plaster manufacturer's specifications, and that if possible all the materials come from the same manufacturer. All modern plaster specifications include metal angle beading.

Plaster coving at the junction of a wall and the ceiling gives a prestige feel to any job, and is available in a number of styles. It is also relatively expensive. As an alternative, polystyrene coving can be installed by the decorator, although it does not give the same class of finish. If you intend to use polystyrene coving do not tell the plasterer!

External render can be given a variety of finishes, and the costs involved also vary. The finish will probably have been detailed in the planning consent, and should be applied to manufacturer's instructions if it is a proprietary material, or in accordance with the best local practice if it is a traditional finish. Aluminium render stop should be used to give a bell finish to the bottom of any rendered surface, and to form a drip over windows and door openings.

A good floor screed can receive fitted carpets direct, without any other surface. The thermoplastic tiles laid in new speculative housing are there as a sales aid, and omitting them will save a useful sum. A first-class screed is worthwhile, and usually the plasterer is the best man to lay it.

All plastering work should be quoted for on a lump sum basis, from the drawings, as the system of measurement employed for 'price per square metre' contracts is extremely complex.

General

All labour and materials subcontractors who are

VAT registered can 'zero rate' VAT, which means that it is not charged to you at all, which is useful. Even more useful is the advice which these tradesmen will give to someone building on his own, and it is always worth asking them how they think the job should be done. Invariably they will be local people, who depend on their local reputation, and provided that normal prudence is exercised in placing a contract, it is rare for difficulties to arise in connection with their work. Insist on all work being carried out to material manufacturer's instructions. This is important because you will be able to refer any problem back to the manufacturer. Use of materials other than to manufacturer's recommendations will void their guarantees.

Labour only subcontractors

'Labour Only' sub-contractors are a mixed bunch, varying from those who are worthy descendants of the wandering medieval masons who were the original journeymen, to the worst of the 'lump' so reviled by the unions. The best introduction to a labour only sub-contractor is a personal recommendation from other tradesmen. It is worth asking anyone who you propose to entrust with your work to take you to see the last job he did, following which you can usually arrange to phone his last employer for a reference.

The labour only tradesmen likely to be met by the selfbuilder are:

Bricklayers

Most bricklayers come in gangs, usually called a '2 and 1' gang of two bricklayers and one labourer. They will have their own tools but will require a mixer, scaffolding, mortar boards, roller, etc. They will lay bricks and blocks, and will usually pour concrete and lay drains. They should be able to set out the building and work to any drawings, and if they are local men who are settled householders, they will probably do it very well. Such gangs like to quote per thousand bricks laid, per metre of blocks laid, and per cube of concrete poured. Unless you are used to all this it is better to insist on a price for the job, to include all the concrete work and drains that they are to be concerned with, with clearly defined stage payments. The only uncertainty should be any excess depth of foundation brickwork required, and this alone can be expressed as a price per thousand bricks. It is usual to get a verbal quote from bricklayers, and it is a good idea to confirm arrangements made in a note with a copy for each party. A typical note is shown on page 165, and should prevent subsequent argument. The arrangement is made with one person only, the foreman of the gang, whose arrangements with his partners are his own affair.

Groundworkers

Some bricklaying gangs specialise in work below dpc (damp proof course), and are known as groundworkers. Treat them like bricklayers. Groundworkers are usual in the south and unusual in the north, where bricklayers expect to work both above and below ground.

Carpenters

Carpenters tend to be more settled than bricklayers, and often have a small workshop with limited machining facilities. A contract for the carpenter's work in a new house is in two stages, and these should be treated quite separately. The carpenter's 'first fix' is the roof, hanging the external doors, setting internal door castings, and fixing window boards and ceiling traps ready for the plasterer. The rest of his work comes later, when the building has been plastered out. A carpenter will usually provide a written quotation.

Casual Labour

Selfbuilders who are already employers will have the experience to decide the basis on which they employ casual labour. For others the choice is to offer short-term formal employment, with all the paperwork involved, or only to employ those who are self-employed workers in the building industry. The latter course is usual, and it is important to establish that they are self employed in order to meet any challenge from the PAYE authorities or involvement with the Employment Acts. This can only be a loose business at best, but a note from a duplicate book confirming the arrangements made will help avoid confusion.

Site management

For some reason selfbuilders run sites that are either obsessively tidy or unbelievably untidy, and it is worthwhile being in the former category. You will have to consider matters like having a site notice for the benefit of those making deliveries, providing fencing if you consider this appropriate, and the disposal of rubbish. In dealing with all of this it is useful to bear in mind that any new neighbours in adjacent houses will be less than pleased when they realise that they are going to live next door to a building site for some months, and you will not be their favourite sort of new neighbour if you make this worse for them than it need be. One minor matter which is often important to neighbours is the use by building workers of transistor radios. Every such situation is different, and if you are paying the workmen concerned you should be able to deal with this.

There are various statutory requirements for those who run building sites. Provision of latrine facilities, a hut for meals, protective clothing, a first aid box and accident register are required by the Factories Inspector, who is unlikely to visit the site.

Unlike the Factories Inspector, the Building Inspector certainly will be seen on site, and it is a condition of building regulations that he is given due notice of the progress of work so that he can make his statutory inspections. Local authorities supply postcards for this purpose and these should always be sent off in good time. As discussed earlier, it is essential that selfbuilders should regard the Building Inspector as a friend and adviser, and avoid the 'him and us' attitude which some sections of the building industry adopt. Remember that the inspector has a statutory duty to make these routine inspections, and if work is covered up before he has checked it, he can require it to be exposed or taken down. He can be a good friend, or a very bad enemy, and if you stick to the rules he will invariably be the former.

Another visitor will be the Custom Build Inspector or the architect, surveyor or valuer who may be appointed to satisfy the bank or building society that all is in order, and to issue routine progress certificates. These gentlemen are as difficult to satisfy as Building Inspectors and not nearly as prompt in getting out to the site when requested. However, as their visits result in the release of funds these disadvantages are happily accepted. Remember - if in doubt, give any official visitor a mug of tea. Don't be ashamed of the shortcomings of either the mug or the tea; if he has been any time in the game your visitor will have seen and tasted much worse.

Selfbuild site foremen

From time to time one comes across selfbuilders who are managing the whole job themselves, but who have engaged a site manager or working foreman for the day to day supervision. Invariably this is a retired professional, and I am usually told that they have welcomed the job to liven up their dull retirement. This has always seemed a very sensible thing to do, and if you find the right man who has spent a lifetime working for a builder or developer he should surely be able to save you enough to cover the cost of employing him, particularly if he is paid on an informal basis.

Someone like this is best found through a personal recommendation, but a small ad in a local newspaper will probably bring interesting replies. As with lonely hearts advertisements, send everyone a nice reply even if they sound hopeless.

Buying materials

If you are building on your own and are not using the services of a package company you will spend at least £30,000 on materials. Buying them 10% more advantageously than the other chap will give you a saving of £3,000, while buying them 10% less well will cost you an extra £3,000. It is worthwhile giving very careful consideration to your buying arrangements.

A major factor in this is the delivery arrangements. A best buy in materials that can only be delivered on a 40 foot trailer at an unspecified time during the week may be less attractive than the same consignment at a slightly higher price on a small lorry that can get to the back of your site with guaranteed delivery on a Saturday morning. It is also worth a lot to have deliveries made on vehicles fitted with cranes, for unlike a builder or developer you are unlikely to have a forklift truck or JCB on the site to unload for you. If the lorry does not have a crane, then you will have to make sure that there are enough people about to unload it by hand. Often you will discover that the delivery arrangements are a key factor in choosing a supplier.

It is well worth while to start thinking where you should buy materials just as soon as you are certain that you are going to build, and you should start to collect leaflets and prices. Most people end up with several box files full of brochures! There appear to be obvious advantages in putting all your business through one builders' merchant or perhaps through one of the DIY superstores, but in practice most selfbuilders use a number of suppliers. Decisions about this depend on how much time you have to shop around and whether you live in an area where there are plenty of sources of supply. With the building industry currently in recession, there are bargains to be sought for all materials, and sometimes a merchant will quote you a lower price than a manufacturer. The best way to go about finding the best prices is to get to know the standard list price for the material or component concerned and then to find the salesman and ask him face to face what is his 'best price'. If this is not possible, ask him on the phone. You are unlikely to be offered very competitive terms by letter if you are only buying in one house quantities.

What always helps in these negotiations is an offer to pay promptly — 'your driver can pick up my cheque' or even offer to pop in to see the supplier and pay on the day before delivery. Skilled negotiators will get the price down to what they hope is the salesman's lowest offer and they will then start to talk about the additional discount for payment on the nail!

Ordering procedures in the building industry are often very casual by most standards, and it is commonplace for established customers to order concrete, timber, blocks or bricks by phone in an informal way. Be careful here: the man at the other end of the phone line may know in some mysterious way that a developers site is taking 4:2:1 concrete pre-mix in large loads this week, and will make appropriate arrangements following a phone call saying '2 on Wednesday and 1 on Thursday Charlie'. You are not part of the building industry and you should always be a great deal more specific. Make sure that every order states quite clearly what you want, when you want it and on what sort of vehicle. Whenever you order anything over the phone repeat like a parrot 'I don't have men permanently on site, so you will have to stick to the delivery time'.

There are some materials which you will only order after having given very careful consideration to samples, particularly items like handmade bricks. In this case make it quite clear that you are ordering 'as per sample' and be sure that you keep the sample safely somewhere. When you take delivery of anything which cannot be checked as it is unloaded, always give a qualified receipt on the delivery note, writing 'not checked' above your signature. This will enormously strengthen your hand in any subsequent debate about whether you got what you ordered, but remember that if there are any such problems you must deal with them immediately.

One of the advantages of buying direct from manufacturers is that you can use the services of their sales representatives, and this can be very useful indeed. Of course, it is rarely cost effective for them to spend much time with a small customer like you, but they will often do so, especially if you go out of your way to welcome them with a cup of tea or a walk down to the pub for lunch. A tile manufacturers rep will not only advise on the ranges of tiles likely to be acceptable to the local planning officer, but he will also talk about features such as modern verge systems, patent flashings around chimneys or the local tiling contractors who are most likely to suit you. Reps are also able to advise where their materials have been used locally so that you can go to see them in situ, which is a good deal better than looking at samples.

The sources of supply used by selfbuilders are generally as follows:

Bricks

Either direct or through merchants. Common bricks usually from a merchant, facings usually direct. Formally quoted at ex works prices plus radial delivery charges from the works plus a crane off-load surcharge. You should ignore this and ask for a delivered price, craned off on your site. If they are special bricks ask for a sample.

Stone

If artificial stone, treat as bricks. If natural stone, be very careful and seek the advice of a professional, perhaps the mason you will use. Essential to take delivery of all stone together so that it all matches. Quantities often impossible to check and taken on trust. Beware.

Blocks

Big manufacturers only supply through merchants although smaller companies may deal with you direct. Obtain 'delivered crane off loaded prices' as with bricks. Packs are usually plastic wrapped which speeds drying out. Enquire about this if it is not offered.

Joinery

Usually from manufacturers or their appointed wholesalers. Always inspect joinery before placing an order, and establish which are stock items and which are made to order: there can be long delays on the latter. Check what window furniture (hinges and latches) are supplied as standard, and enquire if your order can be fitted with furniture of a better grade. Go into the business of weather seals, reflex hinges and glazing platforms with care. The informative literature from the national manufacturers will tell you all about these subjects.

Roof trusses

Direct from manufacturers, who should be asked to quote from your drawings. Place an order for a prefabricated roof to suit drawing reference xxx, so that if there are problems they have to solve them. Also ensure that they know your delivery requirements.

Timber

From timber merchants, who may also be roof truss manufacturers. You may get better prices if you arrange to buy the timber with the trusses, although it is unlikely that it will be convenient for you to have it all delivered together unless you have room on site to stack the trusses: they are required after the joists.

Tiling

Many selfbuilders get a 'supply and fix' price from a tiling contractor, which saves arguments about how many tiles are going to be required, and who is responsible for the broken tiles and the consequent delay when waiting for replacements. You can make a start by contacting the manufacturers, whose reps will advise on contractors. Alternatively you can go to a contractor direct if he is well recommended.

Glass

Many joinery manufacturers will now supply double glazing units to suit their windows, but most selfbuilders buy them separately as they find this more advantageous. If you are ordering glass make sure that the measurements that you give to the supplier are clearly marked as either 'rebate size' or 'tight glass size'. Better still, ask his rep to come and measure up for you, and then if something doesn't fit it is his fault.

Plumbing materials

This is an area where there are huge discounts available, often associated with special promotions of bathroom suites, etc. that have been on display in showrooms. Plumbers' merchants or builders' merchants are the place for these, but very advantageous prices are also offered by the big DIY superstores.

Kitchen fittings

This is a jungle in which a 40% discount is only the start of the haggle! Look for special offers in your local paper.

Insulation materials

Insulation quilt for roofs and insulation slabs for cavity walling are often best bought from specialist insulation suppliers, who you will find listed in the Yellow Pages. However, builders' merchants often have special offers at huge discounts.

Electrical goods

It is probable that your electrician will prefer to work on a supply and fix basis. Even if he does

not suggest this, he will be able to advise you on where you can get the best buys for wire, switch gear, etc.

Plasterboard and plastering materials

Again, you will almost certainly find that it is going to be more convenient to ask your sub-contractors to quote on a labour and materials basis.

Plant and scaffold hire

Every selfbuilder hires plant or equipment at some stage in building a new home, and arranging your hiring in the most effective way is an important part of your project planning. The plant hire scene is changing rapidly, and, as with all the other services used by selfbuilders, it pays to look carefully at all your options.

You want a simple arrangement to hire the heavy duty machines used by professionals, with them being delivered to your site in the most convenient way, complete with full instructions and any appropriate safety leaflets. Machines must start easily, do what you want them to do, having cost you as little as possible. The actual hire charge is not important: what you are concerned about is the total cost of getting a job done. This may mean that well maintained equipment delivered to your site at an appropriate charge is more cost effective than driving to pick up an older machine from a cheaper source. Whichever you choose, the real savings will come from making the best use of the equipment on site, so that you have it on hire for as short a time as possible.

If you want an angle grinder to cut reinforcing for your foundations, you want it to come with goggles, armoured gloves and a supply of cutting discs on sale or return. If you are picking it up you do not want it to leak oil into the boot of your car. You will do the job cheaply if you have marked up the reinforcing ready for cutting in advance, so that you get it back and off hire on the same day. Saving the odd pound in the actual hire charge may not be important to the total cost of the job.

You need to plan your tool hire arrangements at the same time that you are deciding on suppliers for your building materials, and in most large towns you will have a choice of both big firms with attractive catalogues and smaller outfits that have just a typed list of what is available. They are all in Yellow Pages, and you should phone half a dozen of them asking them to send you details of their service. Ask about delivery, the hours when tools can be returned, and arrangements for deposits. If a firm seems keen to have your business they are likely to look after your requirements. If they are casual about explaining their service to you, then they are probably only interested in their established trade customers

Much will depend on where a hire depot is situated, and a small local operation may suit you better than a plant hire super-store. Remember that the firm that you choose is going to be very important to you, and take time to go to see what is on offer, and judge the reliability of both the equipment and the delivery promises that will be made. Most of the leading hire companies belong to a trade association called Hire Association Europe. They subscribe to a national code of practice and have a common form of contract. In recent years this has done much to raise standards generally and particularly to promote safety. HAE also provides an arbitration service when required.

When you have decided on a hire company, consider how you are going to pay them. You may get a better discount as an account customer, or cash terms may be cheaper. The smaller the hire company, the more you can negotiate on this. Remember that they will all want deposits, and two separate forms of identification. A driving licence and a credit card are usually all that is required, although some depots that hire out expensive machines like excavators will ask you to pose for a polaroid photo which stays with them until the machine is returned. All good fun. You may be able to use your credit card for the deposit, and usually the card voucher stays in the till and is destroyed when you return the equipment. This is really a form of 'no deposit' hire, and very useful.

It is important to remember that the plant which you hire is probably the most dangerous equipment that you will have on your site, and

that a juicy accident would really set back your building programme. All HAE hirers are committed to providing proper instruction on the use of power tools, emphasising safety, and have appropriate protective clothing for sale. This instruction is very important and the standard of it varies. The HSS hire shops have specially trained staff responsible for this, and take it so seriously that instructors are required to refuse to hire dangerous tools to unsuitable people, and I am told that this actually happens. *(Sorry chum, you look an accident prone idiot to me. Go away.' —Does it really happen?)* If you hire one of their mini excavators the instructor will come to the site with the machine to teach you how to use it.

This leads us to consider insurance for plant. The standard Norwich Union selfbuilder's policy through DMS Services gives cover for up to £2000 worth of plant on the site at any one time, whether it is owned, borrowed or hired. This limit can be extended for the whole of life of the policy for an extra premium of £17-60 per £1000 of plant value, or high value plant hired in on a short term basis is covered at a premium of £5-19 per £100 of the hire charge for 14 days. HSS hire shops also offer an indemnity to cover damage done to a hired machine by an inexperienced operator for an extra 10% on the hire charge. Otherwise you pay for repairs. Most selfbuilders pay the extra, but remember that the best safeguard is to know how to use the tackle properly, to wear the right safety kit, and go about things in an unhurried way. It is also important to remember security, and to immobilise, chain up, or take home vulnerable equipment at night.

There are two items of hired in plant that need special consideration. The first is your mixer. Small mixers with less than 4 cu ft capacity are unpopular with bricklayers and should be avoided. Larger diesel mixers are ruggedly built, and are often offered for sale very cheaply through small ads in local papers or on notice boards in builders merchants. If you are a fair judge of used machinery it will be cheaper to buy than to hire, particularly as you should be able to sell on your mixer when you finish. Some mixers at Milton Keynes plots have been owned by dozens of selfbuilders, and, incidentally, so have many site caravans at this selfbuilders' mecca.

Scaffolding needs thinking about carefully. Most hire companies have scaffold towers, but these do not usually suit selfbuild projects as they need firm and level ground to stand on, so you will have to find a scaffolding specialist. If you have no experience of erecting a scaffold, get it on a 'hire and erect' basis. Make sure that the rails and the kicking boards required by law are provided. If your sub-contractors offer to erect the scaffold themselves, make sure that they do so according to the book, and as far as possible write this into your arrangements with them. Remember that if they erect a hazardous scaffold it will be used by the other trades including the joiner fixing the roof, the tilers, and the plumber fixing the flashings, and it is *you* who will face a criminal summons if there is a serious accident.

Finally, electric tools. These should always be connected through an RCD contact breaker. The hire companies have suitable plug-in units available, but they do not provide them unless you ask for them because most hirers are planning to do jobs at houses that have contact breakers in the fuse box. If your temporary site supply is not RCD protected, hire a plug-in unit.

A good start when planning your plant hire is to ring the Hire Association Europe information office on 0121 377 7707 and ask for their data pack for those hiring plant, and then ring 01345 345222 for the 170 page HSS catalogue which is full of useful general information together with details, including prices, of the equipment available at nearly 200 of their branches.

Caravans

Caravans are often used as site huts, and in this role they are not subject to planning controls. If you are worried about how safe they will be it may be a good idea to remove the wheels and take them home so that the van cannot be towed away.

If you are going to live in it you will be following in the footsteps of thousands of other selfbuilders, and although the authorities may challenge you, you are most unlikely to need planning consent. This is discussed on page 118. However, the sight of your caravan may make your neighbours uneasy, so it is a good idea to reassure them that it will not be a permanent feature of the landscape. Another factor is that there may be a covenant in the title to the plot which prohibits caravans on the land. If it is an old covenant you can probably ignore it, knowing that it would take longer to get anything done about it than you will take to build the home. However, if the person selling the land to you has imposed a covenant of this sort you should get him to agree in writing to your having a caravan while you are building before you exchange contracts.

Most residential caravans on selfbuild sites are connected up to the services at an early stage, which may give rise to a demand for council tax. You will have to play this by ear. If children are involved it is essential to fence the caravan off from the building work so that they are not technically living on a building site, which would break Health and Safety legislation. A fence will also provide privacy which will be ever more welcome as the job progresses.

Programming

Finally, a job programme is essential. This can be an elaborate bar chart showing critical paths or simply a list written down in a note book, but it must exist, be realistic, and be used. The value of a job programme is not simply that you know when a particular task is to be done, but to act as an aid to ensuring that all the tools, services and other requirements for the job are ready when needed. Every job programme is different, but a typical simple work sequence for a bungalow on strip foundations with a solid floor is set out in 40 stages on page 178. The programme can be elaborated, or it can be used as it is. Starting on page 179 is a sequence of photographs showing what all this means on the ground, in this case a four bedroom family house.

Moving in

Selfbuilders are often concerned about just when they are considered to have occupied their new home if they decide to move in with the building work still going on around them. There are no hard and fast rules for this, and for most purposes you can decide what suits you as the date when you became a householder. Indeed, there will probably be different dates for different purposes, such as:

* Council Tax. You are unlikely to have any problems if you state firmly when you moved in as a householder (as opposed to being a selfbuild family camping in the house to provide security), particularly if everything has moved along steadily. However be very positive. Some councils still issue completion certificates for Building Regulation purposes, which you can quote, or not, as appropriate. If you have been living in a caravan on the site and have paid appropriate council tax, make sure that you do not find yourself being billed twice when you move into the house.

* Mortgage. You will want to pay the mortgage as a home owner as soon as possible, but the bank or building society will want to delay this until their surveyor, or the NHBC or someone else has issued a final certificate. If they delay, you can use this if you get involved with an argument over the Council Tax situation.

* Insurances. Insurance companies which are used to selfbuilders accept that selfbuild policies and home owner's policies usually overlap. Your belongings while you are 'camping' in the house are not covered on a selfbuilder's policy, but can be separately covered as the first stage of a home owner's policy. If your selfbuilder's policy runs out before you finish all the non-structural work, you will be considered to be a home owner who is doing normal DIY work on his property. DMS Services or other specialist brokers can advise on this.

* Visits from in-laws. Technical difficulties over connecting up the heating in the spare room can make visits impracticable for years.

Teamwork

Later in this book you will find a series of case histories giving the stories of a wide range of selfbuilt homes. In the selfbuild magazines you will find many more. In 90% of them a key feature is that they were built by couples who worked together and who enjoyed the challenge of it all. Their mutual moral support was at least as important as their management skills.

Selfbuilding is tough. It invariably means no leisure activities for up to a year, no family outings, no relaxed evenings. It often means no holidays although this may not be a good idea. Setbacks on site do not improve tempers, and frustrations are invariably taken out on one's nearest and dearest. If you can build a team approach that can cope with all of this it will also be fun.

This is not to say that solo selfbuilding is not a practicable proposition for either men or women, but it is important to reflect that family teamwork is a major element in nearly all success stories.

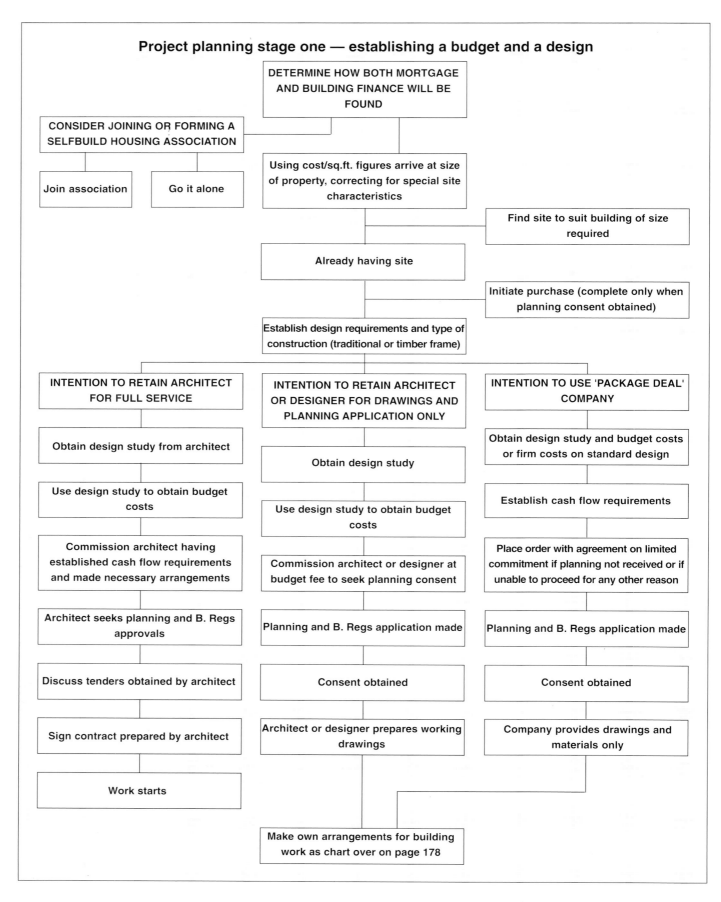

Project planning stage one — establishing a budget and a design

DETERMINE HOW BOTH MORTGAGE AND BUILDING FINANCE WILL BE FOUND

CONSIDER JOINING OR FORMING A SELFBUILD HOUSING ASSOCIATION

Join association

Go it alone

Using cost/sq.ft. figures arrive at size of property, correcting for special site characteristics

Find site to suit building of size required

Already having site

Initiate purchase (complete only when planning consent obtained)

Establish design requirements and type of construction (traditional or timber frame)

INTENTION TO RETAIN ARCHITECT FOR FULL SERVICE

Obtain design study from architect

Use design study to obtain budget costs

Commission architect having established cash flow requirements and made necessary arrangements

Architect seeks planning and B. Regs approvals

Discuss tenders obtained by architect

Sign contract prepared by architect

Work starts

INTENTION TO RETAIN ARCHITECT OR DESIGNER FOR DRAWINGS AND PLANNING APPLICATION ONLY

Obtain design study

Use design study to obtain budget costs

Commission architect or designer at budget fee to seek planning consent

Planning and B. Regs application made

Consent obtained

Architect or designer prepares working drawings

INTENTION TO USE 'PACKAGE DEAL' COMPANY

Obtain design study and budget costs or firm costs on standard design

Establish cash flow requirements

Place order with agreement on limited commitment if planning not received or if unable to proceed for any other reason

Planning and B. Regs application made

Consent obtained

Company provides drawings and materials only

Make own arrangements for building work as chart over on page 178

Project planning stage two — do you use a builder?

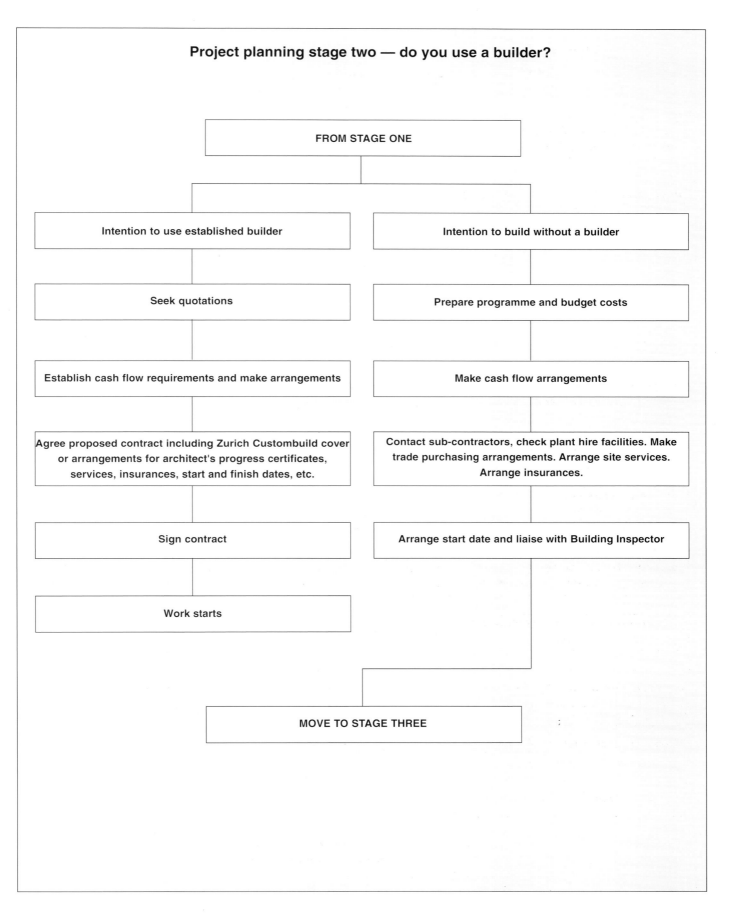

FROM STAGE ONE

Intention to use established builder	Intention to build without a builder
Seek quotations	Prepare programme and budget costs
Establish cash flow requirements and make arrangements	Make cash flow arrangements
Agree proposed contract including Zurich Custombuild cover or arrangements for architect's progress certificates, services, insurances, start and finish dates, etc.	Contact sub-contractors, check plant hire facilities. Make trade purchasing arrangements. Arrange site services. Arrange insurances.
Sign contract	Arrange start date and liaise with Building Inspector
Work starts	

MOVE TO STAGE THREE

Project planning stage three — 40 stage programme for a traditionally built bungalow with solid floors

	Action	Requirements
1	Apply for all services	
2	Apply for building water supply	
3	Arrange insurances	
4	Provide access, hard standing, site storage, notice for deliveries	Hardcore, site hut, wire, notice
5	Strip top soil and stack for future use	Digger hire
6	Excavate foundations, service trenches and drive and spread hardcore	Digger and hardcore
7	Building inspector to inspect footings	
8	Pour foundation concrete	Concrete, possibly reinforcing materials
9	Building Inspector to inspect foundations	
10	Build foundation brickwork	Bricks, cement, sand, ties
11	Fill foundations, blind, lay membrane	Hardcore, roller, membrane
12	Building inspector to inspect foundation brickwork and membrane	
13	Cast slab, leaving ducts for services	Concrete, ducting for services
14	Lay drains, fill soakaways, lay services If possible	Drainage materials, materials for inspection chambers
15	Building Inspector to inspect drains Water Board and others may wish to inspect services	
16	Build off dpc	Cement, sand, bricks, blocks, tie irons, insulation slab
17	Building Inspector to inspect dpc	
18	Build up to square	Walling materials including joinery and lintels scaffold
19	Carpenter to scarf joint wallplate. Bed wallplate Rear end trusses as template for gables	Wallplate and roof trusses
20	Build gables. Build In gable ladders. Build chimney up through roof	Chimney liners
21	Fix roof ready for tilers with barge and fascia fixed to suit type of tiles. Board valleys and make provision for flashings	Balance of roof materials
22	Sub-contract felting, laths, tiling. Plumber to be in attendance to attend to flashings and valley, or more usually tiles left unclipped for him to follow	Tiling contractor
23	Fix guttering required	Rainwater goods
24	Carpenters first fix - door linings, window boards, studding partitioning roof access. Hang external doors	Carpenters first fix materials
25	Glaze all windows, making provision for ventilation	Glazier or double glazing units
26	Plumbers first fix	Appropriate materials
27	Heating Installers first fix	Appropriate materials
28	Electricians first fix, including telephone ducts	Appropriate materials
29	Plaster out	Ceiling board, plaster, angle beading
30	Lay floor screed	Insulation for piping under floor screed
31	Complete drains, external works, paths, drive	Appropriate materials
32	Plumbers second fix and water on	Appropriate materials
33	Build fire-place	Appropriate materials
34	Electricians second fix, including TV aerials, and electricity on	Appropriate materials
35	Carpenters second fix, including kitchen and all fitted furniture.	Appropriate materials, kitchen units and fitted furniture
36	Telephone Installation	
37	Decorations	All decorating materials
38	Lay tiled floors	Tiling contractor or tiles, adhesive, grout
39	Wall tiling	Tiling contractor or tiles, adhesive, grout
40	Clean through. Arrange for householder's Insurance to take over from site insurance	

Making it all happen — one real story

Publisher Colin Spooner had been involved with the publication of *Building Your Own Home* for 13 years when he decided to look into actually building his own house. This is his story in his own words. Pearl is Mrs. Spooner. On the following pages are a series of photographs which illustrate his narrative. This is selfbuild through the eyes of a selfbuilder:

"Pearl and I had just finished renovating a 400 year old cottage in two acres of garden, and it occurred to us that the lawn beside the stream at the bottom of the garden would be a beautiful setting for a new house. Involvement with Building Your Own Home *since 1978 may, or may not, have played some part in our thinking.*

Everything indicated that the chances of obtaining planning permission were remote, and that to have any sort of chance we would have to prepare our case thoroughly and present it efficiently. Local enquiries found the best person to handle this for us. He was invited round and told us that our application would fail, but that it might just succeed at appeal. We should plan for the appeal from the start, and submit an application for detailed planning consent to show that we were intending to build for ourselves, and not just hoping for outline consent in order to sell the site at an enhanced value.

Accepting this advice we set about choosing a design. Regular phone calls to local estate agents gave us a value for the cottage which became the basis for our building budget. We allowed for a slight increase in the ultimate mortgage and a 10% error in the budgeted build figure. This enabled us to decide that we could afford the Potton Grantchester. We liked this design because it had all the features of a cottage, was spacious and was timber framed, which intrigued me. After discussions with Potton and visiting a nearby completed Grantchester (built by an architect!), we bought a set of planning drawings from Potton. We then visited every

neighbour who bordered or overlooked our land, showed them the plans and asked them to support our application. They all agreed and supplied us with letters. Preliminary talks with both the bank and the building society enabled us to arrange funding. Insurances and a Zurich Custombuild warranty were also checked out.

The application went in and, as anticipated, was rejected by the planning committee on a number of counts. We now had the specific grounds against which we would have to argue our case with the Department of the Environment. Those relating to the look of the building were discussed with the planning officers and we persuaded them to advise what features would be considered acceptable, and got them to confirm this in writing. Appropriate changes were then made to the drawing to meet their requirements in every respect. This left the issue of whether the site should be developed at all, and this is what the D of E had to resolve.

We prepared the documents for the appeal and sent it in. The inspector came and went, and we waited. We had gambled about a thousand pounds in fees of one sort or another. Would it be in vain? Four weeks later the letter with the Bristol postmark arrived, and the appeal had been granted subject to conditions regarding the landscaping, etc. We were euphoric.

Now we had to decide whether to push ahead or sit tight because there were definite signs that the housing market was getting even worse. We thought we would take the building to the slab stage and then re-assess the situation. We checked with Potton that their insurance scheme would operate even though we had not yet placed an order and paid the premium for the Zurich Custombuild warranty.

First a services trench down the narrow driveway was dug and services installed, and backfilled. The site was then cleared and levelled, the trenches were dug and the footings poured, and the movement of the lorries served to compact the new drive. The site was on the bank of a stream and the foundations were very complex, involving a

179

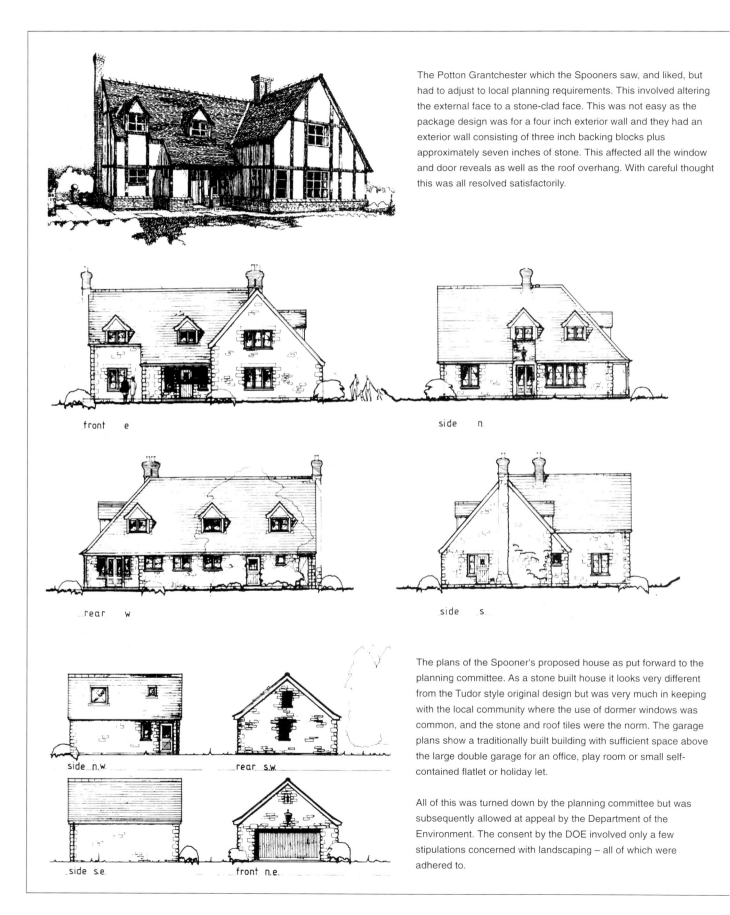

The Potton Grantchester which the Spooners saw, and liked, but had to adjust to local planning requirements. This involved altering the external face to a stone-clad face. This was not easy as the package design was for a four inch exterior wall and they had an exterior wall consisting of three inch backing blocks plus approximately seven inches of stone. This affected all the window and door reveals as well as the roof overhang. With careful thought this was all resolved satisfactorily.

front   e

side   n

rear   w

side   s

side   n.w.

rear   s.w.

side   s.e.

front   n.e.

The plans of the Spooner's proposed house as put forward to the planning committee. As a stone built house it looks very different from the Tudor style original design but was very much in keeping with the local community where the use of dormer windows was common, and the stone and roof tiles were the norm. The garage plans show a traditionally built building with sufficient space above the large double garage for an office, play room or small self-contained flatlet or holiday let.

All of this was turned down by the planning committee but was subsequently allowed at appeal by the Department of the Environment. The consent by the DOE involved only a few stipulations concerned with landscaping – all of which were adhered to.

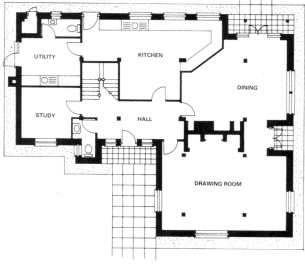

Top left: The back garden of the cottage which the Spooners bought, as it was when they acquired the cottage and land. They subsequently thought they might build on the land on the nearer side of the stream which is overgrown in this picture.

Top right: The same view some three years later when planning consent had been granted and the finishing touches were being put to the foundations and raft. Note the somewhat eccentric, but picturesque, bridge that Colin Spooner built to cross the stream, and which allowed him to drive his mini tractor over to mow the large expanse of lawn!

Above right: The finished house from a slightly different perspective with some work still to be done on the drive and in the close proximity of the house.

Left: The adapted floor plans for the original Potton Grantchester. When finished the kitchen no longer had an angular wall but was squared off and the main bedroom (overlooking the garden) – usually a secondary bedroom in this design – was increased in size by moving the bathroom forward into the main bedroom above the lounge. This also allowed the bathroom to be designed more satisfactorily. Also the wardrobe in the large front bedroom was felt to be unnecessarily large so a toilet and hand basin was fitted in one half of it.

reinforced slab with three massive beams of mind-boggling complexity. In all we calculated we made 7,000 welds to the reinforcing, many by the light of car headlights. Finally the slab was poured. The master mason who we persuaded to run the site for us suggested that although we did not yet want to order the frame, we should go ahead with the accommodation works, and in particular the paths and surface drainage. This we did, and the site was never muddy from the day we started. But we, and many of our friends, had calloused hands, tired backs and suntans!

We had already met a serious problem as we had cleared away far more soil (and buried petrol pumps, beds, plastic sheets and scrap metal) than we had anticipated and then found that the sewer was 18 inches higher than our drains were going to be. This involved installing a mini sewage pumping unit at an unbudgeted cost of £2,500.

The property market had not improved and we were still unable to sell our cottage. We wondered if this was because prospective purchasers could not visualise a house on this massive slab of concrete next door, so we decided to press ahead, confident that the new house would be the best advertisement for the cottage. The kit was ordered and while we awaited its arrival we pushed on with the garage, which was to be built traditionally.

On November 16th the first of five lorries with the frame arrived, together with three Yorkshiremen who were to put it together for us. Eight days later the structure was up and the roofers moved in. They had the house weatherproof by Christmas, but not without drama. Our planning approval had stipulated clay double Roman tiles and the planning authorities were insistent on their use. The significance of this at the planning stage had passed me by due to my inexperience: these are very expensive tiles indeed, and quite outside our budget. We had still not resolved the problem a week before the roofers were due to arrive when we saw an advertisement '6,000 clay double Romans for sale - all from one roof'. Within 24 hours we had a sample,

within 48 hours the planners had approved it and within 72 hours we had the quantity we needed plus matching ridge tiles, all for 40p a tile and delivered at 6 a.m. on a Monday morning by a driver who had instructions that not a tile was to be unloaded without cash. I was expected to have £2,000 in notes lying around!

Building with newly quarried stone also created problems for us as we could not find stone masons who were willing to quote for the entire job. In part this was due to the fact that we were sheathing a timber frame, which was something new to them. Our budget for this part of the work was inadequate, but the overspend was balanced by savings because we could buy other materials very keenly due to the recession.

While the stone walls were being built and the local workmen were solving the practical problems of cladding a timber frame with stone laid in the local style, the first fix tradesmen all arrived. Electricians, plumbers, plasterers, the central vacuum system and heat recovery and recycling systems installers all vied with each other for space and for our attention. Pearl and I now became cleaners and tidiers rather than diggers, shovellers and levellers, (as well as problem solvers, decision makers, and, necessarily, money earners). I made it a rule to completely clean out the house every Friday night so that we could work in comfort over the weekend and the trades people could come back to a reasonable environment on the Monday morning.

In went the insulation (not only in the external walls and roof but also in all the internal walls and floors to reduce the sound transference from room to room), the vapour barriers, the plasterboard, and we could start the second fix. At times it seemed chaotic with one tradesman waiting for another to complete a job but everything came together in due course. The water and electricity was switched on and the kitchen installed. Then, best of all, the cottage was sold with immediate completion, although we had to drop the price by 20% from the original valuation. The timing had *worked out just right. It became a race to get as much done as possible before we moved in.*

Looking back at it now, and we still had much to do, it was really worthwhile although we had a bigger mortgage than we intended. It was fun, very hard work, a test of management and budgetary skills, perseverance and nerve. We had started with a semi-derelict cottage with two acres of overgrown, rubbish strewn ground and we ended up with a lovely cottage style four bedroom modern house with two acres of landscaped gardens, stream, pond and paddocks and we are rightly proud of it.

Colin and Pearl Spooner's costs

Item	£		
Digger	1,566.58	Blocks	2,530.94
Concrete	4,626.47	Plasterboard	888.82
Cement	993.07	Timber flooring	545.72
Sand	902.96	Paint - emulsion	231.60
Aggregate	1,425.00	Paint - wood primer	11.61
Mash	1,557.65	Paint - finishing	33.06
Welding	209.50	Electrics	2,507.01
Timber	1,997.25	Vents, slabs, etc	86.96
Scaffolding	1,252.63	Oddments, inc dpc, lead, mastic,	
Stone	7,401.33	nails	3,115.05
Facing bricks	1,443.49	Central vacuum	898.60
Common bricks/slabs	438.59	Insulation	2,274.84
Stocks/engineering	176.57	Heat recovery system	2,205.45
Potton kit	39,700.30	Kitchen/utility	4,007.50
First fix joinery	1,272.00	Services	2,256.55
Second fix joinery	1,530.00	Fees	4,383.03
Sewage station	2,992.79	Artexing labour	1,567.40
Ground pump	1,633.39	Labour	17,920.00
Rainwater grids	327.89		
Glazing	935.49	**Total**	**135,327.28**
Roofing tiles	2,191.00	**Less VAT.**	**9,224.40**
Roofing	2,891.25		
Flooring lino	217.80	**Total**	**126,102.88**
Beams, lintels, sills	1067.23	Garage	488.84 sq. ft.
Internal plumbing	9,254.94	House	2,600.00 sq. ft.
Chimney	1,432.16	**Total**	**3,088.84**
Plaster	315.81	**Cost per sq. ft. £**	**40.83**

1. Colin and Pearl Spooner, right, discuss the programme with Elsie Woolven and Brent Ackerman of Potton Limited.

2. First the rough outline of the building was marked out in lime as a guide for the digger

3. . . . which dug the trenches

4. . . . to the grudging approval of the Building Inspector and his custom build colleague.

5. Profile boards 'set out' the building ready for the concrete.

6. When it comes there is a typical selfbuilders' concrete pouring fiesta

7. . . . with everyone trying hard not to get it down their wellies

8. . . . under the laconic eye of the concrete vomiting machine operator.

9. After the blocks arrived the foundation walls were built

10. Pipes were set out for the underfloor air supply to the fireplaces.

11. The foundations were filled with crushed stone and had to be watered . . .

12. . . . and compacted with a vibrating plate.

13. The specially designed foundations involved a reinforced slab

14. . . . for which the reinforcing was welded up by, quite literally, moonlighters!

15. The finished job was literally a work of art, all too soon to be covered up . . .

16. . . . with concrete, which involved the return of conscripted friends who demonstrated

17. . . . enthusiasm for the job in direct ratio to the age of the individual concerned.

18. The slab had to have a tamped surface, working accurately to levels.

19. The frame arrives together with the Potton erection gang.

20. They quickly erect the first of the aisle frames

21. . . . followed by the ground floor walling panels

22. . . . and then the first floor panels.

23. In no time at all it began to look like a house

24. . . . with a downstairs

25. . . . and an upstairs.

26. The roof was quickly felted to keep out the weather.

27. The second hand tiles arrived at 6am to be unloaded by hand. (Pearl second from the right)

28. The plumber fixed the lead valleys ready for the tilers

29. . . . and Pearl stained the exposed timber to the dormers.

30. The tiles went on the roof very quickly, and from their perches up aloft the tilers could look down on

31. . . . lesser men laying drains

32. . . . while the masons' labourer sorted the stone

33. . . . or faced quoins inside when it was wet.

34. When the first stone was laid Colin decided to take over brushing out the raked joints.

35. The threat of frost at night meant that the stonework had to be covered up in the evenings.

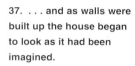
36. Colin's next job was fixing double glazing units which proved relatively simple

37. . . . and as walls were built up the house began to look as it had been imagined.

38. The ground floor was now plaster boarded and a start could be made on the inglenook fireplace

39. . . . which quickly took shape.

40. Meanwhile upstairs Colin tackled the vapour barrier ready for the plasterboard.

41. Meanwhile the masons had reached the top of the chimney.

42. Inside Colin treated the ground floor surface

43. . . . while the fitting out proceeded on the first floor.

44. In the roof the heat exchanger/ventilation unit was installed.

45. Paths were laid, and at this side a retaining wall provided a raised flower bed

46. . . . and after moving piles of rubbish out of the way the house began to look better and better.

47. The kitchen was fitted and equipped

48. . . . and one by one the rooms were finished and furniture carried down from the cottage

49 . . . until the Spooners could move in, although with a great deal still to do.

Moving in!

18 Selfbuild paperwork

First of all, what follows is pretty basic stuff for the selfbuilders who are concerned with management services in their every day work. If you are in this category, it is not for you. It is written for the couple who are a little apprehensive about coping with the paperwork that is involved in building their new home, and hopefully it will help them to realise that given a little thought it really needs to be no bother at all.

The records that you will need to keep will consist of your correspondence files, your orders for materials and your accounts, and your collection of brochures and leaflets. You will also need to know the best way to write letters to the various authorities who will be concerned, to know how to place orders with suppliers and others and to make payments to them, and what you will need in order to reclaim VAT when you move in.

Correspondence records first. You will have to keep eight different correspondence files dealing with the following subjects.

Finding and buying the plot.

Services - water, gas, electricity, phones, etc.

Financial arrangements.

Planning and Building Regulations, etc.

Package companies, architects or others providing professional services.

Sub contractors, builders and others who will work on site.

Suppliers and those who will hire plant.

Other matters.

This will involve you in eight ring binders which cost about a pound each at a stationers, each for a different subject, and in them you keep all the letters on a subject that you receive and a copy of every letter that you write about it. This leads to the problem of how you make the copies of your own letters, because it is really important that you arrange for this. If there is a problem about anything in the future you must be able to say, 'look — this is the letter that I wrote to you', and without a copy you cannot do this. You can make a carbon copy, or a photocopy, but unless you have access to a copier at work this means that you will have to take your letters to a shop that advertises a photocopying service before you post them. It is a nuisance, but well worthwhile. Don't think of buying a copier for the job: good ones are expensive and cheap ones are unreliable.

How do you set about writing letters to the town hall, the water board, and all the others? First of all, do not let them intimidate you with their fancy letterheads, beautifully set out typed letters or pompous style. Above all, do not imitate the way in which they write to you. Remember, as a tax payer you employ these people. Their impressive letters are supposed to show you how professional they are. Your simple hand written letters in ordinary language will show that you are a straight forward direct type who does not need to impress anyone, and who they should deal with promptly and properly as one of the honest tax payers on whom they rely for their salaries.

In the last century officials used to end letters *'I remain, Sir, your obedient servant'*, and the proper way to end your reply was *'You will remain, Sir, my obedient servant'*. Alas these days officials merely conclude with 'Yours faithfully' or 'Yours sincerely', with precious little intention of being either faithful or sincere. But I digress.

All letters to large organisations should be addressed to the office concerned, marked for the attention of the person who you want to deal with it if you know who he or she is. If a letter is signed by someone who has got his name typed under his signature, you reply to him by name, but you address the letter to his office marked 'for the attention of xxx'. This ensures that someone else will deal with it if he or she is away. Your letter should always be dated and start with a reference to what it is all about, but otherwise you should use the same ordinary words that you would use if you were talking directly to the person concerned, and at all costs you should avoid official jargon. You are not a civil servant, and do not try to write letters like one.

There are a couple of examples on the opposite page.

If you receive this:

LOAMSHIRE WATER AUTHORITY
HIGH STREET
ANYTOWN

Mr. A. Selfbuilder,
Plot 14,
Grove Crescent,
Anytown.   22nd June 1996
                        Ref: JC/4/211739/bcd/2

Dear Sir,

Public Drains Act 1882 -
Drain connections by non-designated undertakers

Persuant to your proposal to make a drain connection into the
authorities main sewer in Windmill Road at a point 28 metres
north of manhole 56a as shown on drawing 2164/8/B(revision 4),
I am instructed to advise you that in addition to your obligations to undertake this work and all reinstatement, strictly
in accordance with the Board's requirements, together with an
obligation to protect and light the works, which obligation
you are deemed to have accepted by payment of the connection
fee, you are also required to demonstrate that you hold
approved public liability insurances for 2.5 million pounds in
respect of this work. This is an essential prerequisite of
any work starting in the highway.

Yours faithfully,

I.M. Portant
Sewage Engineer

And if you receive this:

THE PLANNING OFFICE
ANYTOWN DISTRICT COUNCIL
ANYTOWN

A Selfbuilder, Esq.,
Plot 14
Grove Crescent   22nd June 1996
Anytown   Reference: C/00/92/4612/6/D

Dear Sir,

LAND AT GROVE CRESCENT -
PLANNING CONSENT 92/4612, CONDITION 12.

I am instructed to draw your attention to the above
condition in the planning consent for the property which
you have constructed at Grove Crescent, which condition
requires that the over-mature sycamore tree which was
removed should be replaced with a tree of an approved
indigenous species, which should be protected until
properly established. I note that you have now occupied
the building but that no tree has been planted, which is
in contravention of the consent. The council must look
for immediate compliance with the condition of the
consent, and I await early advice of the species to be
planted and your confirmation that the necessary work will
be put in hand forthwith.

Yours sincerely,

for Chief Planning Officer

Your reply could well be:

PLOT 14
GROVE CRESCEN
ANYTOWN

25 JUNE 1996

THE SEWAGE ENGINEER
LOAMSHIRE WATER AUTHORITY

FOR THE ATTENTION OF MR I M PORTANT

DEAR MR PORTANT
DRAIN CONNECTION - WINDMILL ROAD

THANK YOU FOR YOUR LETTER JC14/211789/bcd/2 OF JUNE 22nd. I
HAVE ARRANGED FOR £2,500,000 PUBLIC LIABILITY COVER FOR
THE TIME THAT I AM MAKING THE DRAIN CONNECTION AS AN
EXTRA TO MY EXISTING INSURANCE POLICY AND ENCLOSE A
LETTER FROM THE INSURANCE COMPANY CONFIRMING THIS.
THE COMPANY TELLS ME THAT THIS LETTER IS ALL THAT YOU
USUALLY REQUIRE. I HOPE TO START WORK ON JULY 20TH

YOURS SINCERELY
A. Selfbuilder

A. SELFBUILDER

A reply might be:

PLOT 14
GROVE CRESCENT
ANYTOWN

30 JUNE 1996

THE CHIEF PLANNING OFFICER
ANYTOWN DISTRICT COUNCIL
TOWN HALL
ANYTOWN.

DEAR SIR   PLANNING CONSENT 92/4612

THE TREE TO WHICH YOU REFER IN YOUR LETTER C/00/92/4612/63/D
OF JUNE 22 1996 WILL BE PLANTED IN NOVEMBER WHICH IS THE
APPROPRIATE MONTH FOR PLANTING TREES.
I INTEND TO PLANT A SILVER LIME (TILIA TORMENTOSA) AND
WILL BE GRATEFUL IF YOU WILL APPROVE MY CHOICE AS
SOON AS POSSIBLE.
YOURS FAITHFULLY
A. Selfbuilder

A. SELFBUILDER

DATE	CHEQUE NO.	ITEMS	RECEIPT NO	PAID	NET	VAT.
6 Nov 95	22715	24 LOFT FLOOR PANELS		43-90	37-38	6-52
"		18 mm PINE BOARD FOR KITCHEN	56	167-20	142-30	24-90
"		140 M. SKIRTING		270-12	229-89	40-23
"		16 SETS ARCHITRAVE		112-00	95-32	16-68
"		NAILS AS INVOICE (all above on invoice 21735)		9-26	7-88	1-38
" 95	22716	BUTANE GAS	57	10-72	9-12	1-60
7 Nov 95	22717	MDF FOR KITCHEN WINDOW CILLS	58	30-20	25-70	4-50
7 Nov 95	22718	STAGE PAYMENT CARPENTER		400-00	400-00	—
8 Nov 95	22719	FINAL PAYMENT SCAFFOLD HIRE — SMITHS HIRE	59	185-00	157-48	27-52
13 Nov 95	22720	TIMBER AS INVOICE	60	92-70	78-89	13-81
13 Nov 95	22721	GUTTERING + DOWN PIPES ETC AS INVOICE	61	122-60	104-34	18-26
13 Nov 95	CASH	MASTIC SEALANT	62	5-40	4-60	0-80
14 Nov 95	22722	LOFT LADDER	63	48-99	41-69	7-30
14 Nov 95	CASH	DADO RAIL FOR HALL	64	27-30	23-23	4-07
14 Nov 95	22723	GARAGE DOORS	65	366-50	311-92	54-58
14 Nov 95		GARAGE DOOR OPENERS		582-00	495-32	86-68
17 Nov 95	22714	GUTTERING PARTS AS INVOICE	66	22-50	19-15	3-35
18 Nov 95	CASH	PAY TONY, 22 hrs @ £7.67		154-00	154-00	—
18 Nov 95	22715	KERBING FOR DRIVE	68	207-42	176-53	30-89
		TOTALS THIS PAGE		2857-78	2514-74	343-04

Your account book can look like this. The columns show:

Cheque No. You will see that some cheques, like numbers 22715 and 22723, cover the payment of more than one invoice.

Receipt No. These are the consecutive numbers which you have written on the suppliers' invoices before you put them in order in the lever arch file.

Paid. This is the sum which you have paid with the cheque.

Net. This is the cost of your purchase without VAT.

VAT. This is the important column because you will get back most of it when everything is finished. If the supplier's invoice is marked 'VAT INCLUDED' but does not break down the figures, use a calculator to divide the total sum by 1.175 to get the net sum, and multiply this by .175 to get the VAT amount.

These are the VAT claim forms which you will have to complete.

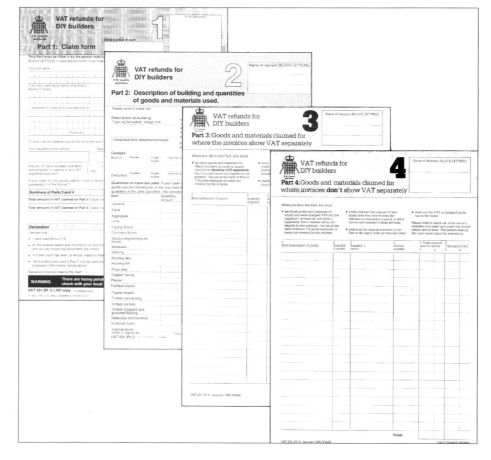

If a letter is very important it is a good idea to send it 'Recorded Delivery', which you can arrange at the post office for a few pence. Stick the receipt for it on your file copy, and then the addressee cannot say they never received it.

So much for letters and letter files. You will also want to keep a whole mass of brochures, leaflets and advertisements about the materials, fixtures, fittings, etc. that you want. For these you need a couple of lever arch files, which come complete with index sheets at about a fiver each. You will require a punch to be able to file your collection of brochures, and when you have got them all in order it will be easy to refer to them.

Accounts are a little more serious. If you are really frightened of what is involved, consider finding a retired professional or a friend to handle it for you. Tell them you will be making between 100 and 200 payments and will want a simple cash book with the VAT itemised, and you will also want all your invoices for which you can reclaim VAT entered on some simple forms. An appropriate fee for this would be £100, and no more. However, it is much better to deal with this yourself, and some selfbuilders or their wives go to evening classes to give themselves the confidence to do this. Good for them, but really there is nothing to it.

To keep perfect selfbuilder's accounts you will need a five column account book like the Guildhall or Collins books available in any office stationers, together with another lever arch file to keep the receipts. Record every payment that you make and the invoice to which it refers. The important thing is to show the total amount paid in one column, the VAT in another, and the figure without the VAT in a third. Another column is for the cheque number and yet another for the number which you have given the invoice which you have put in the lever arch file. There is absolutely no need to keep separate account pages for each supplier, and you need not keep their monthly statements but simply pay on the invoices. Cash payments are simply entered as 'cash', and payments in respect of services which do not carry VAT should show this by putting a dash in the VAT column. A page in your account book will end up looking like the example illustrated.

VAT invoices have to be entered on your VAT claim form and these are shown opposite. Remember that you cannot reclaim VAT on some items, as explained in chapter 22, and VAT invoices *have* to have the supplier's VAT number printed on them. If the invoice does not have the VAT shown separately it is a VAT inclusive invoice and is listed on a separate form. To calculate the VAT in the total use a pocket calculator to divide by 1.175, which gives the net sum, and then multiply that by .175 for the VAT.

When you send a cheque to a supplier you should attach either his statement or a note of your own saying which invoices you are paying.

When you come to reclaim the VAT all that you have to do is to take your claim forms and your file of invoices to the VAT office, carefully get a receipt for them, and expect the congratulations of the clerk for having presented everything in such good order. You will get the receipts back in due course.

The last type of record that you need is a simple duplicate book for making a note of orders which you issue to suppliers, and perhaps of arrangements made with subcontractors or casual labour. You may, or may not, give the original to the supplier or subbie involved, but the important thing is that you have a record. When you pay for the item or service concerned you make a note of this on the duplicate page in the book, and at any time you can tell what you have ordered and paid for, and what you have ordered which has still got to be paid for.

Finally, a word about computers. If you have a home computer which you use regularly, then obviously you will use it. However, if you have a computer which you have not used for anything as complicated as this before, for goodness sake leave it alone. Proper selfbuild accounts are too important for learning-as-you-go with a keyboard and screen, and writing them out by hand is going to take less time.

19 Avoiding Problems

First of all, let us get one thing straight. There are very few great selfbuild disasters, and a higher proportion of all builders and developers go bust than the proportion of selfbuilders who are unable to complete their new homes. One reason for this is that in order to be able to make a start on site at all the selfbuilder has to have everything worked out and checked and re-checked in order to obtain his finance. However, things can go wrong. Sometimes it is the fault of the individual builder, sometimes someone else's fault. In this chapter we look at how you avoid being unlucky.

Twelve of the most common ways in which people building on their own sometimes get themselves into trouble are set out below. They are not in any sort of priority order, and some unfortunate souls manage to get involved in two or three of these difficulties at the same time. One gets the impression that these people are accident prone anyway. At any rate, if you can avoid any of the twelve situations below, your chances of causing yourself any serious problems are remote.

1. Domestic problems

One of the commonest reasons why work comes to a stop on a selfbuild site is that the couple building have split up. This has a disastrous effect on both the finances of the project and on the enthusiasm to get the job done. These are always sad stories, and sometimes I get the impression that the selfbuild operation was a last ditch attempt to save a marriage that was in deep trouble anyway. If your relationship is going through a sticky patch then avoid selfbuild until things are better: it is far too stressful for couples who are not working together as a team.

On the other hand, many successful selfbuilders will tell you that building their new home gave them an opportunity to tackle something together which made them realise just what an effective team they were, and that they really enjoyed working together.

Moral: do not get involved with selfbuild as a family unit unless you are quite sure that relationships all round can stand the strain.

2. Selling the existing home

Until the end of the property boom in 1989 you could build your new house and be confident that you were going to sell the old one at an appropriate stage. Inflation in house values meant that there was a good chance of you getting more for it than you had planned when you first put the project together. This is no longer possible, and the advice of the banks and building societies today is that you should sell the old house before entering into a commitment to build the new one.

Problems with selling an existing house have led many selfbuilders into difficulties, particularly as the interest clock keeps ticking on the building loan for the new house while you are trying to sell the old one.

Moral: until the property market recovers do not gamble on being able to sell an existing house.

3. Site problems

Unforeseen site problems that are discovered after you have bought your site can be a disaster. Examples are discovering that there is a ransom strip which prevents you having access to the site, or that the main drain is not where you believed it to be, or that the highway authority requires a sight line at your entrance and this involves the co-operation of an unfriendly neighbour. These are all the sort of things which you should check out before you sign a contract to buy the land. Remember your solicitor will concern himself with whether your title to the land is good and solid. It is **your** job to make sure that you are going to be able to build on it in the way that you want.

Moral: check every aspect of the site before you buy it.

4. Sub-contractors

Most selfbuilders get on well with their sub-contractors. Some make long term friends of them. However, there are a few who manage to employ the wrong subbies, or who simply find that the arrangements that they made are not working.

When this happens you have to manage the situation, and not let it drift from bad to worse.

You may have to cut your losses, get rid of the person who is not doing what you want, and start again. If you are not used to sacking people you may find this difficult, and if this is the case the other chap is likely to exploit the situation. Get advice from someone who is used to dealing with this sort of thing all the time. Ignore the advice of anyone who does not have relevant experience. And remember that every day that you let a situation like this continue, the worse it is likely to get.

Moral: find the right sub-contractors who have been recommended by other selfbuilders, build the right relationship with them, and if things go wrong manage the situation firmly and effectively.

5. Depending on others
Never, never let a selfbuild project depend on others. If Grandma is going to let you have the last £10,000 of your finances, she should hand it over before work starts. If your Uncle offers the stone slates on his derelict barn for you to use on your new home, then make sure they are removed from the roof and stacked before work starts. If a neighbour has agreed to let you have an easement to connect your drain into his septic tank, then get it signed at the same time that you sign the contract to buy the land.

Remember that Grandma, Uncle and your neighbour can all be involved in the same car crash, and their executors may not be nearly as helpful. Or, they may simply change their minds.
Moral: ensure that key elements in your proposals are always properly tied up before you enter into financial commitments.

6. Ignoring planning conditions
Planning consents are fairly concise documents written in plain English on very bad quality paper. The words on them are specific and precise, and they are meant to be read, understood, and the complied with. This seems straightforward, but it is surprising how many selfbuilders choose to ignore the conditions of their planning consent. They do not follow the building line shown on the approved drawing, or they decide to move a tree which has a protection

order on it, or they simply do not bother to get the formal approval for the bricks and tiles that they choose to use.

Those who do this may decide very lightly that there is no need to bother with what they consider to be unnecessary formalities. Then they find that they are involved in a dispute with the authority that will take a long time to resolve, and may delay all of the work on site until it is settled.
Moral: if your planning consent carries a large number of conditions, get to know them by heart and stick to them.

7. Forgetting about insurance
Many building societies will insist that you take out appropriate site insurances, but others will suggest you take out an ordinary householder's policy giving fire, flood cover etc. This does **not** protect against all selfbuilder's risks. You need public liability cover, employers liability cover and contractors all risk cover. The chapter on insurances in this book which starts on page 206 tells you how to arrange your insurances.
Moral: if you do not take out insurances, you are gambling. Do you really want to gamble with your new home?

8. Coping with death
If you walk under a bus while you are building the new house, your estate will still have the building finance available to get the job finished. However, in order to arrange for this they will probably want up to a further £20,000 of additional money to be able to employ the best builder in town to finish it quickly, perhaps so that it can be sold. If something happens to you, will this money be available in your estate? If there is any doubt about this at all you should take out an appropriate short term life insurance policy. It will not cost very much, and is a prudent thing to arrange.
Moral: remember you are not immortal.

9. Accidents
The building industry has a bad accident record. Selfbuilders are more at risk than professionals. There is a real risk that you will suffer an accident while you are building. Guard against

this by learning of all the hazards, and taking common sense precautions.

In particular, beware cement burns, and do not let small children play with mortar on your site.

Moral: *accidents can happen to you.*

10. Disputes

Once you have started to build a new house it is important to get it finished on time otherwise the interest charges on the building finance will get out of hand. If you do become involved in a dispute with others, settle it quickly so that the whole project is not held up. This may involve making a pragmatic decision to let others get away with things that you think are wrong. If you agree to a neighbours version of where a boundary post should be put, or pay a few pounds more than agreed for materials or services, it may be the right decision if it saves time and keeps up the momentum of the whole job.

Moral: *in any dispute do not cut off your nose to spite your face.*

11. Beware cheap materials

Bargains in building materials are for those who know what they are buying, and if you are not an expert do not use anything to build your new house about which you are not absolutely certain.

This is particularly important with materials that can affect the structural stability of the whole building. In particular, beware hardcore or other fill in the foundations which is not as shown on the drawing and has not been approved by the Building Inspector. Sometimes selfbuilders use cheap industrial clinker ash as fill. The sulphur content of this material can be very high, and it will attack concrete that is not made with sulphate resisting cement. Those selling clinker ash at a fraction of the cost of graded stone will not bother to tell you this! You face the same hazards if you take risks with cheap bricks, lintels that are not to specification, and timber that is not stress graded if stress graded timber is specified.

Of course, this does not apply to bargains in fixtures and fittings which can be easily changed. A cheap central heating boiler can easily be swopped for a better one if it is a bad buy. No real

risk here. On the other hand, a cut price load of concrete that is surplus from another job can be a disaster.

Moral: *don't take risks with structural materials.*

12. Setting off in the wrong direction

When you start a selfbuild project its whole shape is determined by the very first decisions that you make. Are you choosing the right site? Is your solicitor really interested in working for a selfbuilder who wants to ask him a lot of unusual questions? Do you have the right designer working for you? Or the right package company? These early decisions should be made after careful consideration of all your options, and you should not drift into them.

In particular, employing the people and firms who will be central to the whole operation is something that should be given a great deal of thought, and you should evaluate all your options. Think how much care will be taken in choosing your bathroom suites. Give just as much consideration to making sure that you are choosing the right person to design your house.

Moral: *Remember the early decisions are the most important ones.*

IF YOU DECIDE TO ADOPT THESE TWELVE RULES FOR AVOIDING DISASTERS YOU WILL BE VERY, VERY UNLUCKY IF YOU GET INTO SERIOUS TROUBLES.

Avoiding Problems Rule One — make it a team effort.

20 Trouble shooting

Every selfbuilder meets problems, and crisis management is part of the project management that this book is all about. Most problems are annoying but not serious — materials do not arrive on time, or arrive when there is nobody on site to unload them, or workmen do not turn up when promised, or materials supplied are not as specified. Nothing that cannot be sorted out. However, a few selfbuilders meet serious problems. How they deal with them can be crucial to the success of the whole job, and they must get it right. As selfbuilders are very superior human material, they usually do.

When dealing with a crisis remember that your objective is to build the right house, within your budget, and to the planned timetable. This determines the way in which you deal with anything that has gone wrong. It may mean that you will spend some of the contingency money in your budget if it keeps the job on schedule, or that you will decide to concede a point in an argument with the council. This is not always easy to accept, but keeping everything moving forward is more important than winning a dispute over who pays for a dropped kerb or a broken double glazing unit.

Always seize the initiative in dealing with a problem. The jargon is that you are pro-active, leaving others to be re-active. You are in charge and set the agenda: they answer your questions and, hopefully, dance to your tune. Let us look at how this can work.

Problems with the authorities
Suppose the Building Inspector says your trench has disclosed bad ground and wants a soil investigation, or the Roads Department want you to move your access by two metres. Rule one: keep your cool — this is not the end of the world. Rule two: take control and establish what is going to happen. It is usually important to get it on record, involving the person who is ultimately responsible. The Building Inspector will probably be helpful on a personal level, but write to his boss saying that you have had new foundations designed and you want to discuss them on site on Thursday so that the work can go forward without delay. Or write to the Chief Roads

Engineer saying that your access is shown on your approved drawing, that you are unable to understand the suggestion that it is inadequate, and that to avoid expensive delays you must ask for a site meeting immediately with someone able to resolve the matter on the spot.

Always define the problem, explain what you want the authority to agree to, set a timetable, and, if appropriate, involve the boss. This is even more important when you are dealing with the services authorities. If you have paid for a water connection, or a gas or electricity supply, the authority has a contract with you. If they let you down they have a duty to do something about it urgently, and these days the Citizen's Charter makes it very difficult for them to hide behind the small print.

Problems with suppliers
First of all, any organisation selling goods or services on any sort of scale sometimes lets down its customers. In the building industry this is a common problem because of the stop/go nature of the market, the fact that deliveries are made to a multitude of different sites, and because often suppliers do not hold buffer stocks. As a result of this, most selfbuilders are likely to meet one or two minor irritating problems with suppliers, and a few have to cope with major problems.

Let us suppose that you have ordered a new type of central heating boiler. You have the technical literature, which shows that the boiler that you want needs a certain sort of flue and a plinth of a certain size. You place the order for the boiler for delivery in three months time, and build your house with the appropriate flue and plinth. You receive a confirmation of your order, which shows that delivery is required at a specific time, and which also has fifty different terms and conditions of sale printed on the back.

Three months later the boiler does not arrive. You ring the supplier's office, and are fobbed off with a junior employee who keeps assuring you that he or she will look into the matter. Fortunately you have kept the card given you by the salesman who called on you to take your order and you manage to get hold of him. He tells you in a very confidential manner that there are problems with the supply of this boiler, but due to his own superhuman efforts he is sure that you will get one very soon. He tells you that he is heart broken that you did not get delivery on time, and that he will do all that he can, etc., etc.

A week later you get a letter from the company saying that the particular type of boiler that you ordered cannot be supplied, but they will supply you with a different and superior model, at no extra cost. They include the specification for the new model, and you note that this will require a different plinth, and, even worse, a larger flue. You ring your friend the salesman, who tells you that he is so upset at the way you have been treated that he would resign from the firm in disgust were it not that he feels he must stay on in order to get your problem sorted out. You don't believe a word of it.

Where do you go from here? Well, first of all, it is useful to sit back and consider why you think all this has happened. There are two alternatives. It is possible that the suppliers are rogues, or so desperately inefficient that trading with the public at their level of competence amounts to a confidence trick. If this is the case, then you must cut your losses and find another supplier.

Have you a claim against the villains who let you down? Possibly, but the real fault is yours. People who buy building materials are considered to be in business, and generally capable of making proper business decisions, like deciding which are reputable firms and which are not. If you have got it wrong there is not a very great deal that you are going to be able to do about it unless you decide to make it a personal crusade. Crusading takes time, and it is unlikely that you will be able to spare the time.

A much more likely situation is that you are a tiny part of huge problems being experienced by a reputable supplier. They may have been let down by the boiler manufacturer. A significant design fault in the boiler may only have become apparent very recently, and until it is sorted out they dare not send any out, and equally they cannot tell you about the design fault because otherwise their product insurance on the defective ones that have already been supplied will be invalidated. Possibly there is nothing wrong with the boiler,

but everything wrong with the supplier's financial situation.. However, unlike the rogues or the incompetents, a reputable supplier with difficulties may offer you the best chance of getting things sorted out.

Now let us think about what you want. First of all, you want a boiler that meets your requirements, possibly with some compensation because of the trouble that you have been put to. You may also feel so aggrieved by all of this that you want to 'warn others' about the supplier, which is simply a way of saying that you want to get your own back. But what good will this do you?

So what do you do? First of all, evaluate your situation very carefully, and do not be in any way unreasonable. Do not threaten the supplier with legal action until you have settled on your plan of campaign in the matter. At this stage the most important thing is to try to find alternative boilers from other suppliers. If you can, then you simply write cancelling your original order because of the supplier's 'non performance of the contract'.

If you threaten the supplier with legal action, you will probably forfeit any chance of him giving special consideration to your problems and trying to do something special to help you. With a large firm it will simply mean that the whole matter will be taken out of the hands of the sales department and dealt with by the company secretary. This does not mean that you should not take legal advice, but you should not shout about it until you are sure that you are going to shout to some effect.

Sending the paperwork to a solicitor will enable him to advise you whether you have a simple contract for the purchase of goods, or whether you have a contract which involves an element of service, and whether 'time is of the essence' in performing the service. He will also tell you what you can hope to achieve by standing on your legal rights, the cost of doing this, and the probable timetable involved. Almost certainly you will decide that the legal route does not make sense unless you are looking at a very serious and very costly problem.

Next, do beware of casually threatening that you will ring a newspaper or a TV show. If you do tell the supplier that you are going to do this, you will probably be told that it is up to you, and you will forfeit any sympathy or special help. If you want to complain to the media, do so by all means, but do not tell the other side!

This does not mean that you cannot use a threat of legal action if you decide that this is really what you want to do. If so, then it is usually far better for you to continue to be the reasonable customer, writing courteous letters explaining your position, while your solicitor writes separately on your behalf explaining that Mr. XXX is suffering such losses that he has had no alternative but to retain Messrs. Nasty & Sue to press for specific performance of the contract and compensation in respect of losses suffered to date.

If you are going to do this, then make sure that you decide *in advance* exactly how far you will go, and at what expense. Remember that only the lawyers profit from the law, and most judges in civil courts spend their time listening to cases where neither party can be a real winner.

It is much better to concentrate on finding out what has really happened, and to try to persuade those who can help you to make a special effort on your behalf and perhaps to give your order some priority. Two ways of doing this are:

1. to be seen as a nice guy who is being very reasonable in very difficult circumstances, and

2. to try to deal with people as near to the top as you can get on a face to face basis.

See if you can reach the sales director or even the managing director by phone and say that your problems are such that you must arrange to call on him to find out exactly what the position is in order that you can make your own plans accordingly. Do not be abusive, and if appropriate you can say that you appreciate the concern that the company has shown to date, but that you must now come to a firm arrangement with him on a face to face basis if you are to minimise the losses that you are suffering.

Now you may think that this is being rather wet, and the chap really needs to be told his fortune. Stop and think: if you were sitting in his chair would you want to help a customer who

was abusive?

What you want is the boiler that was originally specified, or an acceptable alternative, and possibly some compensation. If you are taking the legal route to claiming this, then leave being nasty to your solicitor. At any rate, what is the point in being nasty and antagonising the other side? Anyone who has worked in a sales office will tell you that this may be a recipe for being put at the end of the queue for the new boilers when they arrive. You think that talking tough will frighten the suppliers? Unless you have a record for GBH you will only make them laugh. On the other hand, if you are seen to be disappointed but philosophical they may try to help.

Another good reason for a personal visit to the supplier is that you may discover what the real problem is, and this may persuade you that under no circumstances do you want the particular type of boiler that you ordered. The more that you can learn about the background to the whole business the better, and the more likely you are to obtain a satisfactory outcome.

If the problem involves a package company it is particularly important to strike the right balance between being a very reasonable fellow and being determined to make sure you get the service you are paying for. Unlike builder's merchants, package companies depend on individual selfbuilders, and advertise that they make things easy for you. If you are not happy, do not hesitate to insist on speaking to the managing director.

Finally, in any dispute of this sort you should always look at the cost of your 'worst case losses' and decide whether or not it would be better simply to walk away from the whole situation. If you are selfbuilding, the target is the right house, on budget, and on time.

Problems with builders

If you have a contract with a builder, and there is no architect as an intermediary, then any problem is dealt with at two levels simultaneously. On a face to face basis your good working relationship will help you to find a solution to the problem. At the same time you make sure that your rights under the contract are pointed out

Individually built house in a
1930s style.

and early warning given that you will hold the
builder responsible for losses. The more formal
the contract, the more important this is. A
solicitor's help may be necessary, particularly if
you are going to withhold any part of a progress
payment. If you reach a point where the contract
has to be terminated you should be advised by

your solicitor about what to do at every stage.
He will probably act for you. In this case your
concern will be to move from the bad contractor
to a new one as soon as possible, with as little
disruption and delay as possible. All situations
like this are different, but you will probably want
to engage a quantity surveyor to give you a report

201

on the work done and the value of materials lying on site, and if there is any question of bad workmanship an independent architect's report may be required. If the NHBC or Zurich are involved remember that they must be consulted.

If your builder fails

Perhaps the most worrying thing that can happen to a selfbuilder is that someone building the whole house for you goes bust, or they simply disappear, or tell you that they are facing bankruptcy and cannot continue with the work on your new home. What do you do?

To start with, you have to act very quickly indeed, and if you work fast enough you may even make a profit out of the situation. What you do will depend on the nature of your arrangements with the builder, and you may have a formal contract with him involving either an NHBC Buildmark or Zurich Newbuild arrangement (not Custombuild).

Your final situation will depend on this, but in either event you must take the following immediate action, at once, without any delay, and ignoring any remonstrations by others.

1 . Secure the site by changing all locks. It is not unusual for a builder's employees and sub-contractors to try to recoup their losses by helping themselves to 'his' valuables on the site, which will be *your* building materials or fixtures. They will convince themselves that they are entitled to do this. So may builder's merchants and other suppliers to whom the builder owes money, and they may arrive with a lorry waving delivery notes saying that ownership of materials supplied does not pass to purchasers until they are paid for. These vultures are trespassing on your site, and although the law is on their side in many respects, they cannot enter your site without your permission. Simply say *'this is my site, the builder supplied these materials to me, and your problem is one for the liquidator. Go away.'* Wire off the entrance to the site and put up a notice saying something like:

'Materials on this site are the property of Joe Selfbuilder, and of no other person. Any attempt at repossession against the debts of others will be treated as theft.'

It may even be worthwhile moving materials elsewhere or engaging a watchman for a week.

2. Next, advise the local police of the situation. They will not help you in an argument with a repossessor, but they will be concerned to prevent a breach of the peace or a criminal trespass. Their interest may also deter the opportunist thief who notices that work has stopped.

3. Finally, consider the insurance situation. If theft, fire, vandalism, etc., were covered by the builder's insurances, and you can get details of these insurances from him, check out with the company concerned that cover still exists on the site and for how long it will be in force. The answer will usually be 'no'. If so you must arrange your own insurance by phone in the next ten minutes. Do not tell the brokers your tale of woe, simply say that you are starting a self-managed selfbuild project with an uncompleted house and want immediate cover. If they do not take down the details, give you a quote and then immediately confirm that you are on cover subject to you sending them a cheque within 72 hours, find another broker!

Having done all that you can to safeguard your property on your site, you must move on to safeguard your legal position. Take whatever contract documents you have to your solicitor and ask him to send an appropriate notice to the builder advising him that by going bust he has voided the contract, and spelling out where he stands. Your solicitor should get off this letter by recorded delivery the same day.

If there is a NHBC or Zurich Newbuild arrangement in the contract, get in touch with the local inspector at once. He will tell you how to proceed, and you will be delighted to discover that you will probably be refunded any deposit with the builder up to 10% of the contract value, and will be given help to transfer the job registration to another NHBC or Zurich registered builder. However, if the new builder takes the job at a price that moves the total cost outside your budget, this is a loss that you will have to stand.

Nor are you covered for any consequential or incidental losses, like having to take time off work to deal with the situation. Sadly this does not apply to the Zurich Custombuild warranty.

Consideration of this explains why no selfbuilder should ever pay anyone working for them anything at all in advance if they can possibly help it, but that if NHBC or Zurich warranties are arranged *by the builder* there is limited cover on advance payments.

While you are doing all of this you may have the builder at your elbow suggesting all sorts of ways of getting the house finished, like 'I will do the work and you pay my wife'. Be very wary, and discuss any such possibility with your solicitor. It would be easy to get yourself in very serious trouble if you are considered to be conniving at something that is inappropriate to his position as a potential bankrupt.

In general, as long as you have not made the mistake of paying the builder more than the value of the work he has done, you may even make money from his sad situation. Do not pay him a penny: he has failed, and any money you owe him is properly due to the liquidator, who will use it to pay preferential creditors, meaning the Inland Revenue and the VAT authorities. When you hear from the liquidator claiming money from you, ask your solicitor or accountant to enter a counter claim to include all your losses, including the additional cost of a contract with another builder, and even compensation for your distress and wasted time, etc., etc. The counter claim will be larger than the liquidator's claim. You will not get any money, but it will be set against the liquidator's claim. You will probably feel genuinely very sorry for the builder, but there is nothing that you can do to get him out of the hole he is in. Give him sympathy, a cup of tea, and nothing else.

Problems with sub-contractors

The more informal contracts with subbies leave only difficulties to be dealt with on a man-to-man basis. Provided that you have not fallen into the trap of making payments in advance, there is little danger of bad workmanship by a sub-contractor causing serious losses provided that the work is regularly inspected. The sub-contractor should be told exactly what is wanted, be told firmly if anything is unacceptable, and his employment terminated if he does not meet your requirements. In particular, pay off at once anyone who you decide is not the sort of character you want working on your new home.

A decision to terminate someone's engagement should not be taken lightly, or used as a threat, but if you have decided that it is the thing to do, never change your mind. Do not bother about the other chap's feelings: any bad subbie has been finished lots of times before. Pay him in full for the work that he has done, and do not deduct the cost of putting right any defective work: your choice of subbie or lack of supervision was at fault, and you will have to pay for your mistake. It is what the contingencies item in the budget is for.

Difficulties with professionals

The professions used to have a very patronising approach to their clients, and many people still feel uneasy when dealing with someone with a string of letters after their name. As a result they sometimes fail to ask the right questions when they first meet, or are not firm and explicit about exactly what they want. This can lead to expensive misunderstandings. As in every other walk of life there are also those who are idle or incompetent.

When you engage a professional, whether an architect, a solicitor, a quantity surveyor, an engineer or a package company providing a professional service, it is essential to make absolutely sure that he or she knows what you want, and that you know how, when and at what price, you will get the service they are going to give you. If you do not know exactly what you *do* want, then it is even more important that the expert explains in detail what he proposes to do for you, offers a timetable, and quotes the fees involved.

The best professionals will always do this, and are punctilious in making sure that both they and their client understand exactly what has been agreed between them. However, this doesn't always happen, and as it is your money that you will be spending, you should take the same amount of time over finding the right professionals to work for you as you will take in choosing your new kitchen, and give just as much care to making the right arrangements with them.

But what happens if you are not satisfied with progress? This can happen, and professionals are not as easily sacked as bricklayers if you are unhappy with their performance. In law your arrangement with them has the force of a contract, but before you start considering your legal rights you should make a very determined effort to sort out the problem in a friendly way, while setting up everything to deal with the difficulty formally. This requires a structured approach with proper records in case the dispute gets out of hand. Avoid 'taking a firm line' on the phone or being abusive. If possible take advice from someone with wide experience of these things, who may tell you that you are being unreasonable and should not make a nuisance of

yourself! However, after trying to take a detached view of the problem, you may decide that something has to be done.

In such circumstances you normally start the ball rolling in a letter that might read:

I am very concerned at the delay in submitting details to the local authority for the revised piling arrangements for the house at Dogswood. As explained, further delay is unacceptable. I will be grateful if you will telephone me to make an appointment to meet this week, when I trust you will let me have a copy of the drawing and sight of your correspondence with the council.

When you meet you can ease the atmosphere by being more friendly than your letter, but the letter is there for you to quote if things get worse — and the professional will be well aware of this. Withholding any progress payment will also concentrate his attention on the matter. No doubt he will have good news for you at the meeting: if so, write to say how relieved you are that he has

Jean and David White receiving the Selfbuilder of the Year award in 1993. They built their house, inset below, after two years of problems with bureaucracy, all overcome against the odds.

the matter in hand, and keep a copy of this letter as well!

Problems like this are rare, but they do occur. If you sack a defaulting professional he will probably react by sending you a huge bill for abortive work, and probably issue a writ if you do not pay it immediately. For this reason any letter terminating a contract with a professional should be written for you by a solicitor.

All of this makes the initial choice of the right experts to act for you very important indeed. Choose with care, and be reassured by remembering that the longer the fellow you have chosen has been in business the more likely he has given a good service to thousands of happy clients!

Trouble over residential caravans

Local Authorities do not like caravans, and they can be a source of trouble. In your negotiations with the planners you may have explained that you intend to live on the site in a caravan while you are building, and obtained the necessary formal consents, which will be for a limited period. However, many selfbuilders do not bother with this, either because it is commonplace for selfbuilders to live on their sites in the local area, or sometimes because they do not want to mention a caravan while negotiating the planning application, as they feared it might adversely affect the outcome. If you are in the latter class, and you have not yet obtained all the consents that you need, avoid putting the caravan on the site until you have, especially if you think your new neighbours may be sensitive about it. When you have actually started work on the site your legal position in most areas is as set out on page 118. If you have problems with the council remember the council knows it will take longer to get an order requiring you to move it than you are going to take to build the house, and will not usually press the matter. If they do, remove the wheels of the caravan and claim that it is the site hut. Silly arguments can go on until you replace the wheels and sell it to some other selfbuilder who will tow it away.

Problems with neighbours

Occasionally neighbours who are less than friendly can cause problems. Usually the difficulty concerns the boundary. There is a saying that possession is nine tenths of the law, and this is very true of boundary disputes. A typical situation arises when you buy a plot with a hedge as the boundary. A neighbour appears and says the hedge was planted in the wrong place, and the real boundary is three feet inside it. What do you do? Well, possession being nine tenths of the law, erect a very substantial row of fence posts at once, string wires between them, and send your troublesome neighbour a letter by recorded delivery saying that you have fenced your property as shown on your title deeds and if he has any doubt about the boundary he should ask his solicitor to get in touch with your solicitor. He is then claiming 'your' land, not vice versa. You have the initiative.

If anyone complains that your building activities are spoiling their peace and quiet, or goes to the town hall to complain, try to establish good relations by making some sort of concession. Diesel mixers are often the problem: offer to change to an electric mixer if they will help you with a temporary electricity supply. This could be the start of a good neighbourly relationship. A marvellous new home is even better with marvellous neighbours, and you may have to work hard at it. It will be worthwhile.

If all this fails you will need legal advice, but remember that although your solicitor will tell you your strict legal position, he may not be the person to give pragmatic advice on how to deal with the total situation. Listen to what he says, but also try to find others with direct relevant experience and listen to them as well. In some cases consider buying off trouble provided that you get some sort of formal withdrawal of the complaint.

Conclusion

Finally, do bear in mind that although all selfbuilders deal with minor difficulties, very few of them meet catastrophic problems. Do not let this chapter worry you. Cigarettes, alcohol or bad company are far more likely to encompass your ruin.

21 Insurances, security and safety

The previous chapter dealt with problems that can arise which you will avoid or overcome by using your management talents. There is another category of possible disasters which may be simply due to bad luck, but unlike the others they can be covered by insurances.

Appropriate insurances for a selfbuild operation are as essential as car insurance is to the motorist, and like car insurance they have to cover claims from others who are affected by an accident of some sort as well as taking care of any losses which you suffer yourself. The difference is that you have to arrange for car insurance before you can licence a vehicle at all, while selfbuild insurance is something you have to remember to do yourself. However, it will probably cost you less to insure the whole of your selfbuild operation than it costs to insure your car, and the premiums are unlikely to total more than a third of one percent of the money that you are spending, or less than you will lay out on light fittings! However, it is in no way incidental and you ignore insurances at your peril.

If you are placing a contract for the whole job with a builder, can you be confident of your builder's insurances? If you are placing the contract with a builder with an architect as an intermediary, and the architect confirms that he has established beyond all doubt that the builder has appropriate comprehensive cover, then you are all right. If you are placing a simple contract with a builder without an architect you must make sure the builder is fully insured in all the work he does on your site, with cover for your materials and all the risks of public and employers liability. If you are sure of this and it is noted in the contract then you are OK. You have to make a careful judgement here. If challenged to prove that he is insured your builder may produce the vehicle policy for his van! Do you feel like challenging him? If not, you may decide to take out your own insurances. Also, if you are covered via your builder's policy and then employ another firm to do any work in the house, like installing the kitchen, and there is a theft or an accident, then you have probably invalidated the cover via the builders policy and you will be left with all sorts of liabilities.

If you are arranging your own insurances you must consider three types of cover. The first is Public Liability Insurance, which is required to cover you against any claim made against you by a third party who suffers a loss as a result of your building operations. This includes injuries to either visitors or trespassers on your site, and also to anyone hurt outside the site because of your activities on the site. Typical claims concern incidents like traffic accidents caused by children moving building materials onto the road, and in 1995 there was a fatality for which a selfbuilder was held to be responsible. Fortunately he was insured.

Other public liability claims result from damage caused to other people's property, as happens when selfbuilders dig up electricity cables or water pipes by accident, even though the services were not where the electricity or water board showed them on their drawings. Other claims can arise from excavations on the selfbuilder's site affecting the stability of adjacent buildings belonging to a neighbour, and such claims can be very expensive indeed. More straightforward claims from neighbours relate to paint spilt by a painter working at first floor level blowing onto a car parked down wind, or to mortar splashes taking the same route.

Most householders have some form of public liability insurance as part of their normal householder's policy which covers them for accidents that happen in their home or garden. However, they do NOT apply to accidents stemming from building work, even if the new property is being constructed in their existing garden, and it is essential to have special cover. The law is definitely not on your side in this matter, and a child trespassing on your site who climbs a ladder left tied to your scaffolding and then falls off has grounds for a successful claim against you.

The next type of insurance is Employers Liability Cover. This is a legal requirement if you are employing anyone. Here it is important to understand what is meant by employment, as most selfbuilders use sub-contractors and do not employ anyone in the general sense of the word. They have no liability to collect PAYE from the people who are working for them, and their arrangements are generally outside the scope of the Employment Acts. However, as far as any accident to someone who is working for you is concerned, you certainly owe him a 'duty of care' to ensure that he does not get injured on your site, and if a sub-contractor is injured on your site his solicitor will try to establish that he has a 'deemed contract of employment'.

Many selfbuilders like to think that they are on such good terms with their sub-contractors that there is no risk of any unpleasantness. However, experience shows that when someone is hurt they tend to look for any advantage in the situation, and they are very quick to go to their solicitor. Remember that self-employed sub-contractors do not get unemployment or sick benefit if they are off work, and they will consider a case against you as a different route to the same financial benefit!

The third type of selfbuilders' insurance is Contract Works Insurance cover, sometimes called site risks insurance. This is the way in which you insure against the more usual

problems, providing cover against theft, vandalism, storm damage, fire, flood, etc., etc. The most common claims are for thefts of materials, plant or tools. Timber, roof insulation and copper piping are always at risk, and concrete mixers that disappear in the middle of the night are another common problem. Claims for this sort of loss are rarely at a figure that would jeopardise the success of the whole project, but fire and storm damage insurance is sometimes all that stands between the selfbuilder finishing his house and losing everything. In November 1987, in January and February 1991 and in February 1995 selfbuilders made hundreds of claims for gale damage, often involving both rebuilding costs and claims for damage to neighbouring property caused by falling brickwork, and the claimants were very glad they had remembered to get insured!

Vandalism is another problem giving rise to claims, and sadly it is growing. Graffiti on bus shelters is cleaned off by the council using

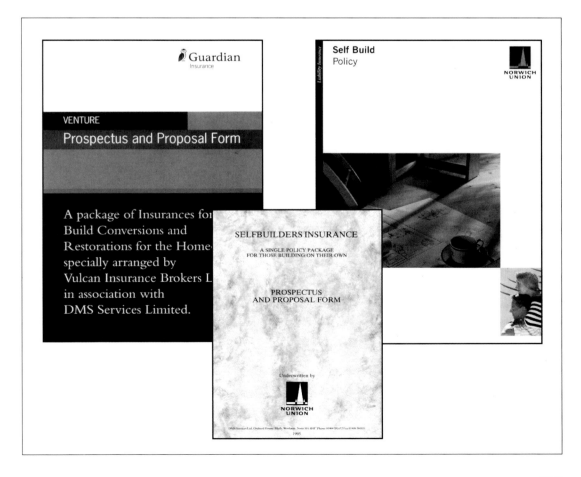

Notes on insurances when a selfbuilt home is completed and occupied

The selfbuild policy insures a project until it is complete. A common sense interpretation is put on 'when it is finished', and this is not tied to a completion certificate issued for some other purpose, or when a selfbuild loan is converted to a mortgage, or when an NHBC or Zurich Custombuild certificate is obtained. It is a matter of when the work that was insured is completed, using the ordinary interpretation of the words, without any legalistic niggling.

If the family moves into the new home before all the building work is complete this does not affect the selfbuild policy, but their furniture, personal effects, etc. which are not relevant to the building work are NOT covered. They should take out householder's insurances for this, and if they decide on a Norwich Union Home Plus policy their possessions can be covered at this stage, together with the other cover extended by a homeowner's policy, and cover on the actual structure need not move from the Selfbuilder's policy to the Home Plus policy until the work is finished. This is quite normal, and is an obvious advantage of having the same insurance company involved in both policies.

In considering your own requirements the following should he kept in mind:

* The selfbuilder's policy covers the building work until it is completed, or for two years, whichever is the sooner. If a selfbuild project is being carried out in stages within the two year period, perhaps with the house being built in year one, occupied at the end of year one, and a garage and drive constructed in year two, the whole project, including the fabric of the house, is covered until the garage and drive are finished. However, as soon as the family move into the house they need a householder's policy to give the cover which is not part of the selfbuild insurance arrangements.

* As stated above, 'completed' is interpreted in a common sense way. It is not recommended that fixing the door knocker should be delayed to keep the selfbuild cover going for the full two years.

* If there are still small items of work to be carried out when the selfbuild policy expires, and they are of the same scope as normal home improvements, then any accidents or losses will be covered by the relevant parts of the Home Plus policy.

* The selfbuild policy assumes that building work moves steadily along to completion and that the new home is then occupied. If work stops for any significant period, or if the new home cannot be occupied for any reason, the selfbuilder's cover is not effective. Such circumstances should be discussed with DMS Services Ltd.

* Finally, if a new home is occupied while building work is still going on, health and safety regulations require that children are not put at risk by the building work.

Selfbuild policies are arranged by DMS Services Ltd who can be contacted on 01909 591652.

Home Plus policies for selfbuilders are handled by Thurcroft Insurance Brokers, who can be contacted on 01709 540348.

Either of these companies will advise on the transition from a selfbuilder's policy to a Home Plus policy.

---

special solvents: the same use of a spray can on your new brickwork will require the damaged bricks to be cut out and replaced, which is a horrendous job. Fortunately the more usual sort of claim involves a small boy with an air gun breaking your double glazing units.

Until the 1970s selfbuilders could only obtain cover against all these risks by taking out three separate builders' policies, and they were extremely complicated. Then the National Selfbuilders Association, which sadly is no more as it was disbanded in 1983 when its grant was discontinued, arranged for a special simple fixed premium policy with the Norwich Union. This was specially tailored for selfbuilders requirements and it covered all three categories of risk. It proved to be so popular that when the association was wound up

Right: Keith Axelby was very careful to avoid an electricity board cable when digging a services trench for his selfbuilt bungalow. Unfortunately it was not where the plans showed it to be and his JCB cut off the power to the local betting shop, making the manager very cross. In law Keith was responsible. Norwich Union handled the claims for him and he is shown here with their local manager.

Below: Craig and Frances Jordan's selfbuilt home was badly damaged in an exceptional gale while under construction. Their insurers approved their estimate for rebuilding without delay and in six weeks the damage had all been made good.

Bottom right: Paul and Lorraine Doran had a series of thefts from their selfbuild home while they were building it. The losses totalled £4000. Their insurers covered the losses, but not the cost of the delays and the frustration.

arrangements were made for it to be carried on by D.M.S. Services Ltd., who offer this same special policy today.

This policy provides the public liability cover, employers liability cover, and contract works insurance to cover all the risks described above, and for most of the United Kingdom there are fixed premiums. These are based on the cost of the construction of the proposed building, and they do not take into account the site value, even though all site risks are covered. The insurance proposal form comes with a table showing the standard premium for different values of construction: for example it currently shows a premium of £499.20 for a house with a declared building cost of £80,000. Cover starts immediately on receipt of the proposal and cheque, or from a forward date if requested, and it runs for two years or until the property is completed. Building societies and banks who are financing a selfbuild project frequently want their interest endorsed on the policy, and this is readily arranged and appropriate confirmation of this sent to the bank or building society concerned.

The standard cover on plant, whether owned or hired, is for £2,000 worth of equipment on the site at any one time. It is meant to cover scaffolding, mixers, site huts, powered tools, etc., and it is not intended to cover excavators or other plant hired in with an operator who is covered by the hirer's insurances. If the anticipated value of plant on site at any one time is likely to be in excess of £2,000, or if the selfbuilder is hiring an excavator on a self-drive basis, there is a small additional premium.

Caravans on the site which are used as site huts are insured as plant. If the selfbuilder lives in a caravan on the site while the building work is in hand there is no extra cost, although if the value of a residential caravan takes the total value of plant on the site at any one time over £2,000 a small additional premium is payable. Personal effects in a caravan on the site are not covered, although tools are.

This selfbuilders' policy is for new dwellings, and not conversions or renovations of existing buildings for which DMS has a special policy available. There is a special case when an existing building has been demolished and a new property is being erected on the old foundations, and usually this is accepted as new construction.

The policy is also limited to individuals who are building a new home for their own occupation, and it is not available to builders. However, if a selfbuilders circumstances change while he is building the new house, and he decides that it has to be sold, the policy can continue until the building work is completed. After that a special policy has to be taken out to cover the unoccupied house until someone moves in.

Insurance when you move in

Your selfbuild insurance is not affected if you move into the house while it is under construction — after all, everything is safer if you are sleeping on the premises. When you do finish you will need to change your selfbuild policy to a homeowner's policy, and the insurers will send you a leaflet about this, together with details of how you can obtain a 15% discount on your homeowner's policy when your selfbuild operation is finished. They give you this discount just because you have been a selfbuilder, which they equate with you being a good risk! The leaflet is reproduced on page 208.

Unoccupied property insurance

If a property is completed but not occupied, possibly because another house cannot be sold, special Unoccupied Property insurance is required. Again, details of a suitable policy are available from DMS Services Ltd.

Legal contingency insurance

It is also worth knowing that it is possible to insure against anyone challenging your title to your land, anyone claiming rights under an old covenant, or anyone disputing your access. This is explained on page 46.

Life and health insurance

There is another category of risk for which you may consider appropriate insurance cover is appropriate. If you are using sub-contractors, or are doing any work yourself, you should consider the effect on the project if you step under a bus, or are taken ill, or are incapacitated due to an accident.

First of all, think about cover to meet additional costs if you are sick, injured or depart this life. If any of these sad circumstances would not affect the progress of the job at all, then you can meet them happily, but if not you must consider health, accident and short term life insurances. The latter is probably the most important, and if you are managing the building work yourself you may consider it prudent to take out a simple short term life policy for a sum that would enable your executors to get it finished. If the financial arrangements for your self managed budget would survive you, and they normally do, then cover of £20,000 would probably suffice. This would enable your executors to put the uncompleted building in the hands of the best builder in town with instructions to get it finished without any bother. The premium for a short term policy to make this possible would be very modest.

Security

Theft and vandalism on a selfbuild site involve financial loss, a setback to a carefully worked out programme and often a serious drain on enthusiasm. Sadly, it is a growing feature of society generally and building sites in particular, but although nothing can be done to guarantee it will not happen to you, it is possible to make it less likely.

First of all, don't lose any sleep worrying about it — the odds are it will never happen. Claims by policy holders who take out Norwich Union Selfbuilder's Insurance indicate that only a small proportion of those covered have any significant losses, about the same proportion of those who make claims on their car insurance every year. You have long ago come to terms with motoring risks, so accept the selfbuild risks in the same way. And, as with motoring, there is a lot that you can do to reduce the risk.

There are three basic rules for this: don't advertise, deter trespassers, and make your property difficulty to steal. They need considering separately.

Don't advertise. Potential thieves may be professional rogues who will use a vehicle, or children who may trespass on the site. The professionals are obviously less likely to take an interest in your property if it is not displayed for their inspection. It is quite impossible to ensure that everything worth stealing is hidden away, but if it is practicable to have the mixer and the site hut at the back of the site it is a very good start. Keep tools and materials in a locked site hut if at all possible, and programme delivery of materials so that they arrive when they are required and not before. This is particularly true of high risk items like roof insulation, timber, copper pipe and plumbing and electrical fittings. If materials which are likely to be stolen must be stored in the open, stack them out of sight as far as possible and cover them up with a tarpaulin.

Secondly, deter trespassers. The first rule here is to keep a tidy site and not to give it the appearance of an interesting scrap yard. This is not always easy, particularly if you are building on a site where there are existing derelict outbuildings which you hope to renovate in due course. However, the tidier and more business like a building operation is seen to be, the less likely it is to be investigated by those with time on their hands.

Consider fencing the site, even with a single strand of wire, as a deterrent to trespassers. If it is already fenced it is likely that there was once a gate: you can buy a farm gate to hang on the old gate posts, and chain it up at night. It may even be worthwhile considering doing any fencing work included in your final plans for the site at a very early stage in order to make it more secure.

If you have electricity on site — and many selfbuilders arrange this so that they can use a electric mixer and electric power tools — consider putting up a floodlight with a proximity switch. If you are going to do this install a thousand watt light, as when this comes on suddenly in the dark it can be very startling. However, make sure that any such arrangement will not upset neighbours, as the detectors are sensitive to dogs and other animals. If it seems plausible put up a 'Beware of the Dog' notice.

Perhaps the most valuable thing of all is to contact neighbours when you first start work

Selfbuild site safety

* Wear a hard hat. It impresses visitors, hides your bald patch and can stop you having a terminal headache.

* Wear protective footwear. Wellies and boots with steel toecaps are readily available - look under 'safety' in Yellow Pages.

* Buy two or three pairs of cheap plastic goggles and always use them with grinding tools, etc. Encourage others working on the site to use them when appropriate. The alternative can be half a day wasted in a hospital out patients department

* If you have no experience of erecting scaffolding, get it on a 'hire and erect' basis. Make sure that the rails and kicking boards that the law requires are provided. If your sub-contractors want to erect the scaffold make sure that they do so according to the book, and as far as possible write this into your arrangement with them. Remember if they erect a hazardous scaffold, it is going to be used by other trades, including the tilers, the joiner fixing the roof and guttering, and the plumber fixing the flashings, and it is you who will face a criminal summons if there is a serious accident.

* With conventional scaffolding the short lengths of scaffold tube that carry the scaffold boards are called putlocks, and project beyond the scaffolding at head level. Building professionals know they are there by instinct: selfbuilders tend to bump into them and need a trip to hospital to get their scalps stitched. It is worthwhile collecting empty plastic bottles to tape these over the putlocks. This looks funny, but it is very effective.

* Whenever you hire equipment from a hire firm ask if instruction leaflets and particularly safety leaflets are available.

You may feel rather self conscious about doing this, but most plant hire firms will welcome your enquiry, and will probably be pleased to give you the benefit of their own experience. They all have stories of the wife returning the tool that put the husband in hospital!

* Keep petrol for mixers in a locked hut, preferably in the type of can that is approved for carrying petrol in the boot of a car, and not in a cheap plastic container. Do not let anyone smoke in a site hut where you keep petrol. Better still use diesel equipment.

* Professional electric power tools from a plant hire company will normally be 110 volts and equipped with the appropriate safety cutouts, etc. If you are using 230 volt DIY power tools, or any other 230 volt equipment, including temporary lighting, take the supply via a RCD contact breaker. These are now readily available as plug in units, and should always be used, whether you have a site supply, a cable from a neighbour or a generator.

* If trenches for services or your foundation trenches are more than a metre deep treat them with respect, and go by the book with shoring. If they show any tendency to collapse, deal with them from above, in company with another person. Never ever work in a deep trench alone on site.

* Packs of bricks and blocks that are crane off-loaded with or without pallets, must always be stacked on a proper base, and never piled more than two high. Take great care when cutting the bands, and re-stack them by hand if packs are in any way unstable. If there is any risk of children climbing on packs of bricks, particularly those which have been opened, it is good practice to sheet them before leaving the site. They will be all the better for protection from the weather anyway.

* Concrete burns are a selfbuild speciality. Bad ones can leave the bone visible and require skin grafts. Never handle concrete or mortar, and particularly do not let it get down your wellies or in your shoes. If it does get in your wellies or shoes, wash out the offending footwear at once and change your socks. Remember cement burns do not hurt until after the damage is done. If you get cement dust in your eyes, flood your face under the tap at once. **Do not let small children play with mortar or concrete, and if they are on the site with your permission warn them of the risk of cement burns.**

* Do not get involved with work on a roof unless you are well used to heights and positively like it. If you are uneasy up there you will not be able to do anything properly anyway. Do not be afraid to tell sub-contractors 'I am not a man for heights'. They will understand, and their probable reaction will be to show off by performing miracles of productivity while standing on one leg on your roof ridge. This will help to keep down costs.

* Selfbuilders regularly fall down stairwells. If they do not their visitors are likely to. Use scrap timber to form a rough balustrade until you fix the real one.

* There are still old type wooden ladders about without a wire under each rung, and these are often owned by selfbuilders. The only place for them is on a bonfire.

* Be obsessive about clearing away any loose boards or noggins with a nail sticking out of them, and in case you miss one, never ever wear shoes with a thin sole on the site.

* Put together a first aid box containing plasters and antiseptic and fasten it on the site hut wall or keep it in your car. You will suffer your fair share of cuts and abrasions, and a poisoned finger is a nuisance.

* Watch your back when unloading a very heavy item, or if you are unloading more weight than you normally handle in a day. This also applies to digging work. The risk of straining yourself is very real, even if you play rugger at the weekends and squash in your lunch hours. The most scrawny and unhealthy professionals can handle heavy weights, or unload and stack 16 tons of building blocks by hand, without any risk of injury. If you try it you are running the risk of putting yourself out of action for a week.

and to leave them your phone number so that they can call you if they see anything amiss. In some situations you may find someone prepared to 'keep an eye on things' for a small sum. If you are building in a built up area, cultivate the acquaintance of the local busybody and encourage him to come to chat to you while you are working. With luck he will identify with you so that he or she will take action if they are suspicious about uninvited visitors to the site.

Make your tools and materials difficult to steal. First of all, take small tools home unless it is totally impracticable, and never leave the originals of planning documents, building regulation correspondence, quotations, invoices or any other paperwork in the site hut. The place for it is in an old case in the back of your car, or safely at your home. The site hut itself should be as secure as possible, with a sturdy locking bar to the door, a good quality lock, and a shutter to the window. Old steel shipping containers are becoming increasingly popular as site huts, although they suffer from not having windows. However, while they are probably secure against children, professional thieves with bolt croppers can gain access to them in a matter of seconds.

Concrete mixers are the tools most commonly stolen from sites. When the one which you have hired or bought is delivered, consider removing the tow bar and possibly one of the wheels, which will make it much more difficult to tow away. Ladders are also a prime target for thieves, and should be hidden away, or chained up, or both. If you take all these precautions you are reducing the risks of direct losses very significantly. You should also guard against indirect losses, the most common of which is damage from water due to theft of sanitary fittings, boilers, or other parts of your plumbing installation. Thieves will disconnect whatever they want with a hacksaw, and if you have not turned the water off, the damage caused by the resultant flooding can cost more than the value of the items stolen. Turn off water at the mains stop cock whenever you leave the site, and if the water board stop cock is not conveniently situated for this, make sure that you have one of your own installed for this purpose. Make sure it isolates the tap which is used for building water so that if small children leave it on it does not turn your site into a quagmire.

If you do have losses, remember to collect all the evidence that you will require to support your insurance claim. Advise the police as soon as possible, asking for the name of the officer to whom you are making your report. This will be required to support the insurance claim, and the sergeant at your police station will not be surprised when you ask him how he spells his name! Take a whole series of photographs of any damage, and then contact the insurers with a coherent story. You will want to hear from them whether they are going to send an assessor along to visit your site, and whether you can start putting things right before he comes. The insurers can only help you if you give them appropriate information.

This can be in a phone call when you will:

* quote your policy number

* explain that it is a contractors risk loss, and that there are no others involved, or that it is an employers liability incident, or involves a third party.

* state simply the extent of your loss (vandals have broken my patio window or someone has stolen £2,000 worth of bathroom suites)

* tell them you are taking action to make the place secure again

* ask them if they are sending an assessor

* and confirm the address to which they should send the claim form.

Dealing with all of this in an efficient way will help to get you back on an even keel, although you will still be concerned to tell everyone about your new found enthusiasm for capital punishment.

Safety

To be a selfbuilder you have to be self confident, and this will lead to your belief that nothing nasty is ever going to happen to you personally. Let us examine this contention.

The building industry has a worse safety

record than coal mining. The building industry professionals who are injured get sick benefits, and possibly compensation. Their absence from work while they are recovering is unlikely to stop the job. Consider that you are not a professional, are unlikely to get sick benefits or compensation if you are injured, and ask yourself what will happen if you cannot deal with the work on your selfbuild site. Add to this the fact that amateurs are always more likely to be injured in any situation than professionals. Perhaps this will convince you that positive safety procedures should be part of your project planning.

First of all, who is covered by what sort of insurances?

Looking at the standard Norwich Union selfbuilders' policy, your employees whose misfortunes are covered by the employers liability section of the insurance are defined as:

* direct employees

* labour only sub-contractors, whether working directly for you or working for someone to whom you have given a sub-contract

* persons hired or borrowed from another employer.

This does not include members of your family who are working for you without any charge for their services, nor does it include friends who are giving you a hand. It is a nice legal point that it does not include other selfbuilders who are helping you in exchange for you helping them on their own job. However, these others who may be hurt on the site are covered under your public liability section of the policy, and this includes those who are on the site in connection with some sort of business arrangement made with you (the architect making a routine inspection), those invited to the site by you (your friends and family), and trespassers on your site (the child who climbs your scaffolding while you are not there).

As far as you are personally concerned, the selfbuilders' policy gives you no help at all if you are injured. For this you have to take out personal accident, death, and permanent injury insurances if you do not already have this insurance cover in some other way. This has been discussed on an earlier page.

Valuable as these insurances are, they should not encourage you to ignore common sense precautions. Not only is the food in a hospital unlikely to be up to the standard that you normally enjoy, your inability to manage the job while you are recovering from your injuries is going to be very expensive, and this loss is not covered by any sort of insurance. Most of the precautions which you should take are common sense matters, but use the site safety check list on pages xx-xx. If all selfbuilders followed its advice, accidents on selfbuild sites would be greatly reduced.

If you have children you will have to do some very careful thinking about the extent to which you are going to let them visit the site. Having your kids help by clearing rubbish is happy family togetherness, and a good thing. The moment one of them is hurt it becomes irresponsible disregard of safety legislation to let them be on the site at all. There is no doubt at all about this: in law they should not be there. New European safety legislation emphasizes this. Unfortunately, in most family situations your children are likely to become involved with what you are doing. You will have to make your own careful decisions, decide what the rules are going to be, and see that everyone sticks to them. If you are living on the site in a caravan this may involve you deciding to fence off the caravan and family area from the building site, but this is not always possible. Remember that besides more obvious hazards, children are at risk from toxic materials on a building site. The worst of these, and certainly the one that gives most trouble, is cement. Cement dust, mixed concrete and wet mortar are very corrosive and lead to concrete burns.

Finally, some sort of first aid box on the wall of the site hut is a good idea as well as being legally required. Probably more important than the bottle of dettol and the plasters in it is a note on the outside of the box 'important' telephone numbers, including the numbers of next of kin. If a brick drops on your head when you walk under the scaffold, and you are taken to hospital by your apologetic bricklayer, it is nice if he can ring to tell your wife what he has done instead of leaving her to notice that you have not appeared for supper.

22 Value Added Tax

Those who are buying a new house from a builder or developer do not have to pay VAT on the property, and the VAT authorities recognise that selfbuilders should not have a VAT liability either. This is arranged through special regulations which are set out in VAT Notice 719, which has the simple and explicit title Refund of VAT to DIY Housebuilders. The regulations themselves are equally simple and straight forward, and all the details are obtainable from local VAT offices, where you can get a claim pack containing leaflets and the claim forms that you will require.

The VAT regulations are administered by the Customs & Excise. Their simple procedure and straightforward way of doing things stem from hundreds of years of experience in clearing ships between tides, and compare very favourably with the ponderous ways of the Inland Revenue, planning offices, services authorities and others who the selfbuilder will get to know. However, their rules require that you are equally business-like, and you should learn what is involved at an early stage.

It is all set out in the leaflet opposite, reproduced by permission of HM Customs & Excise, and if anything is not clear you can ring your local VAT office for advice. If you stick to the procedures, submit your claim on time with all your invoices properly listed, and answer any questions promptly you should receive your refund in well under a month. Typically it will pay for all the carpets and curtains in the new home. The most difficult part of the whole business is finding out where your local VAT office is: it is listed in the telephone directory under C for Customs & Excise and not under V for VAT!

A few points that are not to be found in the leaflet:

* If you use a package company, ask them for an itemised invoice in the form acceptable to the VAT authorities. Ensure that any last minute extras are included. The package companies know all about doing this, but as it involves a lot of typing they may wait to be asked.

* As this book is going to print there are proposed new arrangements for VAT to be reclaimable for some types of conversions. Conversions are outside the scope of this book, but there are selfbuild projects where a new home is being built on the site of an old building with some small part of the original building retained for some pur-pose, usually at the insistence of the planners. In such circumstances the project is classed as a new structure and VAT is reclaimed as if it was 100% new work providing that only one existing wall (and any supporting buttresses or returns) of the old building will be retained above ground.

* A VAT registered builder or sub-contractor will not charge you VAT for their work on the building, or for the materials which they supply. The size of your VAT claim will be dictated by the proportion of materials that you buy yourself.

* You make one claim only, and it must be made within three months of receiving a certificate that the building is completed. It can include VAT paid in respect of boundary walls, drives, patios, a garage, etc., etc. If you leave this ancillary work until after you move in, you may not find it convenient to delay the VAT claim while you wait for the bill from the tarmac contractor who is laying the drive. Take this into account when drawing up your programme.

* When you make your claim, package up all the precious invoices that you have collected so carefully, and either take them to the VAT office and get a receipt for them there and then, or send them off by registered post and make enquiries if you do not receive an acknowledgement within 14 days.

VAT notice 719

The main provisions of Notice 719 are reproduced below, but it is essential to obtain an up-to-date edition of the leaflet and to study all of it.

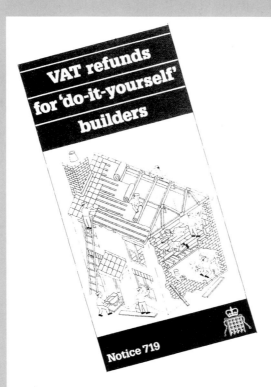

VAT refunds for 'do-it-yourself' builders

Notice 719

What if the building is being constructed in the course or furtherance of a business?

You cannot use this scheme if you are constructing the building to sell, let or for some other business reason. But, if you are registered for VAT, you may be able to reclaim the VAT paid on materials etc in the normal way. If you are not registered for VAT, you should ask your local VAT enquiries office for VAT Leaflet 700/1 *Should I be registered for VAT?*

Who can use the scheme?

You can use the scheme if:

• you have constructed a new building (paragraph 3):

and

• the building was not constructed in the course or furtherance of any business. For the purpose of this notice, 'business' includes the letting of accommodation.

If you are registered for VAT and are constructing a building for non-business purposes, you still have to use this scheme to get back the VAT you pay on goods and materials. You must not reclaim it on your VAT returns.

If you make bulk purchases of some or all of the materials under an informal agreement with others who are buying these materials, you can still use the scheme. But you should ask your local VAT office for advice before making your claim.

What kind of building is covered?

Private dwellings:

The building must be **intended for ordinary residence and entirely new**. It must not be a conversion, reconstruction, alteration or enlargement of an existing building. Any existing building on the site must first be demolished to foundation level – although you can retain a single external wall. The new dwelling must **not** have internal access to any existing building: **nor** may the planning permission forbid its separate disposal or use. Most small additions built onto an existing house – "grannie annexes" – would **not** qualify for a refund.

Provided they are built at the same time, a 'dwelling' includes:

- a built-in garage;

- a detached garage, if it is intended for use with the dwelling;

- other works on the site, such as paths, patios, drives, boundary walls and fences.

Buildings such as greenhouses or garden sheds are **not** included.

Other buildings:

The building must be **entirely new** and:

- **intended for use as a communal residential building;** or

- **intended for use by a charity for non-business purposes;** or

- **constructed by a charity and intended for use as a village, community, church or similar hall** providing social or recreational facilities for the good of the local community. (This includes cricket pavilions and changing rooms for charitable playing field and recreational ground associations, but not civil engineering works such as tennis courts or running tracks.)

You can find more detailed descriptions of qualifying buildings in VAT Leaflet 708/2 *Construction industry.*

Provided they are built at the same time, these buildings include other works on the site such as paths, patios, drives, car parks, boundary walls and fences.

Must I construct the building myself?

You can employ specialist help such as a bricklayer, plasterer, plumber or electrician. If they are not registered for VAT they must not charge tax on the services or materials they supply, so there will be no VAT to reclaim. On the other hand, if you employ a contractor who is registered for VAT, any construction services and most materials he supplies will be zero-rated – by certificate for a building other than a dwelling. Because no VAT is charged, there will be none to reclaim.

If you use a VAT-registered builder to do the structural work for you, you can still claim for the goods you use to 'fit-out' or 'finish off' the building – but this must involve a significant amount of building work. You can't use the scheme to reclaim VAT on extra fittings you incorporate into a complete building that has been constructed for you by a builder or sold to you by a developer.

What can I claim for?

You can use the scheme to reclaim the VAT you pay on most goods, including prefabricated building kits, that you

- import or buy from a VAT-registered supplier

and

- incorporate in the building or its site.

Examples of the kind of goods that are covered by the scheme are given at the back of this notice.

You can't reclaim VAT on

- professional and supervisory services, for example, architects and surveyors' fees

- other services such as hire of equipment or transport of materials

- tools or other goods used for constructing the building.

Examples of the kind of goods that you can't reclaim VAT on are given at the back of this notice.

In addition to these items you can't reclaim any VAT which has been charged in error. This includes VAT wrongly charged by a builder, and VAT on "supply and fit" invoices. You will therefore have to take up the matter with your supplier.

Must I have proof of the VAT I have paid?

Yes. You must have an invoice for any goods which you claim for. The invoice must show:

- the supplier's VAT registration number

- the quantity and description of the goods

- the price of each item

If the value of the goods is more than £50 – including VAT – it must also show your name and address. You must have paid for the goods and, if asked, you must produce proof of payment.

If you have imported the goods and intend to claim a VAT refund, you must have the VAT copy of the import entry showing the amount of VAT you have paid.

You must provide the originals of all invoices and import documents – copies will not do.

If you have received a credit note from your supplier in respect of any invoice you must provide it, together with the relevant invoice and deduct the credited VAT amount from your claim.

If you have any doubts about an invoice or credit note, ask your local VAT enquiries office for advice.

What if VAT isn't shown on the invoice?

If VAT is not shown separately you can still claim it by working out how much of the price was VAT using the VAT fraction. The notes on the cover of the claim pack tell you how to do this.

When can I claim?

You must make your claim within three months of the date when the work on the building is completed. Only one claim can be made for each building.

Please make sure that you make your claim in good time or it may be refused. If you don't think you will be able to make your claim within three months you should contact your local VAT office immediately to tell them why. When you send your claim in you should enclose a letter explaining the reasons for the delay.

How do I claim?

Your local VAT office will give you the claim forms. You must fill them in and return them to your local office with:

- a copy of the planning permission – including the plans; **and**

- proof of the VAT you have paid (paragraph 6), **and**

- evidence that the building is completed. If a detached garage has also been built this must be stated on the document you provide or in a separate letter. The evidence can be:

- a certificate or letter of completion from the local authority; **or**

- a certificate or letter of habitation from the local authority (or in Scotland, a temporary certificate of habitation); **or**

- a rating valuation proposal from the District Valuer; **or**

- a certificate of completion signed by an **independent** architect or surveyor; **or**

- a certificate from the building society stating:

"This is to certify that the .... Society released on .... (date) the last instalment of its loan secured on the dwelling (and the detached garage)* at

....

because it then regarded that dwelling (and that detached garage)* as substantially complete."

* delete if not applicable

To check if your claim is complete, please refer to the checklist on the last page.

What happens next?

A few days after making your claim, you will be sent an acknowledgement. If you don't receive it within 14 days of sending your claim, please contact your local VAT office. You may also be asked to provide additional information – for example, written evidence that your organisation is a non-business charity.

When Customs and Excise are satisfied that your claim is valid, they will send you the refund by payable order and return all the supporting documents to you. If your claim is rejected you will be sent a letter explaining why.

Can I appeal?

Yes. If Customs and Excise refuse to grant a refund or if you do not agree with the amount paid, you can ask them to reconsider your claim. If you are still not satisfied, you can appeal to an independent VAT Tribunal. There are time limits for doing this. Information about the appeals procedure is given in the leaflet *Appeals and applications to the Tribunals*, issued by the President of the VAT Tribunals. This leaflet is available from any local VAT enquiries office or VAT Tribunal Centre.

23 The local authorities

The attitudes of local authorities to DIY builders are diverse and reflect special circumstances rather than any overall policy. The official position is that councils have been urged by successive Governments to consider helping selfbuild Housing Associations, and a Department of the Environment circular lays a duty on councils to examine whether they can find land for local selfbuild associations, and consider whether they should publicise that they have land available for this purpose. To say the least their enthusiasm for doing this varies, but they do make hundreds of plots available to those who want to build for themselves every year.

Some councils have a long tradition of supporting selfbuild groups of all types. Some favour self managed groups and others prefer to have professional managers set up schemes. The role of a council can vary from simply selling land to an association at market value, as an ordinary disposal of land made in the usual way, to making special arrangements to sell land to an association which would not be available to others, and giving the association advice and support at all stages. This sometimes extends to the Chairman of the council's housing committee taking the chair at a public meeting to form an association, and this sort of support is invaluable.

Other councils appear to be wary of groups, but have an active policy of releasing serviced plots to individuals, many of whom they know will build for themselves. Some local authorities will make a point of advertising that such plots are suitable for selfbuilders, and this simple recognition of the selfbuild approach is usually enough to ensure that 75% of those who buy the plots will build on their own, while the rest will employ builders. The way in which such plots are allocated varies, with many councils giving local ratepayers the first chance to buy, and only selling to outsiders when the local demand has been met. Sometimes these plots are offered by tender, sometimes allocated at fixed prices from a waiting list, and occasionally allocated on a first come first served basis on a particular day, which before the collapse of the housing boom used to lead to applicants queuing for a week. Oddly enough the queue system was popular with

purchasers, who turned them into week long pavement parties, and found them a good way to get to know each other. In Milton Keynes, where they were an established feature until 1989, both architects and builders merchants could be seen discreetly soliciting business among the queuers!

Whether or not a council sells land in this way, and with what enthusiasm, has little to do with political attitudes. To start with the council has to have the land to sell. It is unusual to hear of an authority buying land simply to make it available for a selfbuild scheme. Generally speaking, councils that sell plots are in areas where development is being actively encouraged,

Some local authorities do more than simply provide land for selfbuilders. Here the Chairman of the South Norfolk District Council cuts the tape to launch a group selfbuild scheme which it sponsored for first time buyers. Look at the ages of the members. Helping young people like this to become owner-occupiers is the best thing that any community can do if it wants to ensure its long term prosperity.

and this is not in the Home Counties or in the South East.

Over the years many local authority schemes have been examined for this book, and in this edition we look at a council in the North East.

The Gateshead Metropolitan Borough Council first became involved in selfbuild in the 1980s, when five sites were sold by the council for orthodox group selfbuild schemes which were initiated and managed by two competing firms of selfbuild management consultants. The initial approaches to the council were made by the consultants, and the Director of Planning was authorised to negotiate with them on the basis

A few of more than a hundred local authority schemes in 1995 which gave local people — and others — an opportunity to build for themselves.

Selfbuild and area rehabilitation

Some councils make provision for selfbuild schemes as part of area rehabilitation programmes. The photograph on the left shows a street of Airey concrete houses in South Shields which the council made available to the Chirton Avenue Selfbuild Association, who replaced them with the small detached houses shown in the photograph on the right.

that the members of the association would be local people, that agreement would be reached on every aspect of design and amenity, that the contract between the consultants and the association would be vetted, and that the land would be transferred directly from the council to the association when it was formed. Arrangements of this sort were common in the North East in the 1980s, and were responsible for many hundreds of low cost owner-occupied houses built to a high standard. The Gateshead group schemes have accounted for a total of 225 homes at a time when council housing was being curtailed.

At the same time the council began to look at the possibility of providing sites for individual selfbuilders, prompted by a high level of local enquiries and recognising the difficulties which private individuals were experiencing in buying plots, as they were invariably outbid by developers. The first scheme was at High Heyworth, where 13 plots for detached dwellings were sold very quickly, and where everything went successfully. This was followed by two similar schemes at Rothbury Gardens and Rowlands Gill, where work is now in progress although at a slower pace than on the earlier schemes due to the recession. All the plots are now sold at Rothbury Gardens, and the council has a rolling programme to open new sites providing that funds are available to install the services.

The council's brochure for prospective purchasers is 17 pages long and deals in a simple and straightforward way with everything that a would-be selfbuilder requires to know, from prices (circa £20,000 a plot) to details of the back fences required, and from the contact names for all the different services to threats of retribution if building materials are stacked on the pavements.

Perhaps the most interesting feature of the scheme is that the purchasers can move on to their plots after paying 10% of the purchase price, and have a whole year in which to build their house and get a mortgage on it before the balance is payable. They have to meet interest charges on the outstanding money, but as they can put caravans on the site the saving in the cost of temporary accommodation while they are building more than balances this.

The sale of plots is restricted to 'those building their own house for owner-occupation' and in order to ensure that this is not circumvented, there is a covenant prohibiting the sale or renting out of a property within two years of completion.

Besides the councils, the development corporations responsible for the new towns are also active in running schemes of this sort. It is uncertain whether their proliferation is due to a desire to help people build their own homes, or whether council and corporation treasurers are insisting that their estate departments turn surplus land into cash to meet a financial crisis! If a council has experience of a successful scheme, a follow up scheme is very likely.

A special situation exists in Inner City Housing Action Areas, where both councils and the special Development Corporations are often deeply involved in special selfbuild schemes which are part of larger redevelopment project. This is more fully discussed in the chapter on Community Selfbuild.

The Mayor of Gateshead with the first group selfbuild family to move in to a new home on one of the Gateshead schemes described earlier in this chapter.

Some of the most innovative schemes promoted by local authorities are in the new towns. These members of the Fullers Slade Association building at Milton Keynes have a site where each member has a licence to build and a thirty year option to buy his plot at 40% of the current value of the property that he builds on it. As a simple way for a council to promote low cost home ownership this has few equals.

24 Selfbuild groups

In the first chapter of this book there is brief mention of those who build new homes for themselves in teams, usually as members of selfbuild housing associations. The way in which they work is very different from the activities of the individual builders described so far, and it is now time to give the group selfbuilders appropriate mention.

A selfbuild group is formed by a dozen or so individuals who form an association to work together to build as many houses as there are members. When the houses are built each member moves into one of the houses, and they then wind up the association. By working together they can pool their skills, and by forming an association they can obtain loan funds and help in finding land. However, ordinary groups do not receive a grant or subsidy, and when the houses are built they have to be paid for by the members at cost, usually through a Building Society mortgage. The attraction of selfbuild is that this cost is very low. The attraction of group membership is that it provides the framework within which to build at the low cost.

Selfbuild housing associations fall into two types, the managed and the self-managed. The difference between the two becomes more marked as one moves closer to the practical business of building. Managed groups are usually formed by an agency to build houses, already designed, on a site already purchased, using building finance already arranged. Members are chosen from the agency's waiting list or are sought by press advertising and they can expect to be working on site within a month or two of joining. The agencies may be philanthropic, in which case the association is expected to provide its own site management, or commercial. Commercial agencies expect to be retained as managers to the association on a fixed fee basis. Philanthropic agencies are discussed at length in the following chapter.

Self-managed groups have to find their own land, arrange their own legal formalities, and seek their own building finance. This is invariably a lengthy process. The delays and frustrations make demands on the enthusiasm of

members and considerably extend the overall life of the association.

However a group is formed, the legal framework in which it operates is the same. It first registers under the Friendly Societies Act, 1965, with a constitution based on the model rules of the National Federation of Housing Societies, and affiliates to the Federation. To do this it has to have at least seven members, all of whom normally take up a £1 share. Sometimes a Local Authority or County Council will also take a share in the association. This gives associations a legal status, the ability to enter into legal obligations as an association, and limits the liability of members. In addition it provides an opportunity to

seek loan finance, and assistance in finding land from the Housing Corporation, Local Authorities and others. Any attempt to build as a group without being a properly registered Housing Association would be most unwise: apart from any other consideration, a legal position is likely to develop where any one member could be responsible for the debts of all members.

When the association has been formed it elects its officers and a committee, and adopts the standard working regulations recommended by the Federation. These regulations set out in straightforward terms exactly how the association is to operate and what the obligations of the members are. They deal with matters such as the

Barrack Road in Northampton. These houses were built in the last century as part of a scheme to enable 'sturdy artizans' to build their own homes. They paid a shilling a week into the building fund! The list of members survives and it is interesting that one member went on to be mayor of Northampton. Their association became a building society which is now part of Abbey National. The scheme was sponsored by the local Liberal party.

hours which members are expected to work, the ordering of materials, keeping accounts and the authority of the committee.

When the working regulations are agreed in detail each member signs a separate contract with the association to signify his agreement to them, and to ensure that the association will transfer a house to him when all the properties are finished. Before signing these contracts the members will usually have allocated the houses among themselves, often by ballot. Naturally not all the houses are finished at once, and at this stage it is also usual to agree the order in which they will be completed. All the new homes remain the property of the association until the last one is finished, and when members move in they do so on a formal 'licence to occupy' issued by the association. This does not constitute a legal tenancy, so that the association retains control over every house until they can all be transferred to members on the same date. This is important in many ways, not least

to ensure that members who have moved into a home have an incentive to keep working on the other houses! Finally, the association is wound up, and the remaining assets are divided between the members, together with a refund of

The Surrey House Selfbuild Housing Association members shown here built a block of 14 flats in London's docklands. The scheme was initiated and managed by Essex Selfbuild Advisory services, sponsored by the Docklands Development Corporation and financed by the National & Provincial Building Society.

the £1 membership share.

Model rules and working regulations will be found at the end of the book. It must be emphasised that these should not be copied for use by an association and that the Federation should always be asked to provide drafts incorporating the changes made necessary by new legislation or other factors.

The financial framework within which the housing association operates is equally specific. It must appoint a treasurer to receive monies and make payments, and to keep proper books of accounts. A professional accountant has to be appointed as auditor. Insurances and a fidelity guarantee have to be arranged. VAT registration has to be organised, and an application made for the scheme to be approved for option mortgage subsidies. All this is set up for a managed association, while the self-managed association must look to its committee to handle this work, using the publications of the Federation to assist it.

Both types of association obtain their finance in the same way as follows:

SHARE CAPITAL is the nominal £1 shares taken up by members.

LOAN CAPITAL is the contribution which members make to get the association going and to demonstrate their intention to support it. Loans are interest free, and are repaid or set against the cost of the house at the end of the scheme. These loans may be a condition of membership, and may be paid as a weekly subscription. Associations sometimes require comparatively high loans from 'unskilled' members and waive loan obligations from members who are skilled tradesmen. Typical loans required are of £500, which is usually the full extent of the members financial commitment.

BUILDING FINANCE This is the money to get the job done. It may be anything from half a million pounds to over a million pounds which has to be borrowed until it can be paid back by the members' individual mortgages when they move into their new homes. Self-managed groups usually get their loans from banks or building societies, or a consortium of two or three banks or building societies. A local authority loan is

another alternative, particularly if the local authority has found the land for the group. This building finance carries interest charges which are a major element in project costs.

These financial arrangements have been the basis of successful group schemes for many years, but they have two disadvantages. One is that members live in their old homes while building the new ones, and they need to be able to sell them readily when the time to move arrives. The other is that group selfbuilders are doubly disadvantaged by any significant rise in interest rates while they are building, for this will affect both the interest on their building loan and also the maximum mortgage they can obtain on their income.

In the second half of 1988, when the housing market collapsed and interest rates went up from 9.5% to 15% or higher, there were about 200 selfbuild associations working on sites throughout the country. All had difficulties with the interest on their association loans, and those in the South were faced with members being unable to sell their own homes on a depressed marked. Over 70 associations failed, and the members lost their loan funds, wasted all their hard work, and had their hopes blighted. For some this led to serious personal problems. Sad as this was, it must now be seen in some perspective. In mid-1988 thousands who bought new homes on the open market with high mortgages were similarly affected, and most of them lost more of their savings than the selfbuilders did, although it would have needed a brave man to point that out in the site hut of one of the groups concerned. The 130 other associations which were building in 1988/89 weathered the storm.

An unfortunate effect of all this was that the National & Provincial Building Society, which had financed the majority of the associations which failed, had to make provision in its accounts for losses in excess of seventy million pounds in respect of its involvement with selfbuild, sinking its ambitions to be the second building society to become a limited company and persuading virtually every building society in the country that selfbuild housing associations were bad news. It has become virtually impossible for new associa-

Tony and Belinda Snelson in front of their new house at Rastrick which they built as members of the Brighouse Selfbuild Housing Association. They both work in the control room of the local fire service, and had no previous experience of building. Their only contribution to the group funds was a £200 loan until they had finished their new home, when they bought it on a mortgage from the association at significantly less than its market value.

tions to obtain loan funds at all in most parts of the country, although 400 associations are still registered with the Federation of Housing Associations and about 20 have obtained finance and are working on sites, mainly in the North of England. When the housing market improves they will all be back in business, so it is appropriate to look at the people who make it happen.

First of all the membership. Here one would expect flexibility, with scope for individuals to get together simply as people with a common aim. Unfortunately this is rarely the case. Housing associations are afforded a remarkable degree of support by various bodies, but in exchange, they must be established in a way which experience has shown to afford them the best chance of success. The ideal membership is between 12 and 20 in any one group. The Federation is specific in this, as is the Housing Corporation and others who provide building finance. They look for a group in which 50% of its members have building skills, to give a professional feel to the operation. They expect the majority of members to be married and to possess a steady outlook that gives confidence that they will stick a year of unrelenting toil. They look for a balance between men over thirty and under thirty. They look for a membership established in their employment and therefore good mortgage risks.

Managed associations carry this to extremes, making up lists of possible members from

applicants as if they were selecting an England football team. Few employers would admit to being as selective, or looking so carefully at all the circumstances of a prospective employee, but this rigorous selection has one justification - it works. Building one's own home means working outside the system, and so it needs special people. Unless everything is right, the odds against success will lengthen. The Federation, and the experienced professional managers, do know what makes success. Their advice should not be lightly discarded.

Individual members of groups contract to provide loan finance as agreed, arrange to buy their house from the association at the end of the scheme, and to work for the association as set out in the working regulations. Typically, members agree to work a 20 hour week in the winter and a 28 hour week in the summer, plus one week of their holiday. Seventy-five per cent of working hours must be worked at weekends. Sometimes experienced bricklayers, joiners and plasterers receive an allowance of two free hours a week, recognising their higher productivity. Special arrangements cover sickness, otherwise absence or lateness results in an automatic fine, often £5 per hour. This is not paid in cash, but is built into the final cost of the member's house. These fines are rigorously imposed by most associations, who feel that only in this way can personal relationships survive resentment against the offender.

All members are expected to specialise in one aspect of the building work as decided by the committee, with everyone participating in tasks such as concreting, path laying and unloading materials. Experienced building tradesmen have their own role, the inexperienced quickly acquire skills, and specialist work is put out to contract when necessary. This is particularly important where unprofessional work carries a risk - gas fitting, electrical work, scaffolding etc.

Occasionally associations lose members due to ill health or other reasons, or expel members who default on their obligations. The rules make provision for this, and associations invariably arrange for a new member to fill the vacancy. The member who leaves takes the money which he

In spite of the recession, members of some selfbuild housing associations in the North of England are still achieving cost/value differentials of nearly 40% on their new homes. Members of the Basilton Homesmiths Association pictured here paid £250 each as start up capital, worked 30 hours a week for 13 months, and then obtained mortgages to buy their new homes from the association for under £50,000, just over half the market value. Part of the success came from the members finishing five months ahead of schedule, resulting in a huge saving on the budgeted interest charges.

has put into the association but he usually forfeits the value of the work which he has done. The Federation advises all associations to arrange life insurance for members so that the association can be paid for their lost labour should they die, and their dependants can then take over the house.

At this point, the roles of the sexes are of interest. There is a clear-cut pattern which is impossible to explain in simple terms. The man building for himself on his own invariably works in close and effective partnership with his wife. Their joint involvement in the new home is complete, and invariably they speak of what 'we' are doing. If the wife plays no part on the site she will be organising materials, chasing sub-contractors, deducting monthly payment discount from the bills. She will discover unsuspected tenacity in dealing with the water authority, gas and electricity boards, and British Telecom. Frequently she provides far more than half the enthusiasm and drive required to carry the job through. Often she will work on the site

with her husband and sometimes the only visible difference between them is that the wife wears gloves. Such wives terrify male architects.

Community Selfbuild Groups often have women members but, surprisingly, wives play virtually no part in mainstream group projects. Sometimes one hears of a woman secretary or treasurer but this is unusual. They seldom visit sites except as visitors or to make tea, and some associations restrict family visits to specific times. The exception is where associations leave interior decoration to members and this is done by the wives while the husbands are working on other houses. I make no attempt to explain these phenomena and simply report the relative roles of the sexes as they are seen on the different types of selfbuild operation.

Selfbuilding as a member of an association has two disadvantages. Firstly, there is the standardisation of house designs. A single design for all the houses in a scheme was once a condition of Housing Corporation support, and although in recent years it has become common

Up-market group selfbuild. Richard and Leslie-Ann Parker built this new home in Northants as members of the Barnwell Selfbuild Housing Association, and are pictured here leaving for a memorable 'end of scheme' party to which the

for there to be two or three different designs used on a site there is an obvious need for them all to involve much the same labour input. Colour of bathroom fixtures and the selection of kitchen units are normally a matter of individual choice, but little else. This standardisation and personal choice is much the same as that on a developer's housing estate, and this may not always suit the independent spirits who want to build for themselves. The other potential disadvantage of building in a group is the constraint of the group discipline and the need to work to a set programme. For some the team spirit which this fosters is one of the attractive features of the whole business, but for others it presents problems, particularly for building tradesmen who sometimes find it difficult to accept that attitudes on a selfbuild site are different from those to which they are used in their everyday work. However, these constraints of philosophy and outlook are really side issues.

So much for the members, and how they organise themselves. Other players on the group scene are the consultants who for about ten years from the late 1970s instigated and managed well over 90% of group schemes. There used to be about 30 consultancies offering services to associations, of which a dozen still exist. Their old pattern of operations in their hey day was to locate a site, obtain an option of some sort on it, obtain architect's draft development proposals, prepare a costed

scheme based on these proposals, obtain the promise of finance for the proposals from a bank or building society, obtain planning consents and finally set up an association and find members for it. The new association was expected to retain the management consultants to manage their labours and provide many other services, and this usually involved fees of between 7.5% and 10% of the value of the finished properties.

These management consultants had their own trade organisation, the Society of Selfbuild Management Consultants, but this no longer exists. It is essential that anyone considering joining a selfbuild housing association, or any other form of group scheme, whether managed or not, should seek the advice of the Federation of Housing Associations whose address will be found on page 294. In particular, no substantial sums of money should be paid to those who are setting up selfbuild groups under any circumstances without the advice of the Federation.

The selfbuild housing association route to working together as a team is tried and tested, and is to be recommended. There are alternatives. Some of these, such as the special arrangements recommended by the Walter Segal Foundation, and the housing co-operatives that are popular in Merseyside, have a track record of success. Others, including those which advocate that people should obtain the finance for new homes as individuals and then co-operate in buying land and meeting building costs jointly, whether as a limited company or otherwise, are fraught with hazards. Not the least of these is the possibility that individual members of such schemes may end up being responsible for the debts of all the other members. There are also potential problems involved in buying land as tenants in common, which is sometimes recommended. The selfbuild scene is currently littered with unhappy situations which are the fall out from such projects, some of them involving litigation. Once again, the only source of impartial advice is the Federation of Selfbuild Housing Associations.

25 Community selfbuild

Community selfbuild is a term used of projects where those in housing need are enabled to build their own homes, and it has a long history. Over the years the definition of those in housing need has changed: for 30 years to the end of the 1970s the phrase was used to describe those who had not enough points to get to the top of a council house waiting list, or who could not find an affordable home in an era of mortgage rationing. They were invariably in employment. Those in housing need today are usually not in employment and often suffer other disadvantages. There is a view that today's specific concern with the problems of the disadvantaged homeless tends to obscure the needs of those who have no other problem than an inability to find an affordable home. More of this later.

The way in which selfbuild has been, and is, relevant to the needs of this changing 'housing need' scene is essential to a full understanding of the contribution which DIY construction can make to our society.

Group selfbuild in its modern form started in Brighton in 1948 with special schemes for exservicemen to build their own homes on land provided by the Borough Council. They were a success, with 1194 homes being built by 58 associations over 20 years. As such schemes proliferated in an era of controls and nationalisation they attracted grudging recognition, limited support, and a concern that they should be fitted into the housing bureaucracy. By the 1970s a pattern had emerged where the building work was usually financed by the Housing Corporation and led to Local Authority mortgages. They used the National Building Agency as consultants, and were expected to register with the Federation of Housing Associations and to adopt its model rules. The land upon which they built was usually made available by a local authority, which often imposed pre-emption clauses in titles to prevent selfbuilders selling their homes to make a quick profit. There were other approaches too; in the 60s and 70s the West Midlands Council of Housing Associations financed 2,000 self built homes in Birmingham, often using funds from Quaker charities. In time all of this became formalised, and so did the Housing Corporation's view of selfbuild associations: they were specifically for those in housing need, defined as being those on a council housing list. Associations with a significant membership not on housing lists were ineligible for Housing Corporation finance. Selfbuild associations were ways of helping those in housing need to help themselves.

In the late 1970s all this started to change. Financial stringency had closed the NBA and severely limited the finance available through the Housing Corporation. The selfbuild management consultants arrived and found sources of private finance for the associations that they managed. Within a decade the typical selfbuild home was a four bedroom, detached unit instead of a three bedroom semi, and 'housing need' was rarely mentioned where selfbuild was at its strongest, which was, and still is, in the outer suburbs and small towns. By the early 1980s group selfbuild had abandoned the inner cities where housing needs were still very real, and had become something for those who were using it as a way of jumping several rungs up the housing ladder at one bound. It had little relevance to those who wanted to get a foothold on the bottom rung unless they were exceptionally determined or able, unless they could be found a place on a scheme which had made special provision for 'housing need' members. There were a few dozen such schemes in the 80s which built perhaps 500 homes in total, usually on the initiative of a local authority which would sell land to mainstream selfbuild groups on condition that they took a percentage of their members from council house waiting lists. This was a growing trend, but when the mainstream SBHA scene crashed in 1989 these opportunities for selfbuilders in housing need disappeared as well.

Meanwhile, as the plight of the homeless became a national issue in the 80s, a number of charitable agencies and others of a like mind began to seek opportunities to promote selfbuild for those who could benefit from it most of all. One pioneer in this was a Dr. Barraclough, a consultant from St. Thomas' Hospital who, having built a striking house of his own from reclaimed materials on the Isle of Dogs, decided to encourage East-Enders dispossessed by the

Docklands Developments to do the same. Among a number of schemes with which he was concerned was the self managed Great Eastern project on the Thameside site where Brunel built his great steamship. Here 46 dockers and their friends built 46 homes with building society finance. The photograph of them above tells its own story more effectively than any words, and helped get inner city selfbuild a lot of publicity. So did a thirty minute TV feature programme about the Great Eastern members.

'Zenzele' and the CSBA

Another scheme at Bristol was to gain an even

higher profile: it was called Zenzele, and was set up as an experiment in breaking the 'no job experience - no job opportunity' trap in the run-down St. Paul's area. This was approached by setting up a selfbuild housing association for twelve members who were described in the press as young, disadvantaged and unemployed, and this was made possible by some forceful management by a local J.P., Mrs Stella Clarke, and a very effective management team which she assembled.

Their first coup was to get the agreement of the D.H.S.S. that association members who were still unemployed when they finished building should have the interest on their mortgages paid by the department as if they had incurred the mortgage before they became unemployed. At that time the ceiling payment of this sort was £22.50 per week, which facilitated a mortgage of £10,800. The Bristol and West Building Society was persuaded to offer such mortgages. Building finance was arranged through the Housing Corporation. All that remained was to see if a scheme could be put together for new homes to be built for £10,800.

A site was found for £15,000, and architects Atkins and Walters designed a simple and cost effective block of one-bedroomed flats that could be built within the budget. The next stage was to form an association, and instead of the usual advertisements and public meetings the first four prospective members were assembled by Tana Adebiyo, another of Mrs Clarke's team, who was a community worker in St. Paul's. Their first task was to find another eight members, all of them unemployed, who had some minimal building industry experience. The association was registered with the name Zenzele, which is Zulu for 'together'. The Federation's model rules were adopted and the association was in business.

Progress was at first quite fast, as members were on site for a full working day. The budget provided for the services of a working site foreman, and the first one appointed was a groundworks specialist as there were site problems with the cellars of old buildings. This involved some anxious moments as this sort of situation can never be costed with certainty, but after various worries everything worked out. The

groundworks foreman handed over to another who was to take the building on to completion, and the walls started to rise.

Construction was entirely traditional, and the members handled all the trades themselves. The only contractors brought in were specialists to lay the concrete slab first floor, which, together with a 3" screed, had been designed to give the maximum sound insulation between flats. There were no building problems, but one very happy organisational problem ... members started to get jobs. The whole scheme was based on the idea that people find it easier to find employment when they are already working and this is just what happened. The members were soon in two groups: those who were still unemployed and could put in a normal working day and those with jobs who worked on the flats in the evenings and at weekends like most other selfbuilders. At the end of the scheme eleven of the twelve members had found work, which was the original intention behind the whole project. By the time the block was completed costs had escalated by 11%. However the Building Society valuation of the individual flats had gone up to £18,500, and rises in the D.H.S.S. mortgage interest ceiling made mortgages of £12,000 possible. This not only covered building costs but also cookers, carpets and full interior decorations.

Zenzele provided the public housing establishment with a clear challenge to provide the opportunity for other selfbuild schemes for the disadvantaged, and this led to the Housing Corporation setting up the Community Selfbuild Agency with Stella Clarke as its Chairman and representatives of a wide spectrum of bodies on its management committee. The Agency has offices in the south and north of England, from which it offers impartial advice issues linked to the development of these schemes. It promotes the sharing of experiences between groups and will act as a broker in introducing them to possible 'partners' which are needed for these schemes to reach fruition. In particular it provides a selfbuild presence at local authority and housing association housing conferences, where its Director, Anna McGettigan, is a regular speaker.

Stella Clarke, JP. with three members of the Zenzele Selfbuild Housing Association. The block of flats that the members built is in the background.

Many attempts were made to replicate the Zenzele scheme, but sadly these coincided with the collapse of the housing market and problems experienced by banks and building societies in their funding of mainstream group schemes, so that they declined to fund the proposed Zenzele clones. They were approached by the CSBA at board level to reconsider this but to no avail, which merely served to delay a realisation that commercial funding of community self build on any significant scale is just not going to happen until the housing market recovers, in spite of proposals that the Housing Corporation could underwrite 40% of the lenders risk.

A further element in this disappointing situation was the practice of the DOE and DHSS of allowing each local office to make their own interpretation of the regulations concerning benefit payable to unemployed selfbuilders. Selfbuild was frequently new to the officials concerned, who did not understand that the selfbuilders were not paid while 'working' on a scheme, and some potential selfbuilders were given the impression that they would lose benefits if started work on site. The CSBA and other bodies have pressed for a firm national directive establishing the concessions negotiated by Stella Clarke for Zenzele, but the most that has been arranged is a statement from the DHSS centrally that 'the government appreciates and would

encourage the efforts of unemployed people to spend their time usefully while continuing their search for work, and part-time involvement in selfbuild is not necessarily incompatible with continued benefit, so long as this does not interfere with the claimants search or availability for work'. This although participation in a selfbuild scheme is demonstrably the most efficient way of getting the long term unemployed into employment! However, at Brighton, home of the birth of selfbuild in the late 1940s, another high profile scheme with different financial arrangements was starting work in 1990.

John Tee was 16 years old and having to make his own way in the world when he read in the Brighton and Hove Herald of a proposed Council sponsored self build scheme for teenagers in housing need, and he took himself down to the housing office to learn more about it. A fortnight later he found himself at the Aberdovey outward bound centre with 15 other self build recruits of his own age, getting to know each other and discovering their potential via the challenges of the Welsh mountains and the Irish sea. This is believed to be the first time that this approach to team building had been linked with self build. On their return they started work on a site at Hollingbury, a suburb on the downs above the town, which happily provided a convenient challenge by being littered with fallen trees which

233

were victims of the 1987 gale. Clearing these was a splendid way to make a start, combining demanding physical work and the discipline of working in close proximity to heavy machinery. By then the self builders were technically part of a YTS scheme involving TEC training to lead to National Vocational Qualifications (NVQs). (The proliferation of initials reminds an older generation of the army, a comparison worth making in the hearing of some concerned with community self build just to witness the horrified reaction).

Once the trees were cleared and the old iron fencing had been removed, contractors moved in to pour the plinths of 8 semi-detached timber framed homes while the trainees started their classroom training and paid a visit to the Meyers factory in Southampton to see the frames and walling panels of their new homes being made.

This description of the Hollingbury story is written from the selfbuilders viewpoint, but behind their sudden involvement in all this activity lay two years of hard work by architect Kenneth Claxton and Brighton Housing Officer Linda Beanland. Mr. Claxton has been involved with innovative low cost housing initiatives since 1970, and Ms Beanland was responsible for packaging a bid to the Department of the Environment for special funding. No doubt the department saw Hollingbury as a useful way to gain experience of such schemes at a time when various proposals to finance community selfbuild via the Housing Corporation were under consid-

eration. These arrangements are now in place, and are likely to be the keystone of selfbuild for those in housing need for the rest of the decade. The future will probably lie with shared ownership schemes, but the Hollingbury houses were built to be part of the Brighton Council Housing stock, with the selfbuilders getting tenancies at rents which reflected their involvement.

Returning to the story on the site at Hollingbury, the timber frames were erected very quickly, walling panels complete with ready glazed windows were fitted into them, and the roofs tiled by the trainees. This was arranged so that the self builders worked in small teams, with two or more teams coming together for the heavy work of erecting the frames. Contractors were brought in to handle the plumbing and electrics, with trainees working with them like old fashioned apprentices. Site supervision, encouragement and firm direction was provided by Mike Bailey, a YTS supervisor who seems to have been an inspired appointment, and John Semple whose practical builders background was invaluable. They worked with a committee on which the trainees were represented.

The first selfbuilders moved into their new flats in July 1991, and all 16 were in their new homes by Christmas. However, there had been many changes in the group and only three of the original sixteen who went to Aberdovey finished the course and took up their leases. This was not surprising: 15 months is a very long time to a 17

Hollingbury trainees and, at the bottom of the opposite page, some of the 16 small maisonettes which they built at Brighton.

year old, and a number of the most enthusiastic trainees were so inspired by their new lives as selfbuilders they went on to other things and moved away. Others moved on due to changed circumstances, and inevitably a few found the hard work through the winter was not to their liking and were expelled. The vacant places were quickly taken by others who took the scheme through to its successful conclusion, and there was a hard core of members for most of the programme.

The Hollingbury scheme and other similar schemes led to the formation of the Young

Builders Trust which is a registered educational charity with its roots in actual projects, and campaigns for and promotes schemes with objectives to help disadvantaged unemployed youth in housing need. Vocational training is a key feature of all that it does and practise guides are published which draw on its unique experience. The YTB is currently involved in planning thirty future schemes, including one for the Inner London Probation Service. More widely it is part of a European network, and is already working in Bulgaria. It works closely with other agencies and is financed by DOE, special grants, charities and fees.

Meanwhile there were various other initiatives under way, of which the most important was the developing influence of the Community Selfbuild Agency, although almost immediately after it was founded the housing market collapsed, the building societies withdrew their promised finance both for schemes and for the agencies direct costs, and the CSBA was left to promote community selfbuild without any guarantee that schemes could be funded. It took two years for it to become obvious that building society finance was not going to be available again, but during that time the agency and others lobbied the Government and the Housing Corporation to find a formula for making funds available for schemes involving those in housing need. It was not until 1990 that a way forward was found by channelling Housing Corporation funds through established mainstream housing

associations who would "adopt" a self build group. This involved Housing Association Grant finance (generally known as HAG funds) being used to fund shared ownership and rented ownership schemes to a limit of 5% of each of the Housing Corporation Regions' budgets. This was adequate for a national total of about 24 schemes a year while the scope for rented schemes was unlimited.

Chisel Housing Co-op played a key role in pioneering the self build for rent breakthrough. For years the idea had been mooted, but there was some doubt about whether self builders could be recruited for rental schemes. However, the demand and the number of schemes, particularly in the south east, have proved that this is a completely viable approach.

Another major influence was the number of young single people who began to building for themselves, as in the Zenzele and Hollingbury schemes already featured. This particularly influenced some charities, particularly Charity Projects, which is the organisation behind Comic Relief, which established a fund of £330,000 to encourage such schemes. One in three of all schemes now involve young unemployed people who are able to improve their job prospects by joining up. Their achievements have influenced other agencies and the concept of involving young people in self build is being explored abroad, particularly in Sweden and Finland.

Another part of the picture is rural selfbuild, with innovative projects in small rural villages where housing is scarce and can be very expensive due to the influx of commuters and retired people from urban areas. A great deal of thought is being paid to ways in which quality homes can be built by local wage earners, and the Rural Development Commission have been involved with various initiatives for this. One such scheme is described in the case history which starts on page 267.

Since 1989, over 50 community schemes have been completed, all involving people in housing need. They involved all three forms of tenure: rent, shared ownership and outright ownership. Due to the effect of the unstable housing market, outright ownership for those in

housing need has not been as practicable as in the 70s and early 80s. The Housing Corporation has introduced a revolving loan fund to meet or guarantee 50% of development finance, but the balance has to be found from lenders who require that members have access to ordinary mortgages when they move in, but currently building societies and banks are reluctant to lend to those who most need their help.

Another problem with all schemes has been that recent changes introduced by the Government to curb benefit expenditure has affected schemes for shared ownership and outright ownership. Self builders who are unemployed cannot claim assistance with the payments on an interest only mortgage for the first nine months, and potential self builders are expected to take out insurance to cover this eventuality. This is not easily arranged by most of those in housing need. To say the least, this is very unfortunate as anyone prepared to undertake the work commitment involved in selfbuild will usually be doing their utmost to remain in work and to realise the full benefit of their sweat equity.

But in spite of these setbacks, community selfbuild continues to expand. Each new scheme inspires many more potential self builders, and more valuably, inspires more of those who are able to initiate schemes. The know how and experience passed from one group to its successors means less waiting to start on site and shorter building times. The early community self builders had to put their lives on hold for quite a long time, but this need not be so now. Housing associations and supporting agencies are now much more familiar with the funding regimes and the practicalities of getting such schemes on site.

There are now four charities which help those in housing need to get started:-

* The Community Self Build Agency, which was inspired by the Zenzele self build scheme, offers impartial advice on all forms of self build.

* The Walter Segal Trust, which came out of the early Lewisham schemes in the 80s, promotes a particular form of timber frame construction, known as the Walter Segal Method.

Some of the information available about community selfbuild. The arrangements for funding schemes of this sort are constamtly changing, and up to date advice from the Community Selfbuild Agency is essential at an early stage.

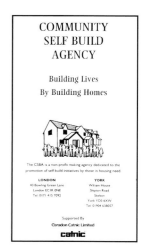

The Lord Mayor of Leeds, Councillor Peggy White D.B.E., with members of the Front Line Self Build Housing Association at the party to celebrate the successful completion of the new homes.

Maltby association members with their new homes. The two women members qualified as carpenters and played their full part in all the work on the site.

* The Young Builders Trust, which came out of the Hollingbury project, is targeting its work at young people.

* Community Self Build (Scotland), which offers information and advice on developing schemes in Scotland.

These charities work together within limits, but inevitably they compete for funding. None of them has any long term security and all rely on sponsorship which has often to be sought by their staff in time which they would rather spend helping to launch a new scheme.

Political interest in community selfbuild schemes is increasing at a pace. The Government's Single Regeneration Budget is taking an holistic approach to inner city regeneration, and requires that people to be helped should be involved in some tangible way. The Housing Corporation is targeting housing schemes of all sorts which they consider to offer more than just housing. Community selfbuild schemes meet both these criteria.

The current approach to community selfbuild housing is demonstrated by two projects in Yorkshire which were completed within a few weeks of each other in late 1995. The Maltby Youth Selfbuild Association has built nine houses in a mining village where the pit closure has led to very high unemployment and resulting social problems. As one of many initiatives to combat this a church based community organisation, the Maltby Rainbow Project, arranged for land for a selfbuild for rent scheme to be made available by the local authority with sponsorship from the Yorkshire Metropolitan Housing Association. The members, all young unemployed single people who did not qualify for tenancies, were enrolled on a training for work package which enabled them to build the two blocks of houses shown on the previous page, leading to NVQ level 2 qualifications and leases on the new homes. The houses are part of the housing stock of the sponsoring association, and the cost savings resulting from the involvement of the selfbuilders are being passed on as cash payments to the members. This is unusual: usually their 'sweat equity' in reflected in lower rents, but this is less important to the young membership than money to be able to furnish and equip their homes.

The impetus and arrangements for the Maltby scheme owe much to Christine Holman of the School of Urban and Regional Studies at Sheffield Hallam University, who had been a selfbuilder in the eighties. Her own experience of the importance to the individual of a selfbuild opportunity led to her concern that the prospective Maltby selfbuilders should not be recruited until it was certain that they would be able to start, with both land and finance in place. As a result, when the scheme was advertised and the members enrolled they were on a ten week training course almost at once, and started digging their foundations a few weeks later.

The experience of the Frontline Selfbuild Housing Association in Leeds has been different, for the impetus came from within the local Afro Caribbean community from which it drew its members. First registered in 1988, the enthusiasm of all concerned was sorely tried while they considered six possible sites and a series of abortive financial arrangements. It took five long years to put together a package to enable work to start in March 1994, but once under way the twelve two and three bedroom houses illustrated were built in a very credible eighteen months. They are occupied on a shared ownership basis, and happily plans for the redevelopment of neighbouring streets should guarantee that they will be a first class investment.

The members, who now have NVQ qualifications, intend to form a building company and work together to handle Housing Association and other contracts. If they are successful for any time it will be the first known example of any selfbuild association demonstrating a common purpose after completing the work that it was formed to do. Given the tenacity shown by the members in their five years in the wilderness, they may well do just that.

THE COMMUNITY SELFBUILD AGENCY

Unit 26, Finsbury Business Centre,
40 Bowling Green Lane,
London EC1R 0NE

Tel: 071-415 7092
071-415 7026

Fax: 071-415 7142

Dear Murray,

Thank you for the opportunity to set out the CSBA message as we see it, and particularly our role in influencing others.

Since the introduction of Housing Corporation funding for community self build schemes, we have seen a variety of schemes promoted mainly by organisations genuinely concerned with the wellbeing of the homeless, and each motivated to break new ground. This has been good in that new ideas have come forward to meet new challenges that have to be faced. However, there are limits to the variations on the theme, and we now need a period where we consolidate and benefit from all these ideas and experiences. The priority is to get many more people to understand that they can build their own homes and to do this in a much shorter time frame.

With over 50 schemes completed in the last five years, there are many examples to impress potential selfbuilders and show the range of homes that can be built. There is now much more scope for individuals to come together and form their own groups. What we need to make sure of is that there are willing "partners" to support their efforts. The Housing Corporation has provided the initial kick start by giving prominence to Housing Plus schemes which cover self build. However, partners have to understand that self builders are motivated by the type of home they want to build, where the site is and the relationship created with its partners - all need to pull together to bring the scheme to fruition.

With all this in mind, the Community Self Build Agency has been promoting self help forums widely. These offer an opportunity to meet those actively involved in building their own homes, to find out what the issues are that they faced and why they are still committed to the idea. Similarly the Agency is now running training courses and developing more training material so that self builders and their supporting agencies can establish good and productive working relationships.

There is much more that can be done. Each of the specialist charities needs to find their own techniques to open minds and hearts to the benefits of community self build. In particular, we need to:

* Influence the Housing Corporation to allocate capital funds early enough for groups to be able to plan well in advance for their work on site

* Reduce the actual building period to as near a year as possible.

* Encourage all self builders to use more ecologically sound and energy efficient materials.

* Ensure that in every county there is at least one scheme involving young people in the planning stage.

* Ensure unemployed self builders are helped to put their newly acquired skills to good use and for these to become assets towards long term employment.

* Encourage the sharing of experiences to keep down costs for everyone.

Many of the early schemes have won major awards from the RIBA, The Times, and Touche Ross Community Enterprise Scheme and the Training and Enterprise Council's Achievement Schemes for trainees. These have provided much needed publicity for a corner of the self build industry which is coming of age. Each scheme though is making its own mark in its own way and helping others to reach out and achieve much greater benefits as part of a team, which is needed to create a more self help world.

We have our own role in this, and welcome approaches from anyone interested in setting up selfbuild schemes or otherwise supporting our work.

Sincerely

Anna McGettigan
Director, C.S.B.A.

Building Lives By Building Homes

26 Energy saving

No one building a new home can ignore consideration of energy costs, nor should they ignore the effects on the environment of the ways in which we use energy. We are not short of advice on the subject. On every side we are urged to adopt new ways of reducing heating bills, and when building a new house there seem to be many opportunities to do this. Unfortunately the choices can seem bewildering. Energy saving is trendy and important. It generates a mass of seemingly authoritarian literature which is sometimes slanted and can mislead someone who simply wants a conventional new home that has the appropriate energy saving features for the next century and who does not want to build an experimental house of the future. Most newspaper and magazine articles are written with one eye to a 'good story'. The same is true of TV programmes. Reflect that a TV feature to demonstrate that something is *not* the energy source of the next century would have few viewers, and beware taking 'Tomorrow's World' too seriously.

Moving on from this sterile scepticism, what can you do? The building regulations require that the design of a new home has various energy-saving features, and these requirements were significantly uprated in July 1995. The requirements are now very high but additional expenditure to save fuel costs can still be cost effective if you make the right decisions. How do you make the right decisions? First of all, if you decide to spend your money to save even more energy than the building regulations require, it will be for one of three reasons and it is very important to differentiate between them. They will be:

* You wish to incur expenditure on any arrangement that is cost effective and which will give you a worthwhile saving in heating costs.

* Or you wish to invest in arrangements that are not cost effective now, but which you anticipate will become so in the future because you expect fuel costs to increase ahead of inflation, and because you think a home with these features will be a better investment.

* Or your interest in the matter transcends sordid financial considerations, and cost effectiveness is not important provided that the arrangement works. For example, you may consider reducing greenhouse gas emissions to be so important that you want the most environmentally friendly heating system that can be found.

Decide where you stand, as different approaches are appropriate to different objectives. You will also have to consider the effect that higher standards of insulation or other energy saving arrangements may have on the appearance and convenience of your new home, and whether you are happy with this. An unconventional home is invariably more difficult to sell than a conventional home, and this will reflect on it's value as an investment. For some this is irrelevant, for others it is critical. A heating system that is so unusual that it makes a house difficult to sell may be the worst possible investment.

If you invest in more energy saving than is mandatory to reduce heating costs, what is a worth while saving? In 1978 the author wrote a book called Heat Pumps and Houses in which he quoted the current price of oil fuel at 18.5p per litre — 1978 pence. As this book is being written the cost is 12.8 of today's pence! Meanwhile the insulation levels which are now required in new houses has reduced their heating requirements to below a third of the 1978 level. Heating costs are not nearly as large an element in the budget of the average household as they were twenty years ago, and fuel prices will have to rise by five times to be as significant to the householder as they were in the 70s. Most people only wish to spend money on additional energy saving features if they are going to get an attractive return on their money, and usually look for the capital cost to be covered by a maximum of ten years savings. This is called the payback period.

Later in this chapter projected annual heating costs for the house illustrated on the front cover of this book are analysed in detail, and vary from £1063 to £829 p.a. depending on the energy saving specification. This would justify capital expenditure of £2300 by someone who wanted to improve the specification and accepted a ten year pay back, and the changes required could probably be made within this budget.

As a bonus, the more energy efficient option reduces the emission of greenhouse gas pollution from the domestic boiler, the fire and at the power station where the electricity used in the house is generated.

After this general appraisal you will be interested in the actual energy costs which you will incur in your new home, and here we consider the new SAP ratings which are now a key element in house design.

SAP Ratings

Since July 1st 1995 all new homes have been given an energy rating assessment to demonstrate the compliance of the dwelling with the Building Regulations. In its most simple form this is known as the SAP rating, although the word simple is relative! It is calculated from taking into account the dimensions of the dwelling and the resultant surface areas, the thermal insulation values of every element in the shell of the building, including its foundations, the ventilation systems and the rate at which they extract air from the building, the draught proofing of external openings, the domestic hot water energy requirements, the internal heat gains, the solar gain which will be experienced by the building, the relative efficiency of the space heating appliances which are proposed, and, most importantly, the price of the fuel which is to be used to heat the dwelling. The resultant assessment produces a figure somewhere between 1 and 100, and the higher the figure the more energy efficient the dwelling is. A house of 80 square metres or less has to have a SAP rating of 80, while higher SAP figures are required for larger houses in accordance with the table below. How the SAP rating is achieved is at the discretion of the designer, and thus of the prospective home owner.

The SAP rating is not concerned with the geographical location of the building — whether it is in Bournemouth or Inverness for example — or any characteristics of the family which will occupy it. It assumes a standard pattern of occupancy of the home, and that standard temperatures are required. How you arrange to obtain an SAP rating is discussed later.

SAP ratings to comply with the Building Regulations

Dwelling floor area m2	AP Energy rating
Under 80 (860 sq ft)	80
80 to 90 (968 sq ft)	81
90 to 100 (1076 sq ft)	82
100 to 110 (1184 sq ft)	83
110 to 120 (1294 sq ft)	84
Over 120	85

The reason for the lower requirement for smaller dwellings is that they have a lower ratio of volume to envelope area than larger houses, and it is not reasonable to expect those building small low cost dwellings to incur higher unit charges for wall insulation than those building larger homes.

The NHER Rating

A more complex rating system which pre-dates the SAP rating was devised in the 1980's by the National Energy Foundation and is known as the National Home Energy Rating or NHER. This takes into account all the features considered in the SAP rating into account and also the geographical location — the Bournemouth v Inverness issue —, the family details of the occupants, the appliances which they will use, and the temperatures at which they like to live. An NHER report includes an SAP rating, and can be used to support a Building Regulation application. It is a more complex analysis of the energy characteristics of a building, and can be considered something for the enthusiast who has a particular interest in the subject.

The NHER rating provides the best way of estimating the total energy costs involved in running a house as it makes allowance for cooking, lights, appliances and standing charges. This allows energy saving options to be judges by the effect they will have on fuel bills.

Examples of both SAP ratings and NHER calculated running costs are on pages 242-246 showing the figures obtained for the house on the front cover with different levels of insulation, different heating systems and other features.

SAP rating - example one

SAP ratings for the same house built with differing standards of insulation and appliances.

The house on the cover built to a typical selfbuilder's specification which just meets Building Regulation requirements.

SAP 85

* 25mm underfloor insulation
* Brick and aerated block walling with 25mm slab insulation in the cavity
* 150mm roof insulation
* 6mm cavity double glazing throughout
* Ordinary gas boiler with zone controls
* Flame effect gas fire
* 110 litre cylinder with 37mm insulation
* Gas cooker
* Ordinary light bulbs

Fuel use and costs

	Energy required Giga Joules (GJ)	Cost per year at 1995 prices	CO2 emission tonnes per year
Central heating system	73.2	£345	4.3
Fireplace appliance	1.9	£9	0.1
Domestic hot water	19.1	£90	1.1
Cooking	5.4	£25	0.3
Lights and appliances	22.0	£508	4.5
Standing charge	0.0	£86	0.0
Totals	**121.6**	**£1063**	**10.3**

Computation by the Energy and Environment Ltd by arrangement with the EEO

Comment: This specification just enables the house to meet the Building Regulation requirements. The following pages show how you can do better if you wish. Even so, this house is very cheap to run compared with the houses most people are used to — this house built to 1970 standards would have a SAP rating of 38 and the annual energy cost would be £2103! The figures for energy usage in Giga Joules demonstrates clearly how much cheaper gas is than the full tariff electricity used for the lighting and appliances. Nearly half the total fuel bill goes on lights and appliances.

SAP rating - example two

SAP ratings for the same house built with differing standards of insulation and appliances.

SAP 91

The house on the cover built with a typical timber frame kit.

* 40mm underfloor insulation
* Typical walling system, 100mm timber
* 150mm roof insulation
* 6mm cavity double glazing throughout
* Ordinary gas boiler with zone controls
* Flame effect gas fire
* 110 litre cylinder with 37mm insulation
* Gas cooker
* Ordinary light bulbs

Fuel use and costs

	Energy required Giga Joules (GJ)	Cost per year at 1995 prices	CO2 emission tonnes per year
Central heating system	60.1	£283	3.5
Fireplace appliance	1.6	£8	0.1
Domestic hot water	19.1	£90	1.1
Cooking	5.4	£25	0.3
Lights and appliances	2.9	£507	4.4
Standing charges	0.0	£86	0.0
Totals	**108.1**	**£999**	**9.5**

Computation by the Energy and Environment Ltd by arrangement with the EEO

Comment: This enables you to make a comparison with the SAP rating for the traditionally built house opposite. However, most people building with a timber frame use 50mm insulation below the floor and 200mm in the roof. This improves their SAP to 92. Top of the market timber frame homes use 150mm timber in the walls, and if used here the additional insulation which this provides further improves the SAP rating to 93.

SAP rating - example three

SAP ratings for the same house built with differing standards of insulation and appliances.

SAP 93

The house on the cover built as example one but with a gas condensing boiler and thermostatic radiator valves.

* 25mm underfloor insulation
* Brick and aerated block walling with 25mm slab insulation in the cavity
* 150mm roof insulation
* 6mm cavity double glazing throughout
* Condensing gas boiler with zone controls
* Flame effect gas fire
* 110 litre cylinder with 37mm insulation
* Gas cooker
* Ordinary light bulbs

Fuel use and costs

	Energy required Giga Joules (GJ)	Cost per year at 1995 prices	CO2 emission tonnes per year
Central heating system	58.5	£276	3.4
Fireplace appliance	1.9	£9	0.1
Domestic hot water	5.3	£72	0.9
Cooking	5.4	£25	0.3
Lights and appliances	22.0	£508	4.5
Standing charges	0.0	£86	0.0
Totals	**103.1**	**£976**	**9.2**

Computation by the Energy and Environment Ltd by arrangement with the EEO

Comment: This shows the difference that a more efficient heating appliance and individual radiator valves can make to overall energy consumption and costs. The comparison is made with example one, but most selfbuilders who install a condensing boiler will want a higher standard of insulation as well, and the results of doing so are shown in example five overleaf. The choice of radiators, or the use of a wet underfloor system makes little difference to energy costs provided they are fitted with appropriate controls, and consideration should be given to the speed of response required to suit your life style.

SAP rating - example four

SAP ratings for the same house built with differing standards of insulation and appliances.

The house on the cover built traditionally as in example one but with an upgraded standard of insulation, still with an ordinary gas boiler.

* **50mm underfloor insulation**
* **Brick and aerated block walling with a 100mm cavity fully filled with insulation**
* **200mm roof insulation**
* **6mm cavity double glazing with K glass**

* **Ordinary gas boiler with zone controls**
* **Flame effect gas fire**
* **110 litre cyclinder with 37mm insulation**
* **Gas cooker**
* **Ordinary light bulbs**

Fuel use and costs

	Energy required Joules (GJ)	Cost per year at 1995 prices	CO2 emission tonnes per year
Central heating system	53.1	£250	3.1
Fireplace appliance	1.5	£7	0.1
Domestic hot water	19.1	£90	1.1
Cooking	5.4	£25	0.3
Lights and appliances	21.9	£506	4.4
Standing charges	0.0	£86	0.0
Totals	**101.0**	**£965**	**9.0**

Computation by the Energy and Environment Ltd by arrangement with the EEO

Comment: This example shows the result of providing a higher level of insulation in the envelope of the house, including Pilkington K glass for the double glazing units. This reflects radiated heat from inside back into the house, but looks just like ordinary glass. With ordinary double glazing units the SAP rating drops to 93.

SAP rating - example five

SAP ratings for the same house built with differing standards of insulation and appliances.

The house on the cover built traditionally with upgraded insulation, a gas condensing boiler, and low energy lights.

* **50 mm underfloor insulation**
* **Brick and aerated block walling with a 100mm cavity fully filled with insulation**
* **200mm roof insulation**
* **6mm cavity double glazing with K glass**
* **Gas condensing boiler with zone controls,**

   and thermostatic radiator valves
* **Flame effect gas fire**
* **110 litre cylinder with 50mm insulation**
* **Gas cooker**
* **Low energy light bulbs**

Fuel use and costs

	Energy required Giga Joules (GJ)	Cost per year at 1995 prices	CO2 emission tonnes per year
Cental heating system	42.6	£200	2.5
Fireplace appliance	1.0	£5	0.1
Domestic hot water	15	£70	0.9
Cooking	5.4	£25	0.3
Lights and appliances	19.1	£441	3.9
Standing charges	0.0	£86	0.0
Totals	**83.1**	**£829**	**7.6**

Computation by the Energy and Environment Ltd by arrangement with the EEO

Comment: These are the features which are recommended by the Department of the Environment's Energy Efficiency Office as appropriate to new homes today, excluding the flame effect fire which is not very energy efficient but which most self-builders will choose on aesthetic grounds. The low energy light bulbs make a saving of £65 on total fuel costs against ordinary bulbs.

SAP rating - example six

SAP ratings for the same house built with differing standards of insulation and appliances.

The house on the cover built traditionally in the country with upgraded insulation, oil fired central heating, a wood fire and electric cooking.

SAP 100

* **50mm underfloor insulation**
* **Brick and aerated block walling with a 100mm cavity fully filled with insulation**
* **200mm roof insulation**
* **6mm cavity double glazing with K glass**
* **Oil fired condensing boiler with zone control**

and thermostatic radiator valves
* **Open log fire**
* **110 litre cylinder with 50mm insulation**
* **Electric cooker**
* **Standard light bulbs**

Fuel use and costs

	Energy required Giga Joules (GJ)	Cost per year at 1995 prices	CO2 emission tonnes per year
Central heating system	43.9	£169	3.5
Fireplace appliance	2.3	£15	0.2
Domestic hot water	15.0	£58	1.2
Cooking	3.1	£71	0.6
Lights and appliances	21.9	£506	4.4
Standing charges	0.0	£46	0.0
Totals	**86.2**	**£860**	**9.9**

Computation by the Energy and Environment Ltd by arrangement with the EEO

Comment: One criticism of the SAP rating system is that because it takes fuel costs into account it favours oil heating. As a result this example, while not having the lowest heating costs and with the highest carbon dioxide figure of all, has the best SAP. Oil prices are at a historic low, and when they rise, which is inevitable, the SAP will fall! However, those living in rural areas usually have no alternative to burning oil. Cutting the logs for their open fire will help them to keep warm.

SAP rating - example seven

SAP ratings for the same house built with differing standards of insulation and appliances.

The house on the cover built with recommended standards of insulation, a gas condensing boiler by a family who decide to experiment with lower than usual temperatures.

SAP 100

* **40mm underfloor insulation**
* **Brick and aerated block with a 100mm cavity fully filled with insulation**
* **200mm roof insulation**
* **6mm cavity double glazing with K glass**
* **A gas condensing boiler with zone controls and thermostatic radiator valves**

* **Flame effect gas fire**
* **110 litre cylinder with 37mm insulation**
* **Gas cooker**
* **Low energy light bulbs**
* *and temperatures 3 degrees lower than usual, ie, 18 degrees C in the living room and 15 degrees C elsewhere.*

Fuel use and costs

	Energy required Giga Joules (GJ)	Cost per year at 1995 prices	CO2 emission tonnes per year
Central heating system	28.1	£132	1.6
Fireplace appliance	1.0	£5	0.1
Domestic hot water	15.0	£70	0.9
Cooking	5.4	£25	0.3
Lights and appliances	19.1	£441	3.9
Standing charges	0.0	£86	0.0
Totals	**68.6**	**£759**	**6.8**

Computation by the Energy and Environment Ltd by arrangement with the EEO

Comment: We heat our homes to much higher temperatures than those which were usual before central heating became common, and for most of us a lower figure would probably be more healthy. However, if you are outdoor people and you decide to try 62F instead of 65F in your lounge, the savings are significant. Set your thermostat to 60F and your annual fuel cost will go down to £713, but this is below the doctors' recommended level for the elderly.

SAP rating - example eight

SAP ratings for the same house built with differing standards of insulation and appliances.

The house built traditionally as an all electric house with electric warm air heating, an all electric kitchen and of course low energy lights.

SAP 100

* 50 film underfloor insulation
* Brick and aerated block walling with a 100mm cavity fully filled with insulation
* 200mm roof insulation
* 6mm cavity double glazing with K glass

* Electric warm air heating using off peak tariff
* No fireplace or fire
* 210 litre cylinder with 50mm insulation
* Electric cooker
* Low energy light bulbs

Fuel use and costs

	Energy required Giga Joules (GJ)	Cost per year at 1995 prices	CO2 emission tonnes per year
Central heating system	49.3*	£401	10.0
Fireplace appliance	11.5	£94	2.3
Domestic hot water	0.7	£18	0.2
Cooking	3.1	£74	0.6
Lights and appliances	15.7+3.9*	£380+£32*	3.2+0.8*
Standing charges	0.0	£60	0.0
Totals	**184.2**	**£1059**	**17.1**

(* off peak tariff)

Computation by the Energy and Environment Ltd by arrangement with the EEO

Comment: The all electric house is widely promoted, and there are many forms of electric heating — storage heaters, underfloor heating, ceiling heating, and electric boilers for use with conventional radiators. There is also electric ducted warm air heating which is analysed here. A dual immersion heater is specified in a very large cylinder, with off peak heating at night and a booster heater for daytime use. The SAP rating (which considers fuel prices) is unfavourable with this standard of insulation, and a much higher standard will have to be specified to conform to Building Regulations.

SAP rating - example nine

SAP ratings for the same house built with differing standards of insulation and appliances.

The house built traditionally with all the recommended energy saving features and with a heat recovery system in the roof and a solar heating system for the domestic hot water on the roof.

SAP 100

* 50mm underfloor insulation
* Brick and aerated block walling with a 100mm cavity fully filled with insulation
* 200m roof insulation
* 6mm cavity double glazing with K glass
* Gas condensing boiler with zone controls

* No fireplace or fire
* 110 litre cylinder with 50mm insulation
* Gas cooker
* Low energy light bulbs
* Heat recovery system
* Hot water solar heating system

Fuel use and costs

	Energy required Giga Joules (GJ)	Cost per year at 1995 prices	CO2 emission tonnes per year
Central heating system	43.0	£203	2.5
Domestic hot water	10.5	£50	0.6
Cooking	5.4	£25	0.3
Lights and appliances	21.3	£492	4.3
Standing charges	0.0	£86	0.0
Totals	**180.2**	**£856**	**7.7**

Computation by the Energy and Environment Ltd by arrangement with the EEO

Comment: The cost saving ascribed to the solar system is £20 per year. Heat recovery systems require that a house is well sealed. The Energy Efficiency Office advise that mechanical ventilation systems with heat recovery may not lead to any improvement in the running cost or rating. This is due to the cost of the electricity used in the fans associated with the system. If the cost of the electricity consumed is greater than the cost of the heat saved then the running costs will increase and the rating will decrease. However, these systems are popular with those concerned with air quality.

Obtaining an energy rating

If you are used to juggling with figures, have plenty of reference books, and want to make energy rating your hobby you can calculate your own ratings and submit them with your plans for the local authority. Disbelievingly they will probably charge you the cost of getting them checked, and it can be good fun. If this interests you, make a start by buying Part L of the Building Regulations.

Most selfbuilders will want their designer to produce a SAP rating and will want him to discuss it with them in detail, demonstrating how the rating can vary with changes to the design, the materials proposed and with different heating systems. It is easy to obtain a rating in this way: there are licensed consultants who provide this service at a nominal fee, and designers usually have established contacts with them. Both the NHBC and Zurich Custombuild offer a service for this, as do most of the leading suppliers of insulating materials such as Thermalite and Tarmac. Some architectural practices are licensed to handle this themselves, probably using one of a number of computer programs which have been approved by the Building Research Establishment.

The need to provide an energy rating with plans submitted for approval under the Building Regulations means that a proposed heating system has to be shown on the drawings, and sometimes a mechanical ventilation system as well. Selfbuilders may not wish to make decisions on these matters at such an early stage, so it is usual to specify notional appliances and fuels that will enable the required SAP figure to be achieved. At a later date revised rating figures are sent along, based on the appliances which will actually be installed and taking into account any other details which will be changed. The local authorities are well used to this.

This is all straight forward: what is more complicated and more interesting is using the system to find out exactly what happens to the rating if you adopt different features — more insulation, a more efficient heating system, an external porch, a conservatory, or a heat recovery system. Either your designer should be able to discuss all this with you and provide the figures you require, or he should put you in touch with a local consultant who can.

All this assumes that this is your particular interest. Do not feel inadequate if it is not: the Building Regulations will ensure that your new home will be several times more energy efficient than the home you grew up in! Also remember that a key element in the heating costs of a home is its pattern of use and the lifestyle of the occupants. A couple who both work, leaving home at 8am and returning at 6pm, have a very different pattern of energy use in their home to a family with very young children. Another factor is the temperature level required. Modern central heating systems are designed to give ambient temperatures of 21 degrees C in living rooms and 18 degrees C in bedrooms. This gives shirt sleeves living. Outdoor types who wear sweaters and set their thermostats 3 degrees C lower will make disproportionately large savings on fuel bills.

Energy saving design

Starting with fundamental design considerations, dwellings can be designed to be energy efficient by ensuring that they make the best use of the sun, and they can also be designed to minimise heat losses. Using the sun involves arranging for the house to be positioned on its plot with as much as possible of the living area and the largest windows facing the South, with the windows on North walls as small as possible. A further development is to have a two storey conservatory or roof height glazed porch to enable the sun to warm up a massive masonry wall called a trombe wall. This helps the whole house to act as a huge storage heater, sometimes using special air circulation arrangements. Such designs are described as being designed for the maximum solar gain, and are examples of passive solar heating. (Active solar heating is a different matter, and will be described later). The limitation of this is that it is assumed that it is convenient for a house to face in one particular direction and that a large glazed area on the front elevation is aesthetically desirable and suits the lifestyle of the occupants. Remember also that net curtains

Milton Keynes Energy Park — A practical demonstration of low energy housing

Many people in this country have a misconception about an energy efficient house, thinking it must be covered with solar panels, with a windmill in the garden, elaborate heat pumps, heat exchangers and heat recovery gadgets. Not so. The experience of Milton Keynes, energy efficiency capital of the UK, shows us that energy efficient homes can be built using existing construction methods and traditional design. All that is required is careful planning and attention to detail, coupled with a good understanding of all the elements of a home that contribute to its energy use.

The following is a brief guide to just a few of the 800 energy efficient homes built in the Milton Keynes Energy Park since it was launched in 1986 at the Energy World Exhibition. Here, you will see a range of examples from large executive style houses to smaller homes to suit different tastes and budgets. Their descriptions highlight the main energy efficiency features of each home but more detailed information is available from the National Energy Foundation which is itself based in Milton Keynes. All the dwellings in the Energy Park have low energy lighting and carefully chosen electrical appliances.

You can go to see the Energy Park at Milton Keynes, where many of the more interesting houses have a notice board at the pavement edge giving details of their rating, etc. However, the houses are privately owned, and the privacy of the occupants must be respected. They are very proud of their houses, but do not expect to be treated like goldfish in a bowl!

The best guide to the houses is the NEF leaflet *Milton Keynes - Energy Capital UK*, which is usually available at the Milton Keynes Information Centre. However, as the Centre is not open at weekends, it is a good idea to write to the NEF before making a visit, enclosing a large stamped addressed envelope, asking them to send their current package of information for the general public including the Energy Capital UK leaflet.

Swedish/British Collaboration
National Home Energy Rating 10.0

The technology used in achieving the outstanding energy efficiency levels in this house is all tried and proven in either this country or Sweden. The principles demonstrated here are a high degree of insulation and air tightness coupled with a very effective ventilation system with heat recovery, double glazing throughout and draught lobbies to front and back doors ensuring a low heat requirement.

The air changes in the house are via a controlled ducted ventilation system. This extracts the air from the moist areas of the home and transfers the heat to the fresh air coming into the drier rooms. Heating is from a wall mounted balanced flue gas boiler with zone control. Domestic hot water can be heated either by the boiler or an electric immersion heater. The hot water tank has 100mm of foam insulation.

Old style modern build
National Home Energy Rating 9.3

Although traditionally styled, this 4 bedroom house was built using modern construction methods which give a high level of energy efficiency.

The timber frame structure incorporates fibre-glass wall insulation and a block fill insulated suspended flooring system. A balanced flue gas-fired boiler with boiler manager and individual thermostatic radiator valves ensure even greater efficiency. All windows and patio doors are low-emissivity double glazed (equivalent to triple-glazing) with a conservatory running the full length of the living room. This minimises heat loss whilst the conservatory provides additional space heating through passive solar gain.

A mini heat recovery system retrieves heat from the kitchen, utility and bathrooms, redistributing conditioned air to all living rooms.

All electric

National Home Energy Rating 7 5

This group of homes is a development of all electric low energy detached homes. Using off-peak electricity, electric boilers heat a central heat store overnight. When heat is needed the store gives up its energy into a conventional wet radiator system.

Other energy efficiency features include high levels of insulation throughout, low emissivity double glazing and orientation of the houses to maximise solar gain.

Passive solar orientation

National Home Energy Rating 10.0

This unusual 3 bedroom bungalow was designed primarily to demonstrate the potential energy efficiency of glazing systems. The key element of the design is the fully-glazed and draught-proofed south-facing wall which maximises passive solar gain.

The glazing uses argon-filled triple glazed sealed units, with an additional outer pane of low emissivity glass which has the equivalent effect of quadruple glazing. An automatic blind system is incorporated into the system to avoid overheating in summer.

Heating is by a high efficiency gas condensing boiler with separate zone control, boiler manager and thermostatic radiator valves. Very high levels of insulation to roof, walls and floors, together with a whole house heat recovery and ventilation system to avoid condensation, complete the main energy saving features.

Zero extra over cost

National Home Energy Rating 9.0

Built for shared ownership, this group of homes demonstrates that houses can be built to a high level of energy efficiency at no extra cost compared to that of building conventional homes.

There are various ways in which this can be done but this particular site has achieved an NHER of 9.0 by the 'room in the roof' technique. This saves a considerable area of cavity wall which is expensive to build. The resulting saving offsets the costs of increased insulation installed throughout.

Modified traditional construction

National Home Energy Rating 10.0

All the major living rooms in this home have a south east or south west aspect while the service rooms (kitchen, utility, cloakroom) are positioned on the north side providing a buffer against heat loss. Energy saving measures have been introduced whilst maintaining traditional construction methods.

The heating system consists of a high efficiency condensing gas boiler connected to thermostatically controlled radiators with pre-insulated pipework. Using an outside temperature sensor the boiler controller calculates optimum heating periods. Hot water is provided from a solar heated system integrated into the garage roof.

The most striking external feature of the house is the octagonal conservatory which acts as a sun trap and forms an integral part of the ventilation system. Heated air can be drawn off into the living area by ducts and a fan located at the top of the conservatory. Air from the conservatory is also passed through the heat exchange system where it absorbs up to two-thirds of the outgoing heat drawn from the 'wet' areas of the house.

eliminate the advantages of passive solar design, so you must be happy with large unobstructed windows.

Designing to avoid heat losses involves making the structure as compact as possible to give the best possible ratio between internal volume and external surface, arranging the room layout to minimise heat loss, especially by having chimney breasts on inside walls, and providing porches at front and back doors as draught lobbies. Design features like this can be very effective, although they are often overlooked as most people immediately think only of insulation when considering what they can do to reduce heating costs.

Energy saving materials

Before the introduction of energy ratings the insulating qualities of the various materials used in the shell of a building were considered very carefully in isolation, but now it is the contribution that they make to the overall rating that is important. However, there are two areas where selfbuilders often seek guidance, and this is with glazing and cavity wall insulation, both of which are advertised very widely in sensational terms. Triple glazing only makes a significant contribution to energy efficiency in modern construction when used for very large glazed areas. All double glazing sealed units are much the same, and increasing the width of the cavity has little overall effect on the heat loss figures for the whole dwelling, although it may provide superior sound insulation and have other advantages. The best window insulation of all is provided by continental double windows with two separate casements, but they are not generally popular here. New types of glass, like Pilkington's K Glass, give double glazing units the performance of triple glazing, and they are gaining in popularity and worth considering.

Advertising for cavity insulation overlooks the fact that the cavity was put into the wall for a purpose, which was to keep the weather out however badly a particular joint may have been pointed, or however porous a bad brick may be. Early cavity insulation based on foams which filled the cavity had a reputation for causing

damp patches. As a result, local authorities in many areas have regulations to control their use in exposed situations. Modern practice is to use blown mineral fibre or polystyrene beads. The NHBC accepts fully filled cavity walls for most areas in the UK — see the *NHBC Good Practice Guide; Thermal Insulation and Ventilation* and the map contained in it. However, if you are building a new wall the same level of insulation can be built into it while still keeping the cavity by using a slab system. This involves using special tie-irons in the wall to hold a slab of insulating material firmly against the inner skin, maintaining the cavity between it and the outer skin. This gives the best of both worlds, although the advocates of fully filled cavity systems claim that their approach ensures there are no gaps in the insulation. In practice both systems are only as good as the care taken by the workmen involved.

All dwellings with a high level of insulation and effective draught proofing could have a built-in condensation problem. To avoid this the roof space must be adequately ventilated, extractor fans must be installed in kitchens and bath-rooms, and this is dealt with at length in the Building Regulations. The NHBC publish an excellent booklet on ventilation and condensation, which all registered builders are supposed to give to their customers. If you have very deep pocket you can make a virtue of this problem by install-ing a whole house ventilation system, with ducts to extract warm stale air, and heat exchangers to save the energy that is locked up in it.

Energy saving heating systems

New and better ways of putting heat into a building are either more efficient ways of using conventional fuels, or the use of alternative fuels, or solar collection systems, or heat pumps. All are on the market. How do you choose between them?

Starting with better ways of using coal, oil or gas, there are new appliances of all sorts which have in common that they put more of the energy from the fuel into the house as heat and less of it up the chimney. The most widely promoted at present are condensing boilers and using them

Energy Efficiency Office
DEPARTMENT OF THE ENVIRONMENT

will boost your SAP rating, but only as part of a professionally designed system with a modern automatic control system.

New control systems are being developed all the time. Some like zone heating arrangements are mandatory under the Building Regulations. Others monitor outside temperatures or use computer chips to learn the pattern of heat demand in a house and set the boiler controls accordingly. One such, the Dataterm by Warmworld, has been installed by one local authority in 195 old peoples' bungalows to replace standard controls and is giving 3 to 4 year payback savings, prompting the author to install one at his own home where it does all that is claimed for it and more. This sort of sophisticated controller costs the same whether it is for a small house or a large one, giving bigger savings and a shorter payback in the larger property. Thermostatic radiator valves (TVRs) are also valuable, and are not expensive. Controls of any sort are the cheapest part of a heating installation but rarely receive the consideration that they deserve, but they must be user friendly to use so that they are used.

In all of this it is important to recognise that there are fashions in heating systems and new ways of heating a house appear all the time. In the sixties and early seventies the fashion was for ducted warm air heating, which is now hardly ever seen. Later came ceiling heating which was unpopular with bald men but promoted widely in its day. The current vogue for underfloor heating has been here before in the sixties. The choices between heating systems are rarely made on wholly rational grounds, and for most people a fireplace has an appeal which has nothing to do with energy efficiency. An open fire, or a modern smokeless appliance in a fireplace, is usually valued as a key feature in the decor of a living room, and may have a deeper psychological appeal. Our cavemen ancestors sat around fires, and we still seem to appreciate a point source of heat more than ceiling or underfloor heating arrangements. The choice in all of this is yours.

Non-fossil fuel systems
Moving to non fossil fuels, (electricity being

While a decision on what to spend on energy saving beyond the Building Regulations requirements is yours, the Energy Efficiency Office currently promotes the following as worthwhile features for any house with more than two bedrooms
* Fully filled cavities in masonry construction
* 140mm insulation between timber studs in timber frame construction
* 200 mm insulation in lofts
* 50-100mm of insulation under floor slabs
* Condensing boilers (gas or oil), or efficient solid fuel boilers.
* A Thermostat situated in a room without TVRs
* TVRs to radiators in all other rooms
* Easy to use programmers
* Careful attention to construction details

considered a fossil fuel while it derives from oil or gas burned in power stations), there is a fashion for new homes in the country to be heated from renewable sources — meaning logs or straw bales. Burning logs has a mystique of its own, and there are those who happily spend every Sunday with a chainsaw cutting up their fuel for the following week. As a spare time activity this is even more hazardous than jogging, and the benefits are equally personal and intangible. Remember that either the supply of timber or your enthusiasm may not last, and it is important that a log burning fire or stove is a multi-fuel appliance so that you can burn coal as well.

Straw is only available to farmers, but new farmhouses are still being built in these hard times for agriculture, and boilers to heat them with straw are a practical proposition. They are efficient, reliable, ugly, dirty and require an outbuilding of their own with a small yard to store the straw bales.

Solar collection systems
Solar collection systems, sometimes called active solar heating, involve liquid filled solar panels on the roof and in this country they can save up to 60% of the annual cost of heating domestic hot water. Reliable installations cost at least £2000

for an ordinary home and are unlikely to save more than £60 a year. They are generally not cost effective unless you use them to warm a swimming pool or have a high summer demand for hot water, for instance if you are a seaside landlady. The panels can be so ugly that they detract from the value of the home unless they can be situated on a south facing roof that is not seen from the ground. If they interest you, the Solar Trades Association publishes an excellent guide to all of this called *Solar Energy — What's in it for you!* which you can obtain by phoning 01208 873518.

Heat pumps

Heat pumps are not new. Heat transfer technology has been with us since the 1890's, and their day came from 1972 to 1982 when oil prices were rising at a rate which made the savings offered by heat pumps seem attractive. A few thousand were installed in large houses in Britain and many tens of thousands on the continent. By and large they have proved reliable and gave the savings promised by the experts, although not the super savings promised by some salesmen. However, oil prices have since fallen and heat pumps have lost their advantage. When oil prices rise they will be back again.

Heat pumps do not 'make heat' by burning anything and instead collect heat from a convenient source and deliver it to where it is wanted.

Above: Solar panels on a south facing roof slope. The total area is 7 square metres and provides 60% of the energy required to heat the domestic hot water in a large house.

This is not easy to understand but it helps if we realise that a domestic refrigerator is a heat pump working in reverse. It collects heat from inside the cabinet and dumps it to waste via the coil at the back. A heat pump collects heat from somewhere outside the house and pumps it into the building. Almost all heat pumps take their heat from the air and work however cold the air is. If the air is at 8 degrees C when it is sucked into the heat pump, it will leave it seconds later at 4 degrees C and the heat extracted will end up as hot water in the radiators in the house.

The appeal of the heat pump lies in the fact that it is more economical to use electricity to collect heat and move it about in this way, than

Left: The author with his heat pump. The wooden building behind contains a heat store. At present fossil fuel prices this sort of technology is not generally cost effective on a small scale, but its day will come.

simply to use electricity to generate heat. Heat and electricity are both forms of energy which are measured in kilowatts. Using 1kW of electricity in an electric fire generates 1kW of heat. Using the same 1kW of energy in a heat pump enables it to collect between 2kW and 3kW of heat and shove it into the radiators. This is a marvellous trick, but it has its price, which is the high capital cost of the machine.

Small heat pumps are sometimes incorporated in modern ventilation systems which reuse the waste heat that ends up in the roof, and are also a feature of full air conditioning systems. The SAP rating will indicate the advantages to be gained from this. Such installations are normally only made in large houses.

Lights and appliances

The major element in all the examples of projected energy costs on pages 243 to 246 is the cost of electricity used for lighting and appliances, and a 20% reduction in usage here is a significant economy. This is outside the scope of this book, but the obvious first step is to install low energy lights. Although they currently restrict your choice of light fittings they are slowly becoming smaller and more elegant. Some washing machines and dishwashers are now promoted as having a low electricity consumption, and there are plenty of automatic switching arrangements available. It all makes sense.

Further information on energy saving

A final word about exhibitions and specialist organisations. If you want to learn more about energy efficiency, you will be able to divide those who are offering help and advice into three categories: those who want to sell you something, the wild enthusiasts, and the scientists. The first two categories are very good at promoting their own particular interests, but naturally will emphasise the advantages of whatever they are pushing, and will tend to gloss over the disadvantages. The scientists take a bigger view, and their views are most effectively promoted through the Building Research Establishment or the National Energy Foundation, which is based at Milton Keynes conveniently close to the Energy Park.*

If this chapter on energy saving appears to be somewhat cynical, this will serve to balance the over-enthusiasm with which the subject is usually approached. The more important a topic is, the more important it is that it should be considered realistically. Energy saving is very important and merits very careful consideration, not simple enthusiasm. Incidentally, the author lived for 15 years in a house with concealed solar panels linked to a heat pump, which was itself linked to a heat store, and it was most effective and usually reliable. The controls involved over thirty switches and dials, but this was mainly so that the performance of the system could be monitored. It rarely required adjustment but this was invariably when its creator was away from home. It was not popular with his wife. But then the author is an energy conservation enthusiast! Which brings us back to where we started: decide why you are making your choices!

* The Energy World Exhibition at Milton Keynes in 1990 involved 40 different homes all built on the same site by many different developers and organisations to demonstrate every conceivable approach to energy saving in domestic construction. At the end of the exhibition the houses were sold and they are now occupied, but can be viewed from the road. Details of them are available from the National Energy Foundation. It is useful to see them if you want to consider the design implications of some energy saving features. Each house has a notice board at the pavement edge giving details of its special features.

Case histories

In each of the fourteen previous editions of Building Your Own Home there have been twenty four case histories of completed self build projects. However, as this edition is being published the monthly selfbuild magazines both feature three or four case histories in every edition, making these stories of selfbuild as it really is very widely available on a month by month basis. For this reason, and because the main text in this book grows in each succeeding edition, there are only ten case histories here — plus of course the Spooner story which started on page 179.

These selfbuild stories demonstrate the many different ways in which people build their own homes, but all have in common that those concerned had to provide all the impetus themselves. They had to make it all happen. This is the essential difference between buying a home off the peg and building for yourself. If you want to buy a house that is already there a whole mafia of estate agents, surveyors, bank managers, building societies and solicitors will make everything happen for you. The families featured on the pages that follow wanted to do better than this, and to have a home to be uniquely their own. This is how they made it happen.

Index of case histories

A selfbuilder of the year

For most selfbuilders their new home is its own reward, but for one couple every year there is an unplanned bonus — someone has to be Selfbuilder of the Year. In 1994 it was Ron and Janet Cox, who received their award at a black tie dinner at the Inn on the Park in September 1994 together with 200 luminaries of the selfbuild industry who were celebrating the most successful ever selfbuild exhibition at Alexandra Palace. The prize was a trip to the USA which they enjoyed to the full in 1995, but their real prize is surely the new home in a Kent village which is a text book example of a successful selfbuild project.

In 1992 Ron was a transport manager with the GPO who was anticipating early retirement after 40 years service and Jan worked in the county council offices. Early retirement offered a challenge, and they had long before decided what it would be — building their dream home. Ron had been trained as a carpenter during his National Service many years before, and was a DIY enthusiast. Ten years earlier he had been looking in the library for a book to help him plan some home improvements when he had found the 7th edition of *Building Your Own Home*. As a

guide to home improvements it was useless, but is sowed the seed of a determination to build a dream home when the opportunity arose. Ten years later early retirement provided the opportunity.

There were various factors to be taken into consideration: they wanted to live near to their family roots in the Medway towns, but wanted to find somewhere to build right in the country. They wanted to do as much of the work themselves as they could, and they wanted to avoid any sort of mortgage. This meant that the budget limit would be the equity in their existing home plus Ron's retirement gratuity, and this would involve living in a caravan on the site between selling their existing home and moving into the new one.

A year before they anticipated that Ron would leave the GPO they started their preparations for the great undertaking in earnest, collecting information, visiting exhibitions and looking for land. At first they concentrated on plots advertised as being for selfbuilders, signed up with the selfbuild land finding agencies, and spent weekends visiting and videoing plots that they were told about. They found none that they liked

Ron and Janet Cox with
Peter Griffiths.

Selfbuilders of the Year!

locally, and were considering Lincolnshire as a low cost option when an estate agent told them of a corner plot in a village on the edge of the Downs. It had roads on two sides, and because of this they were not particularly interested but the agent pressed them to go to look at it. They quickly found that the traffic flow along the roads was measured in cars per week rather than cars per hour, and that it had all the features that they wanted. It had been for sale for a year at £80,000, and the owners were very anxious for offers as they wanted to complete their own purchase of a property. By then Ron had retired, and his gratuity was sitting in the bank. £45,000 with immediate settlement was offered, and accepted. The plot was theirs and the old home was put on the market.

For plans and help with the planning and Building Regulation applications they looked to Peter Griffiths, an old friend who was a chartered building surveyor. There was existing detailed planning consent for a house which was definitely NOT what they wanted, particularly as it did not take advantage of the sun. They prepared sketches, they were amended until no one could think of any more changes to be made, and quickly approved by the planners. The Building Regulations application resulted in the usual list of queries but nothing that presented any difficulties, and trench fill foundations down to the underlying chalk were approved.

Dealing with the services authorities was not so straightforward. The water board wanted £2,100 for a connection and a few yards of pipe, and could not be budged*. Even more seriously, it seemed that the Rivers Authority might require a cesspit as there was some doubt about whether there would be room for an appropriate area of herringbone drainage from a septic tank. Peter Griffiths proposed a special soakaway made from concrete rings filled with shingle as an alternative, and after conducting percolation tests this was agreed for use in conjunction with a Klargester septic tank.

By the end of April the old house was sold, a

* These high prices charged in 1994/95 have now been curbed by new legislation.

new access made into the site, and a new mobile home was bought in a sale at a bargain price and moved onto the site. Ron and Jan moved in, and immediately there was the worst thunderstorm the area had experienced for 50 years, with hailstones the size of marbles playing a deafening tattoo on the caravan roof. Was it a welcome or a warning? Happily it turned out to be the former.

Work started immediately with a hired JCB fitted with a narrow bucket digging the trenches for the special foundations. Ron involved himself in the work of every trade, and was quickly joined by Jan who had resigned from her job to play her part full-time. The garage was built first to house the family furniture which was then in store, and then work continued on the house. As no borrowed finance was involved there was no pressure from interest charges, which enabled Ron and Jan to take things at their own pace, involving themselves fully in everything that was going on with some of the sub-contractors hired to advise and help them in their work. The first bricklayer had to leave the work after injuring his back, but by then Ron had the confidence to carry on himself with another bricklayer coming in for the difficult parts of the walling for just two days a week. He went on to do all the carpentry, including raising the roof, piped out for the plumbing and central heating, and installed the electrics, accepting help and advice from professionals at every stage. The electricity authority charged £40 to check his amateur work before connecting up, and it passed every test!

Jan helped to board the ceilings, supporting the heavy plasterboards with a broom in a way familiar to many hands-on selfbuilders. This is a common cause of matrimonial disharmony between those who are building for themselves, but in this case there was not a word of complaint! She also handled all the paperwork, made a home of the caravan and swept out the new house every evening so that they had the cleanest building under construction that the professionals had ever seen.

Ten months after starting they suddenly received an offer for the caravan that was too good to miss, and moved into two rooms that were just habitable. The grandchildren were due to arrive for a visit six weeks later, and by working the clock round somehow everything was ready for them. This was just ten months after the job was started. The costs were right on target at the extraordinary low figure of £28 a square foot.

Long term subscribers to *Build It* magazine,

Ron and Jan decided to enter for the Self-builder of the Year competition and sent along their diary of events and some photographs. Six months later they were collecting their award at the Inn on the Park, receiving a splendid plaque for their mantelpiece and a trip to the USA which was particularly welcome as they have grandchildren in America.

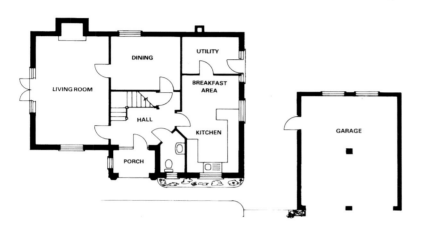

Costings and quantities

LAND	£45,000
PRE-BUILD COSTS:	
Drawings and surveyor fees:	£1,400
Site insurance:	£560
Planning/Building Regulations	£195
National Rivers Authority fees for sewerage certificate	£84
Water, electricity, telephone	£2,681
TOTAL	£4,920
BUILDING COSTS: LABOUR	
Bricklayers	£7,110
Plasterers, screeding	£2,070
Carpenter	£410
Plumbing/Electrical	£1,020
Other labour	£455
TOTAL	£11,065
SUPPLY AND FIX:	
Roof tiling (13,000 reclaimed clay tiles)	
For house	£4,181
For garage	£1,563
VAT	NIL
TOTAL	£5,744
MATERIALS:	
Ready mixed concrete for oversite and footings (39.5-3m)	£1,635
Bricks (Ryash footings, 5,000)	£646

Bricks (Ibstock multi stocks, reds, 15,650)	£3,290
Celcon blocks (five, four and three ins, 335 sq metres)	£1,860
Cement (140 bags)	£646
Sand (52.5 tons)	£652
Ballast (9.05 tons)	£216
Lintels and arch formers (35 plus eight arches)	£1,100
Door and window frames (22)	£1,936
Doors, including garage doors	£1,840
Stairs (softwood)	£468
Double glazing and other glass	£1,082
Carcassing timber for roof and floor joists	£2,528
Window boards, architraves, skirting in MDF	£968
Reclaimed oak beam treated for fireplace	£104
Plasterboard (94 sheets)	£316
Septic tank and drainage materials	£1,766
Wood burning stove and flue linings	£1,339
Electrics	£1,000
Central heating system, including Combi boiler	£1,867
Kitchen and utility room, accessories and pipework	£4,046
Two bathroom suites, cloakroom, including pipework	£2,441

Burglar alarm	£429
Door furniture and locks	£295
Plaster (1.75 tons)	£181
Flooring (first floor, 76 8x2 tongue and groove boards)	£415
Loft insulation (75 square metres, 6 inches)	£142
Paving blocks (26 square metres)	£472
Emulsion paint (100 litres)	£136
Ceramic tiles (floor and wall)	£821
Other miscellaneous items	£4,555
TOTAL MATERIALS INC VAT	£39,192
LESS reclaimed VAT on materials	5,277
TOTAL	£33,915
MACHINERY AND PLANT COSTS:	
JCB with driver hire	£414
Lorry hire for moving soil, etc	£764
Small tool hire	£138
Scaffold hire	£550
Total hire costs	£1,866
ACTUAL BUILD COST:	£52,590
TOTAL COST OF PROJECT:	£102,510
Total floor area of house and detached garage 1,867 square feet	
Construction cost per sq ft	£28

Jan with a time capsule that she built into the chimney breast for the benefit of posterity. This is becoming quite popular, and it is fun to list appropriate contents that will be of interest to the finder in the year 2500. Jan and Ron chose:

Photo of the house being built and the budget.

Photo of themselves, their families and the family tree.

Daily papers, and sets of postage stamps and coins.

Seeds of various sorts which might germinate in the distant future.

Pictures and contemporary details of the Royal Family.

Topping out. The chimney pots were moved from the old house.

The house on the cover

The author first met Kevin Choules at a selfbuild seminar at Alexandra Palace, where Kevin kept an audience of 200 enthralled with his story of how he built for himself, and in particular the account of how he found and bought his building plot. There can be few selfbuilders who went to such pains to find the land on which to build.

It all started when a friend from Scotland visited Kevin and his wife Christine at their London home. They were thinking of moving, and their friend had just built his own new home. As selfbuilders do, he talked about it with great enthusiasm and suggested that they should go to see the Potton show houses in Bedfordshire which, he promised, would result in them stopping looking for a new house and instead starting to look for a building plot. This is exactly what happened!

Kevin is a detective sergeant in the Metropolitan Police, and understands the importance of a planned and methodical approach to any undertaking. They bought all the books, and subscribed to all the magazines, collected piles of brochures, and when they felt they knew

something about it all, they made the firm decision to build for themselves. They decided to do three things simultaneously: to sell their existing home and to move into temporary accommodation, to find a plot, and to carry on researching the features that they wanted for the house they would build. A good start was made when a cash buyer was found for their house which released £90,000 equity to buy a plot, and the family moved in with Kevin's mother. Christine is a teacher, and the Woolwich Building Society offered a loan based on their joint salaries which enabled them to set an overall budget of £225,000 — enough for a very nice house off the peg, but giving the prospect of a really splendid selfbuilt home if they could find somewhere to build it. Unfortunately it had to be in Surrey, which is the most difficult part of Britain in which to find a building plot.

A start was made in all the obvious ways. They subscribed to all the land registers, they approached the estate departments of all the Surrey district councils, they got their name down with every estate agent and rang them all twice a week. They advertised regularly in all

Kevin and Christine Choules with their children.

the local papers and put notices in shop windows. They wrote to every builder, to every developer and every architect in the yellow pages. This produced enough leads to keep them busy looking at potential plots in every spare minute, but few of them were right, and the ones that were could not be bought at a sensible figure. Most building land that does come on the market south of London is bought privately, and they soon realised that they were outsiders in a closed world of builders and developers. Twice they were encouraged to think they had a deal and then found the vendor was being less than straight forward. Once they were definitely used as pawns to beat up the price on a plot already promised to another, costing them £450 in surveyor's and solicitor's fees. It was all very discouraging.

They kept going, but Christine felt that this could not go on forever and began to talk about looking for the right house instead of the right plot. Kevin kept plugging away, and in desperation promised all his friends that he would give a thousand pounds to anyone who told him about a plot that he liked and which he would be able to buy. News of this offer spread to the local pub, and only two days later he got a phone call out of the blue about a house for sale with OPP as part of a large garden. The consent was due to expire and the owner might sell the plot separately for a quick sale, especially as he was wanting to move abroad.

It all turned out to be genuine. Kevin signed a contract subject to the planning consent being renewed and the phone caller got a

thousand pounds.

The existing consent was for a house 'of around 2,500 sq. ft.' with its own access, and Kevin got it renewed but with one unfortunate changed condition, which was that access should be shared with a neighbour. He bought the land on this basis, confident that the access shown in the original consent would be reinstated on appeal. It was.

All the energy that had been directed to finding the land was then applied to deciding on a design, and every spare minute was spent inspecting show houses on up-market estate developments, taking photos and cutting out pictures from magazines. Chris wrote down the ideal size for every room, what the outlook should be, the time that the sun should come in through the window, and all the special features that it should have. It was a good start, and Kevin drew it up. It came to nearly 5000 sq.ft! Two months later, after dozens of redraws, the plan for a 2515 sq. ft. of house, plus a conservatory, plus a detached garage was professionally drawn up and submitted to the council.

The design concept was well received, but the planners wanted it turned round so that the chimney was further from a line of large trees on the boundary which had preservation orders. Arranging for this was not simply a matter of redrawing the design to the other hand: it involved completely redesigning the interior layout and adding 200 sq. ft. to the overall size. Given the choice of the delay involved in another appeal or making the alterations, Ken and Christine decided to bite the bullet, change everything round and get cracking.

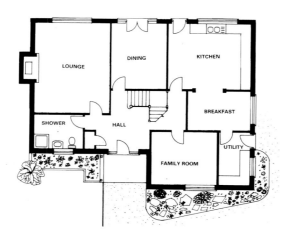

An early decision had been made to build timber frame, and tenders were invited from companies who made frames to clients' requirements. The most attractive quotation had been obtained from Herefordshire Timber Frame. Quite out of the blue the company also offered to build the whole thing, and suggested a price which was so attractive that the family travelled to Hereford to meet the proprietor, Mike Jones, and to see other houses which he had built. His quotation was so low by London standards that it just had to be taken seriously, especially when he said the extra 200 sq.ft. that the planners required would not cost them anything extra. He had recently built in Kent using his own west country tradesmen for the shell and local sub contractors to finish the building, giving the whole job his personal supervision. It sounded right, everything checked out, and a contract was signed.

The construction period was ten months, which Christine in particular found an interminable time. She visited the site every day, where the workmen were living in caravans. Except for trouble with one sub contractor everything went splendidly, and the finished house has all the features which were admired in the expensive show houses that the family had visited, and the materials are far superior. All the timber in the house is english oak. But, nice as it is, there are other selfbuild houses that are equally well built. However, there can be very few where such dogged perseverance in looking for the right plot has been so splendidly justified.

The drawing room.

A Tudor home in Herefordshire

Hillary, Christopher and Ron Stokes with their new home.

The lovely house in the photographs is the result of three selfbuild ventures over 25 years by Ron and Hillary Stokes, who have made building their homes their hobby. Ron drives Bulmers Cider delivery vehicles all over the country, and knows more of regional design trends, and what is being built these days in every part of the country, than many architects. Hillary now uses her expertise as a sales negotiator for a housing developer. It all started when they bought a plot in the early 70s for £4,100 and built a bungalow on it for £10,000. Four years later they doubled their money by selling it for £28,000, and bought a derelict cottage for £15,000. They spent £10,000 on it, lived in it for 12 years, and made it so attractive that in 1989 they sold it for a £185,000. This was reinvested in a three acre small holding with a cottage and planning consent for a further dwelling. This was where they intended to set about building their ultimate home, and they knew exactly what they wanted.

The Stokes had always lived near Kingsland in Herefordshire where the unique Border Oak homes are made, and building one had always been an ambition. They are distinctive and unusual and Ron and Hillary not only knew all

about them, but had friends, working for the company. However, they had first to consider the outline planning permission, which was a conditional consent that had a standard 'agricultural use' clause restricting occupancy of the building to someone working on the land, retired from such work, or the widow of such a person. This type of consent is intended to prevent

The Border Oak frame being assembled.

Part of the hall showing the effective use of polished stone flooring.

outsiders taking over rural homes, but can sometimes be varied in favour of others who, while not strictly engaged in farming, are part of a rural community. This is a good definition of the Stokes family, and they got detailed planning permission for their Border Oak home with relatively little difficulty.

They then decided to add to it a replica of a traditional Herefordshire oast house to provide a teenager flat for their son Christopher, who promised to swop with his parents one day when he takes over the main part of the house! This took further negotiation with the planners, but after some concessions over the design all the planning and building regulation documentation was safely tucked away with the title deeds.

Thanks to the margins from trading up from previous selfbuilt homes only a small mortgage was required, and was quickly arranged. Border Oak found an architect for the progress certificates which the building society required, and quoted £56,000 for the foundations, erecting the frame, fixing the infill panels and tiling the roof. This did not include the windows, which were made in hardwood to a very high standard by a local joiner and fitted with traditional leaded

lights. Local tradesmen were employed for plumbing, electrics and the gas fired central heating installation with its radiators. Special plastering was required between the exposed beams, and a plasterer who worked on ceilings with stilts was found as some of the rooms have exposed rafters. This is unusual in the UK although often found abroad, and saves internal scaffolding.

Rob and Hillary were living in the cottage on the holding while the house was being built, and did a lot of the work themselves. Among the jobs they tackled were laying the stone slabs, building the hearth for the wood burning stove, waxing the exposed beams — which was a very demanding job — staining all the joinery, and doing all the wall tiling and decorating. A feature of building a 16th century home is that the ceiling boards are fixed between the ceiling beams from above, before the first floor boarding is fixed, providing a natural trap for all the debris created by the other work. Hillary was determined to get completely rid of it, so an industrial vacuum cleaner was hired and used to clean out the voids every evening.

The final stage was the oast house, which is

linked to the main house with a study. At the time that oak framed homes were originally built the Englishman drank ale. Bitter beer with the hops to make it and the Oasts to dry the hops were still in the future. The Oast house was therefore built in masonry in the proper 19th century manner, although modern Velux windows were set in the sloping roof. Small dormer windows would have been preferred, but the planners insisted in keeping the profile of a windowless 19th century oast. When the building is viewed from a distance he is probably right, but viewed close up he was certainly wrong — though this is to debate the fine details of a building that fits splendidly into its setting, and which is greatly admired locally.

Only the quarter acre surrounding the house is being landscaped, leaving the balance as a paddock with post and rail fencing. The grazing is let to a livery stable, so that horses are always part of the view, completing the total effect of what must be the ultimate self build story.

It is inevitable that modern community selfbuild has developed as an urban activity, meeting the acute and much publicised housing and social problems of large towns. It has largely ignored the fact that rural areas have equally acute low cost housing problems, not only for the unemployed but also for wage earners who find that they are priced out of the housing market by the combination of planning restrictions in villages and the influx of relatively wealthy retired people and commuters. The adverse effect of this on rural society is now widely appreciated, and the need for low cost homes for local people in villages is well established. On the continent this situation is dealt with in many countries by planning legislation which favours those with local connections, and by heavy local property taxes on incomers. In Britain our concept of democracy made this favoured treatment of genuine local people unacceptable until recently, when the first signs of a pragmatic approach arrived with the issue of planning policy guideline paper PPG3 which has now been adopted by the majority of district councils in areas where it is relevant.

In simple terms it permits the building of low cost homes outside the 'development area envelope' in villages provided that arrangements are made to restrict ownership of the houses to

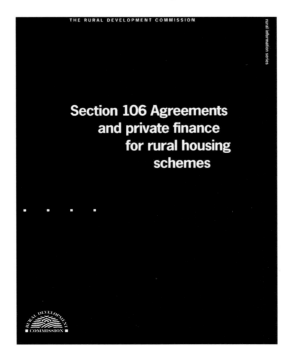

THE RURAL DEVELOPMENT COMMISSION

rural information series

Section 106 Agreements and private finance for rural housing schemes

RURAL DEVELOPMENT COMMISSION

Details of Section 106 Agreements are available from the Rural Development Commission.

local people in order to satisfy a demand for low cost homes and to secure the viability of the local community. A condition is that arrangements must be made to ensure the accommodation is affordable in perpetuity.

If such 'exception sites' outside the village envelope can be found, and can satisfy other planning criteria, the owner of the site may be willing to sell to a selfbuild housing association at a price below normal residential building land price but still in excess of agricultural land price. Combined with the economies of selfbuild this can produce homes that can cost up to 40% less than their open market value. Arrangements can then be made for the house to be affordable in perpetuity by what is called a Section 106 Agreement, requiring that the title contains covenants restricting the resale to local people at a discount of 20% to value. This has the effect of keeping the homes affordable in the future, which is the basis for the planning consent where approval would not normally be granted.

The Alpine Selfbuild Housing Association at Mildenhall in Wiltshire is an example of a scheme of this sort which was started spontaneously by two of its members in 1989. They looked for possible exception sites and found one belonging to a local farmer who was the employer of one of the members. With the support of both the Parish Council and the Kennet District Council, qualified planning consent was granted and a purchase negotiated at £19,500 for land for six plots. A Section 106 Agreement was drawn up to limit subsequent sale to certain local families at 80% of the market value at the time of the sale.

At this point the district council advised the association to find a consultant to assist them to take everything forward, and Wadsworth Landmark Limited of Bristol were appointed as scheme managers. Wadsworths have thirty years experience of providing services to selfbuild housing associations, and provided comprehensive management services on their standard contract terms.

Planning consent was obtained for three different house types, of which one with three bedrooms was budgeted to be built at a cost of

Members of the Alpine Selfbuild Association in their site hut, which after fifteen months or working on the site had begun to feel like home.

Alpine Association members and the houses they have built to a budget of £39,690.

Association member Arthur Cook working on his new home.

£39,690 with an anticipated value of £76,500. The plan is shown below. The total budget was only £226,817 to build six new houses, all destined to be sold by the association to local wage earners in steady employment who had pre-arranged mortgages. The social value of this was obvious, but it proved extremely difficult to obtain the necessary loan finance. Eventually the Housing Corporation provided a loan of £97,540, the Kennet District Council loaned £40,000 and the members themselves raised £90,000 in individual loans of different amounts. It must be emphasised that these were loans, to be repaid when the houses were finished using the members' mortgages to buy their own homes from the association.

Work started in March 1994 and was approaching completion when the photograph was taken in June 1995, when the members were on programme for all of them to move in before the end of the year. By any standards it is a remarkable success.

This sort of arrangement is only one of a large number of approaches to affordable rural housing. It is complex, and requires a level of co-operation between those initiating the schemes, local authorities and lenders which is more evident in policy documents than in real life. The example of Alpine may prove to be very important in getting projects like this the practical support that they require, and the achievements of the members deserves wide publicity.

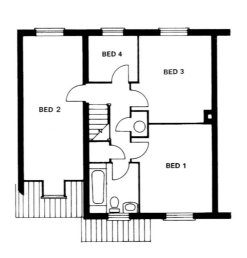
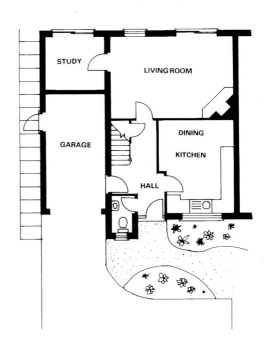

Making it happen

Selfbuilders have to make it all happen. If they are people who are extraordinarily good at making things happen, and spend their days starting up and then selling on entrepreneurial high tech companies, they can turn their talents to development opportunities which are usually too complex to interest developers. This can result in really splendid houses, and the photograph is of one such built near Huntingdon by Mike and Jean Hall. Mike runs his own computer software operation at the cutting edge of new technology. His house results from putting his drive and enthusiasm to work in the selfbuild world.

In 1991 the Halls decided to look for a site on which to build a large family home with an annexe for Mike's parents, and two years was spent looking for an opportunity to do this. Many options were considered, and among them were some large plots available through a company called Technique, whose houses have been featured in earlier editions of *Building Your Own Home*. Beyond these plots, and with

access through them, was another which had all the features they were looking for — and every problem they could imagine. It had outline planning consent and the planning officer was in sympathy with Mike's ambitions, but there was also a difficult access situation involving both a road adoption and obligations to others, a ransom strip situation over a ditch which was essential to provide surface water drainage, a need for pumped foul drains, and many other uncertainties, all to be solved through complex negotiations with adjoining land owners. This sort of thing was the breath of life to Mike, who had known David Barker, managing director of Technique for some time, and who intended to retain him to build the house when, and if, work could start. Technique introduced a designer, Mervyn Rossin, and in early 1994 the arrangements to buy the land had reached a stage where it was time to draw some plans. Not surprisingly Mike had drawn his own — reproduced below — which were discussed at length, and abandoned. Mike

FRONT ELEVATION - B.

The first design study, drawn by Mike.

and Jean's son Matthew, then in his last year at school, was involved in this debate and produced a computer generated mark two version using the KeyCAD programme, shown below, and this gradually evolved into the final version. This had the granny flat wing cranked through 40% to soften the appearance of the whole building and also to further distance it from the main part of the house. Detailed planning consent was obtained very quickly, rather to the surprise of the designer.

It had been hoped that work could begin in March 1994, but it was not until July 24th that Technique could start work. This was before all the legal details of the land purchase were tied up, but Mike, and David Barker who was also totally involved, felt they had the whole situation under control and could proceed on an assumed licence.

Technique quoted for a full builders contract for the whole job, including the outside work and handling the road and services, and everything was tied up with a JCT formal contract. Mike says he debated every detail, including a 22 week construction programme which David Barker wrote out with him by hand, and then when it was signed put the contract in a drawer and forgot about it. The family sold their old home and moved into rented accommodation to be able

to visit the new house every day — and are still close friends with the Barker family! They could have moved in for Christmas, but chose to delay until February 24th.

The granny flat situation is interesting: Theo and Anne Hall who were to live in it were very much involved in its design and all the other arrangements, and explained how their concern that it should be completely self contained had to be balanced against the need to demonstrate that it was a subordinate part of the main house to establish that it was not a separate dwelling, at any rate as far as the planners were concerned. Their investment in the flat was also a carefully thought out part of their Inheritance Tax planning. Among other arrangements made to emphasis this was a shared heating system, common 'service rooms', and even making sure that the granny flat did not have its own letter box. At 800 square feet, the rooms in the flat have the feel of a much larger building, particularly the L-shaped lounge.

The whole venture is an example of just what can be achieved with a selfbuilder with wide business experience working closely with a builder who has local knowledge, local contacts and a good track record in one off housing.

Matthew's Mk II computer drawn plan.

Three generations of the Hall family outside their new home.

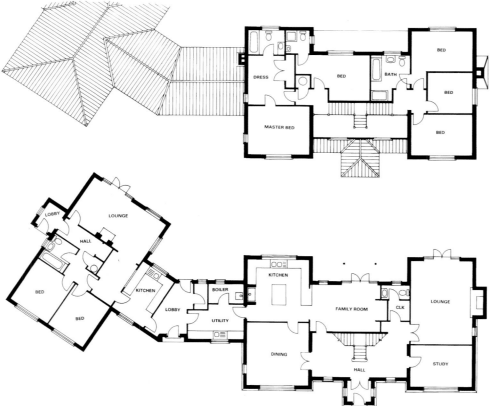

Opposite

Top: The Halls in their new kichen.
Bottom left: Theo and Anne Hall in the lounge of their annexe.
Bottom right: Mike and Jean hall on their impressive staircase.

The final agreed floor plan.

Building a retirement bungalow

In 1993 Alan and Norma Price were living in North Yorkshire but had long term plans to retire to East Yorkshire where they had their roots, and where they wanted to live in a house like one that had caught Norma's eye in the Ideal Home Magazine. This was a Caxton half timbered house by Potton, and they had sent for the Potton literature. All of this was very much something for the future until they suddenly saw a sign advertising plots for sale near to their daughter's house in just the area where they wanted to live. One of the plots, off a private road and backing onto farmland with distant views, was just what they hoped to find some day. They are good at making up their minds quickly, and in no time at all it was reserved for them while arrangements were made to sell their business in Whitby.

The next step was to go to see the Potton showhouses at Sandy which they thought fully lived up to the advertisement in the magazine. Unfortunately the Caxton design that they so liked was on two floors, and planning consent for the plot which they were buying was for a bungalow. However, Potton assured them that most of the features of the Caxton design could be incorporated in a large bungalow, and sent them plans together with details of local builders who specialised in building Potton homes. The plans were marked up with alterations that were required and sent back to be redrawn, and the list of builders studied with care. They noticed that one, David Jackson (Hull) Ltd., also advertised in Build It Magazine so they made a start by calling at the company office. This was as far as they got with interviewing builders for they struck up an immediate rapport with David Jackson himself, and when they had seen some of the other homes he had built they decided that he was the man for them.

A guideline contract price was agreed and Jacksons submitted the planning and building regs applications with drawings supplied by Potton. The consents were obtained in six weeks, and a contract signed. This was for the builder to handle every detail of the job, including supplying the Potton timber frame kit, so that he was able to arrange to provide an NHBC certificate and to 'zero rate' the VAT in his invoices to them.

Meanwhile Alan and Norma had been choosing bricks and tiles, and the more they looked into the options, the more they realised just how wide their choice was. The Yorkshire Brick Company provided a list of homes built

The conservatory looks out over open farmland.

Alan and Norma Price in front of their new home.

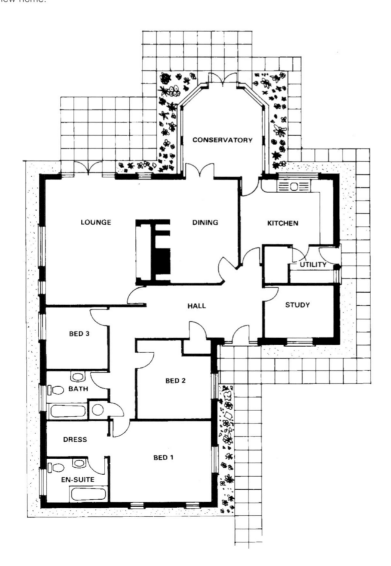

with their handmade bricks, and after many miles travelling to see them the choice was the company's Galtress brick, and so the bungalow is called Galtress Cottage. Then the whole business was repeated to choose the right tile, so that Alan says that he now always looks at the roof of any house that catches his eye.

Work started on May 6th 1994, and when the floor slab had been cast the Potton frame was delivered and erected in just three days. (This was a wall panel frame — the Potton aisle frames are only used for houses.) The bricklaying was then finished in two weeks, and the house was tiled, weatherproof and locked up in less than a month. The rest of the trades followed one after the other without any serious difficulties or delays, although there were the usual minor problems. All were overcome amicably and in the first week of August, after just 12 weeks, the Prices moved in.

As keen gardeners they appreciated the free garden design service from David Stephens, the Gold medal Chelsea Garden designer, that came with the service from Potton as a special promotion in 1994, but of course they modified this as well. Within a year the garden looked as if it had been there for ever. The whole project is a triumph of a package company/builder/client co-operation, and Alan, now retired after his business career, emphasizes that fostering a team approach is what made building his own home so enjoyable.

Building at Milton Keynes

For many years each succeeding edition of *Building Your Own Home* has had a case history about a new home built in Milton Keynes, which is the largest centre of selfbuild activity in Britain. All the building land in the new town is owned by the Commission for the New Towns, who have dozens of plots reserved for owner-builders available all the time, all with roads completed, services ready to connect and a great deal of help and guidance available. This year the house featured was built by Mike and Christine Evans who in 1993 were living in Leighton Buzzard in their fifth house. All the homes that they had lived in had been bought new, and been fitted out by the developer to the family's requirements. Building for themselves was a natural next step.

Mike is a project manager with an international communications company, and Christine was a training manager with the same company before she retired to manage building their own new house. Not surprisingly, the new home was a text book example of carefully structured project management. A start was made by subscribing to all the magazines, reading all the books, and listening to the author speaking in seminars at the Alexandra Palace exhibition in 1993! They methodically searched for the right plot in the local area, and eventually decided that one of the Milton Keynes plots was best suited their requirements, reserving one which overlooked the equestrian centre.

They knew from the New Towns Commission that they would have to employ an architect, and so the next step was to choose one. A list of five questions for possible contenders was drawn up:

* Can you show us what you have designed for other plots like ours?

* What will it cost for you to design something similar and what will it cost to build?

* What will the time scale be?

* What different levels of help and supervision can you offer, and at what cost?

* What advice have you on our different options?

Five architects were chosen from the list provided by the CNT, and all five were interviewed. Two were obviously not on their wavelength at all, two came across as possibles, but one, David Sim RIBA, impressed them at their first meeting as being the obvious choice. He gave them a long list of houses in Milton Keynes which he had designed which they could go to see, he was specific about construction costs, he set out the different levels of professional involvement that he could offer them and his fees, and convinced them he was the man for the job.

Mike and Christine then produced a four page list of their requirements and discussed it in detail. David Sim took it away and within a week returned with four possible 'solutions' — contemporary architect speak for 'design studies'. All of this is illustrated on page 278.

The final design incorporated features from all of the studies, and is shown opposite. It was submitted to the planning authority in January 1994 and consent was obtained in March. The New Towns have their own planning procedures, and Mr. Sim has provided a note on these which is on page 279.

A builder for the shell was selected with great care, and arrangements made for a NHBC 'shell only' certificate as described in chapter 2. Sub contractors for the follow on trades were found, mainly at the suggestion of the builder who was handling the shell. The house in Leighton Buzzard was sold and the family moved into rented accommodation, and Christine left her job to devote all her time to the new home and to making sure that three young children were not left out of things. Two banks — Lloyds and Barclays — were induced to bid against each other for the privilege of providing the finance, and in July 1994 work started. Christine had a fixed meeting with the team at 10 am every Monday morning, visited at some time every day, loved every minute of it all, and made sure everything went according to plan. In February 1995 the family moved in. There is 3156 sq. ft. of accommodation on two floors, with another 900 sq. ft. available for a loft conversion when required. It was hoped that the double garage could have had a playroom above, but this was not acceptable to the planners.

The Evans family and their new home.

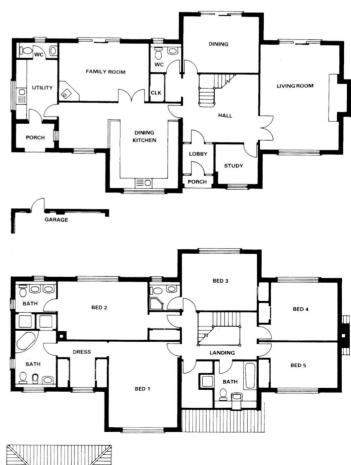

The final plans. The design studies from which they evolved are shown overleaf.

Solution 1

The project brief . . .

Solution 2

Solution 3

Solution 4

. . . and the four alternative solutions. The broken line shows the area within which the house had to be contained to comply with the MKDC design brief.

ARCHITECTS

The Studio   55 High Street   Great Linford   Milton Keynes   MK14 5AX
Telephone (01908) 201501   Fax (01908) 201502   Car (0860) 879082

Dear Murray,

I am interested to read that you intend to feature Mike and Christine Evans' now home in the next edition of your book: they were text book selfbuilders, and text book clients as far as I was concerned because they knew exactly what they wanted.

I think it important that in the next edition of Building Your Own Home you should emphasise the changes that have been made in the planning procedures at Milton Keynes,. My practice as been closely involved with selfbuilders in Milton Keynes for many years, and we have always considered it important that clients should understand the special requirements for obtaining consent to build in the new towns.

Following the disbanding of the Development Corporation (MKDC), the Commission for the New Towns (CNT) now has responsibility for continuing the development of Milton Keynes. They continue to sell plots to private individuals who wish to build for themselves, or who wish to have a builder to construct a house for them, and they continue to provide roads, services and detailed design and landscaping briefs. However, there has been a very real change in the way in which planning consents are obtained. Previously, under MKDC control, the system was quite informal. Once a design had been agreed with the selfbuilder and the MKDC officer, formal approval was almost by return of post following submission of an application.

Nowadays the selfbuilder is still offered a plot with outline planning permission obtained under the New Towns Act, and his or her architect still need to go through the process of obtaining the informal approval of CNT's planning officer before a formal application is made and approved. Only then does he buy the plot.

However, CNT's site constraints are now interpreted very strictly, and the formal application drawing must contain much more information than is required in a normal planning application to a local authority. The proposed external materials must be shown in colour on both the plan and the elevations to give a good representation of the complete house, and each submission must also include a NHER certificate. All this has to be done before the purchaser can buy.

Following receipt of the application the CNT -- unlike MKDC before it -- consults with the local authority and with any neighbours before issuing an approval. This takes time, and taking into account the time required for initial consultations a selfbuilder should assume that the planning procedures are going to take him at least eight weeks.

Once approval is granted, the plot is sold to the selfbuilder with permission to build the approved scheme. Any external alterations, no matter how small, need the further approval of the CNT, which cannot be taken for granted.

I suppose all this is the price of having an opportunity to build in this exciting new community, and of course there are other advantages for the selfbuilder here. The banks and building societies are well used to the selfbuild scene, and are able to give a great deal of relevant advice. There are lots of builders and sub contractors who specialise in working for selfbuild clients, and, perhaps most valuable of all, building in an area where so many of the homes have been built by their owners helps to keep enthusiasm going. Selfbuild is a well accepted way of having a new house in Milton Keynes, and once people have obtained consent to built what they want, it is very easy to make everything happen.

Yours sincerely,

David Sim Dip Arch ARIBA Chartered Architect

Hands-on selfbuild

In June 1995 the secretary of the Self Builders Association was asked to nominate one of the members to be approached to provide a case history for this edition of *Building Your Own Home*, and immediately suggested Steve Haywood. The author contacted Mr. Haywood by phone, and was told that reading earlier editions had given them the idea that they might build their own home, and that the book was responsible for all that followed. Fortunately for the author's piece of mind, what followed has been a great success.

Steve, full time biochemist and part time bass guitar player in a band, and Chris, full time mother of two small daughters, were needing to move from their three bedroom semi to a larger property when Steve picked up a copy of *Building Your Own Home* while on a business trip to Luton. The idea of selfbuild appealed to him, and he mentioned it to a friend who is a builder. Rather to his surprise his friend did not scoff at the idea, and even suggested that he should go to look at a site in a nearby village.

It proved to be a disused farm yard opposite an ancient lime kiln which is now a Site of Historic Interest. This is not only beautifully maintained, but will never be developed, guaranteeing attractive surroundings in perpetuity.

Three houses had already been built on the site, and there was room for a fourth for which there was existing detailed planning consent. Although not advertised, the plot was for sale at a silly price, so in January Steve and Chris put in a silly offer. In June, after six months of bargaining, they bought it at a sensible compromise figure of £19,000.

During these negotiations the Haywood's went to the Individual Homes Show at the National Exhibition Centre where they joined the Self Builders Association. They also attended the seminars, and took advantage of all the other opportunities to learn about their options. This gave them the confidence to press ahead quickly, starting with retaining the architect who had drawn the plans which had already been approved. They arrange for him to obtain consent for various minor alterations, and to provide architects's progress certificates for the new

building. With the plans settled, Steve drew up a business plan and then asked a friend who is a financial consultant to arrange the finances. This led to the Leeds Building Society offering a loan for 75 % of the land price followed by a stage mortgage on the building as work progressed. With this all settled the existing house was sold and the family moved into rented accommodation, the only snag being that it was 20 miles from the site.

A start was made in July 1994 using ground workers who were recommended by the architect. The first excavation work disclosed that they had to construct a complex reinforced raft foundation at an additional cost. A further problem arose when it was found that the suppliers of the steel reinforcing had not followed the bending schedule which had been sent to them, but with compromises the difficulties were quickly resolved. After these early setbacks things went more happily. A friendly local site agent found the Haywoods a brick laying gang to put up the shell, and Chris' father, a retired carpenter, took over all the timber work and thoroughly enjoyed himself. He had encouraged Steve and Chris from the very start of the project. A plumber to work on a labour only basis was found by word of mouth, and local electricians were invited to tender on a supply and fix basis. A plasterer lived in the village, and after seeing his work elsewhere was the obvious man for the job. Steve and Chris worked on the house themselves every weekend, and moved in on May 25th 1995 after ten months which they had thoroughly enjoyed.

The total cost including the land was under £65,000, and the construction figure was £40 a square foot, including the £8,000 cost of the raft foundation. This latter expenditure had not been anticipated when drawing up the budget, which had allowed for strip foundations at £3,000. Fortunately the overspend was covered by a contingency provision. The value of the completed house is £85,000, giving a cost value differential of 30%.

Steve and Chris Haywood in the last stages of building their new home.

The project costs as set out for *Building Your Own Home* by Steve Haywood.

BUILDING COSTS

1. Fees / Services

Solicitor	400
Architect (drawings)	881
Architect (certificates/oversee)	881
Council inspections	195
Building Regs Fees	73
Structural Engineer (footings)	440
Site Insurance	476
Connections (Foul)	200
Connections (Water)	200
Connections (Electricity)	417
Connections ('phone)	116
Connections (Gas)	292

2. Land

Purchase Price	19,000

3. Works (External)

- Groundworks

Labour only	5,000
Soil Materials	568
Concrete (Readymix)	2,173
Steel	1,386

- Blockwork/Brick Plinth

Labour only	3,500
Materials (Blocks, Rockwool, reclaimed bricks, Sand, cement etc)	5,085
Joists	457
Scaffold Hire	460

- Roofing

Labour only	1,100
Tiles / Felt	940
Preform Trusses / Bracing	1,018

External Boards + erection	429
External Rendering + materials	1,900

4. Works (Internal Fitting)

Stairs + erection	850
Ground + 1st floor chipboard	310
Stud Wall / Ceiling Material	555

- Plastering

All internal rendering / Skimming + Artex ceilings + coving + utility Screed (inc. materials)	1,700

- Electrics

Supply / Fit	1,570

- Plumbing

Labour only	1,800
Soil / Rainwater materials	395
Central Heating Materials	1,076
Copperware	405
Sanitaryware / Shower	893

- Carpentry

2nd Fix materials	487
Double Glazing units	2150
Fitted Kitchen	2360
TOTAL	**£62,138.**

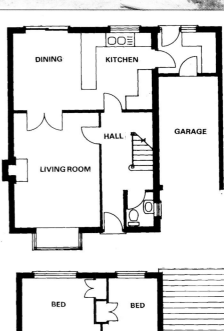

281

Building in the country

Tony and Carolyn Hole first learned of the selfbuild option in the early 80s when they saw a Potton show house at the Earls Court Ideal Homes Exhibition. Tony's career as an accountant involved moving to a different part of the country every few years, and with each move they had grown more demanding in their choice of a new home: building the next one themselves sounded attractive. With this in mind they read all the books and attended exhibitions, listened to lecturers — including the author — and visited showhouses. A possible opportunity to put all this to good use came when they were posted to Essex in 1987, but after some time searching for land they were discouraged by the difficulty of finding a plot at a sensible price at that time, and so bought a house off the peg.

1992 saw another move, this time to Wiltshire. Land prices were now depressed, and as they had lived in the Cotswolds before this seemed the time to build. In particular it provided a way of using the reduced value of the existing house in Essex to build a larger home at the same figure. Years of thinking and learning about selfbuild could be put into practice, and they started to look for a plot.

The ideal was to live in a rural setting within a few miles of a market town, and they decided to base their search on Malmesbury. An estate agent directed them to a plot for sale in a small hamlet a few miles from the town. It was so small that when they got there they assumed the first 'for sale' sign that they saw was what they were looking for. In fact there were three plots for sale within a few hundred yards of each other, which were the first to come on the market locally for 50 years. By the time they had sorted out the confusion they not only knew which one they wanted, but also knew all about the other two, which was later to their advantage.

The site they wanted was a disused half acre paddock adjacent to a pub, and they made an offer for it subject to obtaining detailed planning consent. This was quickly accepted.

The next step was to determine the right house for the plot.

The stages in this are shown opposite — notes made in the design requirements checklist in an earlier edition of this book, a page from another of the books in this series which illustrates a style that was thought appropriate, and the first sketch of a layout to suit. All of this was taken to the planning officer, who was generally encouraging but made various suggestions. In particular he wanted the house to be part stone, part ochre coloured render to match the mix of stone and render on neighbouring properties.

At the same time Tony's business travels took him to Yorkshire, and he took the opportunity to call at the offices of Design & Materials who he knew of from their stands at selfbuild exhibitions. He was quoted for the service he required, accepted the quotation, and D & M took over negotiations with the planners. Three months later the Holes received their planning consent and bought the land.

By then they had sold the Essex house and rented a cottage a few miles from the site. Mortgage finance was negotiated with the Abbey National Building Society with arrangements for a local chartered surveyor to provide progress certificates. The Nat West bank was also keen to provide a loan, but although the banks interest rates were comparable, their high arrangement fees tipped the scales in favour of the building society.

With a site, planning consent and building finance, work could start, and the Holes decided to clear the site and make the entrance themselves. Building insurance was arranged, and an excavator and dumper hired — Tony to drive the former and Carolyn the latter. They learned as they went, and still talk about it! They had already obtained builders prices for the rest of the work, none of which were at less than £50 a square foot, which confirmed them in their feel that they should use sub-contractors. They had kept closely in touch with events on the two other plots, where work on homes of a similar size was in advance of their programme. One was being built by a local builder who quickly became a friend, and who suggested that they engage the sub-contractors who were working for him. Although prices were obtained from other subbies, and their work carefully inspected, the men who were working for their new friend

A design questionaire from an earlier edition of this book marked up by the Holes' with their own requirements, a design from another of the authors books which was to be the basis of the new home, and Tony Hole's own sketches which he took to D & M.

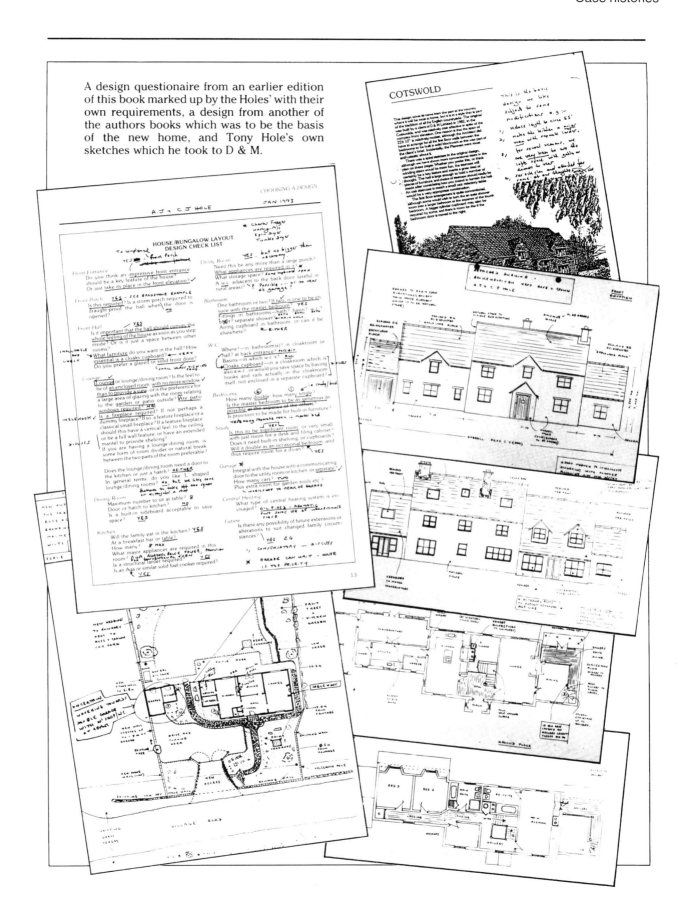

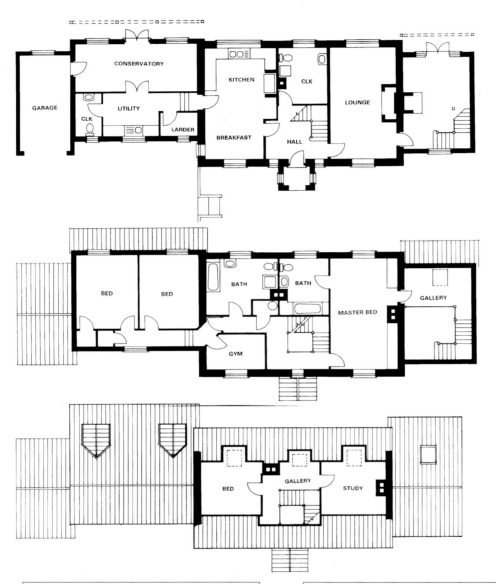

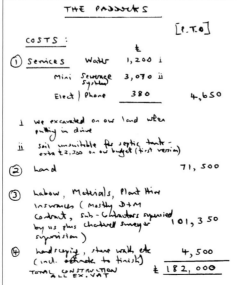

Tony and Carolyn Hole as the new house approaches completion.

seemed definitely the best buy, were engaged, and did a first class job. The bricklayers, who were a 4 + 2 gang, quoted £9,000 to do their part of the work in six weeks: the next quote was from a 2 + I gang who wanted £15,000 to finish in 15 weeks! Similar margins were experienced for quotes for other trades. Tony and Carolyn visited the site every day, and found themselves a great deal of practical work to do besides arranging for deliveries, co-ordinating everything, and dealing with a thousand and one problems as they arose. Among much else they installed the site water supply, stained joinery before it was built in, installed the bridging between floor joists, laid floorboards, handled the floor and wall tiling and much else. They were helped at times by their son, an instrument engineer who fortunately lived only 15 miles away.

After nine months they moved in with the decorating and other finishing work going on around them. The land had cost £75,000 with the services adding a further £3,000 to this. The house, includeing features like the Aga cooker and an expensive wood burning stove, cost £102,000 after an £11,000 VAT refund — significantly less than their £35 a square ft target. Landscaping costs still have to come, and

this will not be skimped. The value is estimated at £215,000.

When asked the usual question about advice for others, Tony suggested that while obtaining technical advice is easy — 'read it up, or ask someone' — the difficult part, and the essential part is getting the management right. This is where the effort must be made. Carolyn wants to warn other wives against being over ambitious in planning their own participation in the heavier work: the hired dumper has left its scars. But throughout they found it a marvellous challenge, and of all the houses that they have lived in this is the ultimate one.

Building as a family

The Morton family have made two new homes a three generation project, building them on a plot that had a derelict bungalow on it. The first of the photographs showing the houses wrapped in scaffolding was taken just one month after the walls started to come out of the ground, and the whole undertaking from first seeing the site to moving in, was programmed to take exactly a year. It has not been uneventful.

Clive Morton was a pilot with the RAF in Germany until he left the forces to join an airline. He and his wife Margot have always been enthusiastic about making improvements to whatever home they happened to be living in, and when living in Germany had developed some very firm views about the features that they would have in a house in England some day. These included ceramic tiled floors, continental windows with shutters, underfloor heating, and above all a continental standard of finish. Once Clive had settled into civilian flying they set about looking for opportunities to make it happen.

Meanwhile Clive's parents, Ted and Ruth Morton, were coming to the conclusion that their large garden was really more than they wanted to cope with in their retirement, and had an idea that they too might build a new home some day.

In September the family found the derelict bungalow for sale with planning consent for two replacement dwellings, and everything moved into top gear. The National Selfbuilders Exhibition was then running at Alexandra Palace and they went along to look at many options which were presented there. This led to a decision to use Design & Materials Limited to provide architectural services and to supply all the structural shell materials for the two houses, and to choosing the Birmingham Midshires Building Society to finance the project. The reasons for their choices were interesting — Midshires because they were more flexible than other building societies and would tailor arrangements especially for them, and D & M because they have a package service for

Three generations of the Mortons six weeks after they had made a start on the site . . .

. . . and when they had just moved in.

traditional construction and have field staff who specialise in working with people building on their own. The Morton's felt that this was well worth paying for.

Airline pilots are great D.I.Y. home builders, probably because their work patterns suits this very well, and Clive was not short of advice from friends. Margot also had a clear view of what would be required: as well as studying the kitchen catalogues she took a book-keeping course and gained a certificate to prove it. An interesting example of the Morton's pragmatic approach was that although they had done their own conveyancing when buying and selling their previous three houses, this time they found a solicitor to handle the purchase of the land as they felt that a D.I.Y. land title for a D.I.Y. house might seem a little too unusual.

D & M have a huge range of standard designs, but the Morton family knew what they wanted and gave the company a very specific brief for their own design. The plans went to the Council in February and in May all the consents were obtained.

In the meanwhile a lot of careful attention was being given to the costings. The original intention had been to employ sub-contractors for such of the work as Ted and Clive did not think they

could tackle themselves, but they then discovered that local builders were so short of work that they could find one to be responsible for everything except the work which they were going to do themselves. This would cost no more than the total cost of employing the different sub-contractors, and in this way they could get NHBC certificates and someone else to deal with co-ordinating the different trades. It was all arranged on this basis, and then less than a week before the start day the builder concerned had personal problems and had to withdraw.

The D & M local representative consulted his little book of builders, and within a week they had another firm to take over. Work started at once with the demolition of the bungalow. Its drainage had been to septic tank, while the new houses were to be connected to the main drains, but the existing electricity, telephone and water supply arrangements could be used. Unfortunately the water board claimed the existing supply was for one house only and levied a charge of £2,000 for connecting up the other with just a few feet of pipe — a story which made mention in the media about the wicked ways of the new privatized water companies.

Margot Morton's own diary of events which is reproduced below tells the story better than any third party could, and the accounts which she has kindly provided show how the construction costs ended up as £35 per square foot. As will be seen, the families moved in on schedule, and when the second of the photographs was taken only the landscaping and the drive remain to be tackled. The septic tank for the old bungalow was being cleaned out and arrangements made to bring it into use for garden water storage, fed from the downpipes from the roof guttering.

Roughly two years after returning to Britain from Germany with a clear idea of the sort of house they wanted, Clive and Margot were living in it. Ted and Ruth live next door in just the house that they wanted, with babysitting simply a matter of walking from one back door to the other. It was simply a matter of making it all happen.

Margot's diary

SEPTEMBER

Visited Self Build Exhibition at Alexandra Palace which rekindled interest in building for ourselves. Mentioned to Clive's parents and they too are keen. Search for plot for two began.

OCTOBER

Found suitable plot with outline planning permission for two houses. Put in offer.

DECEMBER

Offer accepted.

Met with David Snell of D & M to discuss preliminary drawings.

JANUARY

11 Jan
First set of drawings from D & M.

21 Jan
Met with David Snell again to discuss amendments.

FEBRUARY

18 Feb
Planning permission and Building Regulation approval applied for by D & M.

Research carried out into heating, lighting, plumbing systems required. Several items purchased in sales!

APRIL

19 Apr
Exchange contracts on land.

MAY
10 May
Completion on land. Started clearing trees and made one part of the garage/outbuildings into an office. Phone and electricity moved into office.

27 May
Moved from our house to rented accommodation.

28 May
Digger arrived on site and demolition of existing bungalow started.

29 May
Planning permission granted at planning meeting.

31 May
Demolition complete! Bungalow and about 170 conifers had gone and the site was bare except for the old garage complex.

JUNE

5 Jun
Building Regulation approval granted.

6 Jun
Last minute change of builder due to original builder's personal problems.

12 Jun
Builder started on site- trenches for foundations for plot 1 dug.

13 Jun
Building Inspector on site early and left chalked message on the digger to say 'OK to concrete'. This was done!

17 Jun
Bricks delivered. Heavy rain made the site slippery for the lorries.

18 Jun
Drains dug. Floor beams delivered.

20 Jun
Brickies started with the foundations of plot 1.

21 Jun
Materials to wallplate delivered. This made for a lot of materials on site and we had a large pile of lintels etc. I didn't know that there were so many bits!

24 Jun
Foundation trenches for plot 2 dug.

25 Jun
Concreting of trenches in the rain.

26 Jun
Site still too wet for delivery of blocks. Beams put on plot 1 floor.

JULY

5 Jul
Beams put on plot 2 floor. Walls on plot 1 were going up. The houses looked really small!

7 Jul
1st lift scaffold on plot 1.

17 Jul
Ground floor lintels on plot 1....and it has been raining for a couple of weeks so the site was pretty awful. Neil (aged 18 months) had problems lifting his clay-caked boots!

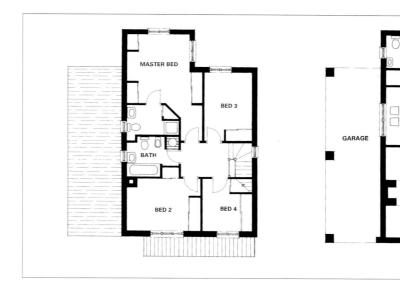

18 Jul

1st lift on the scaffold on plot 2 which joined up with that on plot 1.

22 Jul

Lintels in on plot 2. Floor joists in on plot 1.

25 Jul

Floor joists in on plot 2.

26 Jul

Electricity connected to plot 1 on a temporary meter so extension leads no longer needed from the garage.

31 Jul

2nd lift of entire scaffold. More deliveries but these were easier as the site has dried out - problem was where to put it all!

AUGUST

2 Aug

I installed the '1991' plaque on our house!

6 Aug

Roof trusses delivered.

7 Aug

3rd lift on scaffold. Wall plate on plot 1.

9 Aug

Wall plate on plot 2.

12 Aug

Roof trusses up on plot 2. Clive and Ted began notching and drilling joists for 1st fix plumbing and electrics.

13 Aug

Roof trusses up on plot 1.

15 Aug

Garage lintel 3" too low for the door. Entire gable end had to come down again.

16 Aug

Further lift on scaffold for the gable ends of the houses to be done.

20 Aug

A little ceremony was performed atop our house as the chimney pot from the old bungalow was put on our new house. Builder wanted to continue with plot 2 first (which is not what we want) so Clive and Ted concentrate on carcassing there.

21 Aug

Upstairs ceiling on plot 2 and a couple of stud walls put up. Clive had almost finished carcassing our home.

23 Aug

Rain. Clive and Ted worked inside on 1st fix. Ruth and I went to buy ventilation units which were £40 cheaper from the builder's merchant than buying direct!

27 Aug

1st floor in on our house (plot 1). The tilers continued with the main roofs.

28 Aug

Plasterboard to the upstairs ceiling in plot 1. Clive and Ted took advantage of scaffold being there and put up the TV/FM aerials.

29 Aug

The Building Inspector raised a query re the buttress walls upstairs in both houses which had to be sorted out by the architect at D & M.

30 Aug

Upstairs plasterboard walls on plot 1 were complete and light circuits in plot 2 were done. Building Society surveyor visited to see the roof on plot 1; cash flow was beginning to get tight so we asked him if he could send his report asap.

SEPTEMBER

1 Sep

Clive and Ted still wiring houses.

4 Sep

Stairs arrived.

5 Sep

Boilers arrived.

6 Sep

Scaffolders arrived to begin removing scaffolding. This took 3 days!

9 Sep

Now the houses were looking like houses with the windows and doors installed today. The plumbers arrived to install the gas. A bricklayer was here to do some of the finishing bits and, with our vehicles, we had 7 vans on site.

11 Sep

Stairs installed in both houses.

13 Sep

Downstairs ceilings up in plot 1.

19 Sep

Plasterers start in plot 2.

SEPTEMBER

25 Sep

Builder started porch. Clive finished wiring in our house. Underfloor heating stuff arrives.

26 Sep

The garage was broken into last night and a few tools were stolen.

27 Sep

Porch on plot 2 put up.

30 Sep

Rafters up for carport.

OCTOBER

4 Oct

Gas contractors on site to connect us. Clive jumped every time the phone on site rang in case it was me to say that the baby's on the way!

Editor's note — The first mention that Margot is doing all of this while expecting a baby!

5 Oct

Brickie starts on carport arches but rain stops play. Plastering starts on plot 1.

7 Oct

2nd fix carpentry started in plot 2.

8 Oct

Artexers in plot 2.

10 Oct

Clive and Ted started installing the underfloor heating pipes in plot 2, they finished both houses in 3 days.

14 Oct

Gary born at lunchtime. Clive returned to the site to sort out problems with the floor screed, which was proving difficult to lay.

17 Oct

Artexers start on plot 1.

19 Oct

Plasterers in to finish the floors.

23 Oct

Burgled again. This time all the tools and the door furniture had gone. Sorted out our second insurance claim.

28 Oct

Water board dug up the road and connected us.

2nd fix carpentry and electrics continued.

NOVEMBER

12 Nov

Phone moved to our house.

13 Nov

Electricity Board connected our new meter.

18 Nov

Gas board put in meter.

20 Nov

Heating on in plot 1! Our builder hadn't turned up for several days despite being well aware that we were due to move in that weekend and the carpet men come on 22 Nov.

22 Nov

Carpets fitted in plot 1. Still painting in the lounge so they had to start upstairs.

23 Nov

Moved in our furniture but still slept at the rented house as the plumbing wasn't done and the house was still rather damp.

25 Nov

Clive sets to to get the plumbing done whilst I headed for my mother's with the boys.

DECEMBER

4 Dec

Bathrooms and half of the kitchen completed so we moved into our new home and I began putting things away. A few days later Ruth and Ted move in on plot 2.

Since moving in Clive has worked hard. The ventilation system went in next, then the shower.

In Jan he had great fun with a mini-digger for a week doing a bit of landscaping and putting a surface on the drive.

We disappeared again in May whilst the ceramic tiles were put down on the ground floor as this constitutes about 60% of the floor area.

There is still a lot to do to get everything as we want ... not least the garden which has yet to be levelled and planting will be an ongoing task.

Still, we have got the house we wanted. Sorted out the total costs for the V.A.T claim and sent a copy to Murray Armor for his book. Will send him a photocopy of this diary as well.

The Mortons' accounts

	Plot 1	Plot 2	
Preliminaries			
Feasibility study	300	900	
Local authority fees	676	676	
Insurance	415	415	Well worth it! (We had burglars.)
Legal fees	1948	1948	Includes stamp duty.
Services: electricity	245	245	
water	175	2145	
sewage	510	510	Including rebuilding path
Demolition	600	600	
Materials			
D&M package	20,285	18,774	
Floor tiles/carpets	1,600	1,524	
Wall tiles	270	213	
Plumbing/sanitary ware	1,484	1,210	
Kitchen	1,931	1,952	
Windows	5,100	5,748	
Electrics	1,057	1,342	Includes light fittings
Wardrobes	310	226	
Doors/door furniture	475	392	
Heating			
Underfloor	980	980	
Radiators	137	128	
Gas fire and surround	590	446	
Boiler	1,027	1,047	Gas condensing boiler
Hot water cylinder	500	500	Mains pressure systems: great showers
Ventilation	727	713	Whole house heat recovery and more to come!
Drives, etc	833	833	
Miscellaneous	108	108	
Labour			
Builder	20,066	21,217	
Plumber	125	125	
Land	61,000	61,000	
Total	121,510	123,353	

The sale of our house just before building began meant that, in the initial stages, we didn't have to rely on the building society advances to progress. Even so, cash flow was tight and we had to be cautious about ordering the fittings etc too early. Mid-project was particularly sticky - D&M wanted payment up front for large quantities of materials before the advances had come from the building society whereas if we'd bought from the local builder's merchants we wouldn't have had to pay until the end of the following month.

Interest on the whole project cost us a staggering £50 a day at the end of the project and the pressure was really on to complete the sale of Clive's parents' house. Our total interest charges were £7,300 and £8,600 for Plots 1 and 2 respectively.

VAT could have caught us out but for a loan from a relative - the reclaim amounted to almost £5,000 per house and took a couple of months to come through after completing.

Selfbuild Housing Association model rules

The National Federation of Housing Associations publishes model legal documents for selfbuild associations. These are based on their 40 years of experience of registering associations and dealing with their legal and financial problems, and the whole of this experience is that the right legal framework for an association is essential. The draft rules that follow are part of this documentation, and are reproduced by permission of the Federation as an example of appropriate rules only. Any new association should seek the current advice of the Federation as a matter of course.

Note that selfbuild matters are dealt with by the Federation at 175 Grays Inn Road, London WC1X 8UP. Telephone 0171 278 6571.

The regulations detailed below under 'Building Programme' have been designed to protect the interests of all members of the Group. It is obviously important that every member knows that he will not be called upon to 'subsidise' lack of effort by others.

These regulations are suggestions only and should be discussed in detail at an early meeting and ratified by the members of the association.

Membership

1 Nothing in these regulations shall apply, or be deemed to apply to any Local Authority or County Council holding a share in the Association.

2 For the purpose of filling vacancies, the committee shall cause to be kept a list of the names of persons desirous of becoming members of the Association. From this list all new members shall be elected by the Committee, who need not take the first on the list, but may take others matters into consideration.

3 The amount of Ordinary Loan Stock required under the rules of the Association to be taken up by members may be provided in instalments if necessary.

4 All payments made by members under the rules of the Association shall be made to the treasurer and to no other officer or person.

Building programme

1 With the exception of the building tradesmen referred to in Paragraph 2, all members will work 14 hours per week December, January and February, 16 hours per week March, October and November and 20 hours per week April to September inclusive, Breaks, when spent on site, are included in these figures. 75% of member working hours must be worked between the hours of 7.00 a.m. and 8.00 p.m. Saturdays and Sundays. Hours worked through the week between 8.30 a.m. and 5.00 p.m. will count as week-end hours.

2 Experienced bricklayers, house joiners and plasterers will receive an allowance of two hours per week, other fully experienced site tradesmen will receive an allowance of one hour per week. These allowances are at the discretion of the Managers and may only be altered by them.

3 All members will, when starting work, sign in at the exact time and sign off when they stop work. Details are to be recorded of the exact work if done at times other than at the week-end. Members must sign off when they leave the site, unless instructed to do so by the site foreman, and members are expected to sign off when they are not working. (Breaks accepted 10 mins morning, 30 mins lunch break, 20 mins tea - when part of the working day and spent on site.)

4 Each member's hours will be added at the end of each month, on the last Sunday in each month, and for each hour below minimum the member will be charged £xx, irrespective of any previous overtime worked, or sick hours granted. In the event of certified sickness during the last week-end of the month the members will be allowed to the end of the following month to make up the hours lost during that week-end. All fines are paid to the Association and are debited to the Members account.

5 Any time lost through site accident will not be subject to a fine unless the Association decide otherwise.

6 The Time-keeper will keep a cumulative total of each member's hours. Irrespective of any fines which may have been made, by the end of the scheme each member must be above an agreed minimum. In the event of certified sickness the fine will still apply but an allowance will be added to the member's total. The effect of this is that hours not worked on site will be fined and will also have to be made up, but although sickness hours are fined they do not have to be made up. The committee would also consider if necessary granting hours in respect of wife's illness. All fines accrue to the Association.

7 The Time-keeper will keep a list, up to date, showing the number of hours members have worked, sickness hours, penalty hours etc., and which members are above and which below the average.

8 Members will work 1 week of the annual holiday as well as their ordinary times, at a time to be agreed by the Committee and for this week the hours will be 8 hours per day. Members will also work one day extra at Spring Bank Holiday and Easter, the day to be decided by the Committee.

9 Extra Holiday Incentive Scheme - In order to help members have a longer holiday, a special bonus will operate as follows. 50% of extra hours worked between March and September (inclusive) may be used to offset extra holiday hours. The Time-keeper must be informed in advance as to when the member proposes to take this bonus.

10 All members will completely specialise in one aspect of the building work as decided by the Committee, and will be expected to become completely proficient in the craft.

11 Members will not do work of other trades unless specially instructed to do so by the site foreman.

12 The Association will insure its property against fire and also insure itself against claims made upon it by third parties or members in respect of accidents for which the Association may be held responsible.

13 A personal accident policy will be taken out by the Association to cover accident of the members. Any benefit from the policy belongs to the Association, but may be given to the injured member or to his dependants. In the event of the death of a member, his next of kin has the same right to occupy and purchase one of the houses. Due to the increase in insurance premiums and because of the earnings related benefits which now apply, there will be no weekly income accident insurance. However, the members are insured for major accidents, loss of life or limb.

14 The Managers shall, prior to the start of building operations, determine the order in which completed houses shall be allocated to the members. The order of building the houses may be varied by the committee. The committee may pass over the allocation of a house to a member who may be in default.

15 Members are requested to open a Building Society account with an approved Society at the beginning of the scheme and save regularly with that Society.

Equipment

1 The Association will purchase or hire certain items of equipment, and these will be maintained by the Association.

2 Members will be expected to completely equip themselves for the job they are required to do.

3 In certain cases, the Association will purchase certain items of equipment and put them into the care of various members, who will be responsible for them

Design

1 The basic design of the houses is as the approved drawings indicate and no other; standardisation is vital for economy. Certain minor internal modifications are permissible if approved by the Managers, the Association's Architects or otherwise, and will be charged to the individual. An additional 100% of the extras will be charged to compensate the Association for the losses which invariably arise when modifications are introduced.

2 If members choose to do personal work on their own houses before completion, this must not delay the main building programme; penalties may be imposed by the Committee if they do so.

3 The Managers reserve the right to amend the design in the interest of improved performance or general economy.

Amendment of regulations

1 In the event of there being any conflict between these regulations and the registered rules of the Association, the latter shall prevail.

2 Ignorance of the registered rules of the Association, and of these regulations, shall not be accepted as a valid reason for noncompliance herewith.

3 Amendment for these regulations may be made at a general meeting or a special general meeting providing two weeks notice of the resolution is sent in writing to the General Managers. A two thirds majority to amend the regulations is required.

Penalties

Any member who fails to observe the working conditions can be expelled from the Association upon a resolution carried by two thirds of the members, called for the specific purpose. If he should hold a licence to occupy, proceedings will be initiated to evict such a tenant.

Further information

For further information call the numbers listed and ask for details of services available for those who are building a new house or bungalow. A detailed enquiry is best made after studying the literature which is available.

BOOKS, MAGAZINES, VIDEOS

HOME PLANS, Murray Armor, J.M.Dent. Companion volume to *Building Your Own Home*, order form on page 285 of this book.

PLANS FOR A DREAM HOME, Murray Armor, Ebury Press. Companion volume to *Building Your Own Home*, order form on page 285 of this book.

PRACTICAL HOUSEBUILDING, Matthews, J.M.Dent. Construction techniques for selfbuilders.

THE SELFBUILD BOOK, Broome and Richardson, Green Books. Largely concerned with Segal construction.

COLLINS COMPLETE DIY MANUAL, Collins. The standard DIY guide.

FIND and BUY A BUILDING PLOT, Speer and Dade, J.M.Dent.

GET PLANNING PERMISSION, Speer and Dade, J.M.Dent. Excellent guides to buying land and planning matters.

FURTHER SELFBUILD BOOK LISTS FROM RYTON BOOKS. 01909 591652

BUILD IT. Monthly. Subscriptions 0181 286 3000.

INDIVIDUAL HOMES. Bi-monthly. Subs 01527 836600.

DIY MAGAZINE. Monthly. Subs 01908 371981

THE HOUSE THAT MUM AND DAD BUILT. Video. 0171 833 4152.

BUILD IT YOURSELF. Video. 0782 638339

AGENCIES PROVIDING SERVICES TO INDIVIDUAL BUILDERS AND SELFBUILDERS

INDIVIDUAL HOUSEBUILDERS ASSOCIATION. 01753 621277. Provide lists of package companies etc.

THE ASSOCIATION OF SELFBUILDERS. 0188 958 4221 or 0116 270 8843. Run by selfbuilders for selfbuilders.

ROYAL INSTITUTE OF BRITISH ARCHITECTS (RIBA). 0171 580 5533. Leaflets and Lists of local architects.

ASSOCIATED SELF BUILD ARCHITECTS. 0800 387310 Details of architects who specialise in selfbuild clients.

ROYAL TOWN PLANNING INSTITUTE. 0171 636 9107. Lists of planning consultants.

PLANNING INSPECTORATE
ENGLAND - 0117 987 8754.
WALES - 0122 282 5493.
SCOTLAND - 0131 244 5657.
N.IRELAND - 01232 244710.
Appeal forms and booklets

THE BRITISH CEMENT ASSOCIATION. 01344 762676. Useful booklets on foundations, cellars etc.

THE GLASS AND GLAZING FEDERATION. 0127 781 0882. Glass options, glazing techniques, conservatories.

CONSERVATORY ASSOCIATION. 01480 458278. Specialist advice on conservatories.

SOLAR TRADES ASSOCIATION. 01208 873518. Solar energy information.

NATIONAL ENERGY FOUNDATION. 01908 501908. Energy saving information and the NER rating.

TIMBER RESEARCH & DEVELOPMENT ASSOCIATION. 001494 563091. Publications on timber frame construction.

TIMBER AND BRICK INFORMATION COUNCIL. 01923 778136. Leaflets on timber frame construction and lists of manufacturers.

THE BUILDING RESEARCH ESTABLISHMENT. 01923 894040. Wide variety of publications on problems of all sorts. Also sell advice. A good place to start if you are looking at a problem site.

DISABLED LIVING FOUNDATION. 0171 289 6111. General advice and contacts for specialist advice.

H M LAND REGISTRY. 0171 917 8888. Will identify owners of registered land titles.

BRICK DEVELOPMENT ASSOCIATION. 01344 885651. Information on bricks and brickwork.

THE CLAY PIPE DEVELOPMENT ASSOCIATION. 01494 791456. Advice on design of drains etc.

NATIONAL RADIOLOGICAL PROTECTION BOARD. 01235 831600. Leaflets on Radon.

THATCHING ADVISORY SERVICES. 01256 880282. Thatched roofs.

COMMISSION FOR THE NEW TOWNS. 01908 692692. Plots in the New Towns, including Milton Keynes.

LANDBANK SERVICES. 01734 618002. Plot finding service.

NATIONAL LAND FINDING AGENCY. 01371 876875. Plot finding service.

BRADFORD & BINGLEY BUILDING SOCIETY 0800 252993. Guide to selfbuild mortgages.

AGENCIES PROVIDING SERVICES TO SELFBUILD GROUPS, INCLUDING COMMUNITY GROUPS

THE COMMUNITY SELFBUILD AGENCY.. 0171 415 7092. The key agency for advice. Useful publications.

THE NATIONAL FEDERATION OF HOUSING ASSOCIATIONS. 0171 278 6571. Advice and formal registration for selfbuild groups.

ROYAL INSTITUTE OF BRITISH ARCHITECTS (RIBA). 0171 580 5533. Information on grants for feasibility studies.

THE HOUSING CORPORATION. 0171 393 2000. Information for those sponsoring community groups.

THE WALTER SEGAL TRUST. 0171 388 9582. Unique timber frame system and special group structures.

THE YOUNG BUILDERS TRUST. 01730 266766. Kenneth Claxton's team featured in chapter 25.

CHISEL HOUSING COOPERATIVE. 0181 768 2036. 'Selfbuild for rent' schemes.

CONSTRUCTIVE INDIVIDUALS. 01904 625300. Consultancy services for low cost group schemes.

EXHIBITIONS AND COURSES

THE BUILDING CENTRE. First class permanent exhibition on six floors at 26 Store St, London, near Goodge St underground station. The Building Bookshop is in the same building.

BUILD IT EXHIBITIONS. 0181 286 3000. Selfbuild exhibitions and roadshows throughout the country, with seminars and advice centres.

THE INDIVIDUAL HOMES SHOW. 01527 836 600. Annual Individual Homes Exhibition at the NEC with seminars and advice centres.

IDEAL HOMES EXHIBITION. Hardy annual at Earls Court every March. Some selfbuild.

CONSTRUCTIVE INDIVIDUALS. 01904 625300. Courses of all sorts for Selfbuilders.

CENTRE FOR ALTERNATIVE TECHNOLOGY. 01654 702400. Ecologically friendly selfbuild courses and exhibition.

FINDHORN FOUNDATION. 01309 690311. Ecologically friendly selfbuild courses etc. near Elgin in Scotland.

BRITISH GYPSUM. 0115 984 4844. DIY courses on dry plastering techniques.

KNAUFF UK LTD. 01795 424499. DIY courses on dry plastering techniques.

INSURANCES AND WARRANTIES

DMS SERVICES. 01909 591652. Insurances for Individual Builders and Selfbuilders for new properties.

VULCAN INSURANCE. 01622 671747. Insurances for renovations and conversions.

THURCROFT INSURANCE. 01709 540348. Arrange to convert selfbuild insurances to householders insurances when the project is finished.

FRASER MILLER. 01483 797948. Insurances for selfbuild groups.

NHBC. 01494 434477. Or for N.IRELAND. 0232 683131 The NHBC building warranty.

ZURICH CUSTOMBUILD. 01252 522000. The Zurich Custombuild warranty.

PACKAGE COMPANIES AND BUILDERS FEATURED IN THIS BOOK

DESIGN AND MATERIALS LIMITED. 01909 730333.

MEDINA-GIMSON LIMITED. 01732 770992.

POTTON HOMES LIMITED. 01767 260348.

PRESTOPLAN LIMITED. 01772 627373.

SCANDIA HUS LIMITED. 01342 327977.

OLIVER HOMES LIMITED. 01506 826218.

TECHNIQUE (UK) LIMITED. 01954 789988

STATUS TIMBER SYSTEMS. 01224 248989.

TAYLOR LANE TIMBER FRAME. 01432 271912.

BORDER OAK LIMITED. 01568 708752.

INVESTMENT HOMES. 01635 521525.

COUNTRY HOMES. 01432 820660.

CUSTOM HOMES. 01293 822898.

ARCHITECTS AND DESIGNERS WHOSE WORK IS FEATURED IN THIS OR THE 14TH EDITION OF THIS BOOK

ASSOCIATED SELFBUILD ARCHITECTS. 0800 387310.

ROBIN ASHLEY, RIBA. 01909 564341.

DAVID SIM, RIBA. 01908 201501.

ANTHONY BUCKLEY. 01203 311496.

WILLIAM WESTALL ARCHITECTS LTD.0462 422440.

PETER GRIFFITHS. ARICS. 01795 890286.

CONSULTANTS FEATURED IN THIS BOOK

ESSEX SELFBUILD. 01268 557332. Selfbuild group managers.

CHARTSERVE LIMITED. 01323 412161. Selfbuild group managers.
ECD ENERGY CONSULTANCY AND DESIGN LTD. 0171 405 3121. Energy saving consultancy.

MANUFACTURERS WITH PARTICULARLY USEFUL LITERATURE OR ADVISORY SERVICES

BUTTERLEY BRICK LIMITED (alias Hanson Brick). 0773 570570. Bricks and brickwork, paver bricks.

REDLAND TILES LIMITED. 01737 242488. Tiles and tiling.

RICHARD BURBIDGE LIMITED. 0691 655131. Staircases and balustrading etc.

PILKINGTON GLASS LIMITED. 01744 28882. Advice on glass, and particularly insulating glass.

FLYGT LIMITED. 0115 940 0444. Pumps to solve drainage problems.

CONDER PRODUCTS LIMITED. 01962 863577. Septic tanks, Cess Pools, Pumps, and advice on them.

CLEARWATER SYSTEMS. 01373 858090. Mini sewage systems for individual homes.

HEPWORTH BUILDING PRODUCTS LIMITED. 01226 763561. Useful booklets on drainage.

CAMAS BUILDING PRODUCTS. 01335 372222. Reconstituted stone products.

WARMWORLD — DATATERM HEATING CONTROLS. 0117 9498800

Acknowledgements

The author wishes to express his appreciation of the help and advice received from many sources in writing this edition of *Building Your Own Home*, and in particular to:

Robin Ashley, RIBA
Roland Ashley
David Barker of Technique
Mr. Michael Barraclough
The Birmingham Midshires Building Society
Phil and Melanie Bixby of Constructive
    Individuals
Keith Burrell of the Association of Selfbuilders
Kevin and Chris Choules
Ken Claxton RIBA, of the Young Builders
    Trust
Rob Cousins
Ron and Janet Cox
Richard Crisp of Potton Ltd
Mike Daligan of the Walter Segal Trust
Jenny De Villiers
Dr. Simon Duncan, London School of Economics
Mike and Christine Evans
Mike Flint, Customs & Excise
M J Gaines, N.R.P.B.
Mike Geelan of the Central Statistical Office
Dan Gillbanks of Gateshead MBC
The Great Eastern Housing Association
John Greene of Border Oak
Nicci Griffiths of the National Energy
    Foundation
Simon Hayes of Hedlunds Ltd
Mike and Jean Hall
Steve and Chris Haywood
Christopher Heath of Custom Homes
Tony and Carolyn Hole
Sue Holliday of Build It
Christine Holman, Sheffield Hallam University
The Individual House Builders Association
Nick Jones of the Building Research
    Establishment
Keith Julier, Essex Self Build Ltd
Michael Kilcommons of Zurich Insurance
Tony Lush of Juvan Courses
The Land Registry
Bruce Macdonald and others of D & M Ltd
Anna McGettigan of the CSBA
All the Morton family

Norwich Union
Julian Owen RIBA, Associated SB Architects
Janet Parker
Beverley Pemberton
Robert Pennicott of Landbank Services
Alan and Norma Price
RL Photography
Barbara Mawer and the staff of R.T.S. Ltd
Phil Rees
Rosalind Renshaw, Editor, Build It
David Sim RIBA
David Snell
Adrian Spawforth RIBA, Associated SB
    Architects
David Spence
Colin and Pearl Spooner
Hilary and Rob Stokes
The Surrey Docks SB Housing Association
Colin Wadsworth
Gunnel Westley of Scandia Hus
Maggie White of Medina Gimson
Steven Whitehill of E.C.D. Ltd
The Zenzele SB Housing Association

Index

Also by Murray Armor

HOME PLANS
and PLANS FOR A DREAM HOME

Britain's best selling books of Home Plans, between them presenting 700 house and bungalow designs for homes of all sizes to suit all budgets, most of them in today's traditional styles. There are also features on everything that you need to know when choosing a new home, the choice of materials and components, heating systems, insulation, and much else – all in the author's inimitable style.

Home Plans has 400 pages, 300 house and bungalow plans, 100 colour photographs, 360 line drawings – a treasure trove of information and ideas for anyone thinking about a new home.

Plans for a Dream Home is a blockbuster of 464 pages with plans of 400 homes that have been built by clients of five leading designers in different styles to suit every sort of site. They vary from five and six bedroom houses to small retirement bungalows, and the plans of all of them give all the dimensions and a wealth of construction detail.

The plans in the two books are all different designs, and the two volumes complement each other.

Full sets of drawings are available for all the designs in the books at competitive prices. Whether you are wanting to choose plans for a new home of your own, or just wanting to know more about the best of today's new houses and bungalows, these are the books you must have.

Home Plans is published by J.M. Dent & Sons Ltd and Plans For A Dream Home by Ebury Press. In bookshops everywhere.

Available either by post using the coupon opposite, or by phone.

 CREDIT CARD ORDERS
– Phone 01909 591652
(Books despatched within 24 hours)

ORDERS BY POST
Ryton Books, 29 Ryton Street, Worksop, Notts. S80 2AY

Please supply books as indicated.

Name ....

Address ....

....

Plans for a Dream
Home – £19.95 inc p&p ☐   (Two books deduct £1)

Home Plans
£19.95 inc p&p ☐

Enclose Cheque or quote Visa or Access No.

☐☐☐☐ ☐☐☐☐ ☐☐☐☐ ☐☐☐☐

expires ☐☐☐☐

BUILDERS RISKS INSURANCES
FOR THOSE BUILDING ON THEIR OWN LAND

NORWICH UNION

The Norwich Union is able to offer an insurance package for those who are building for their own occupation private dwellings of traditional construction with or without the help of builders or sub-contractors. It does not apply to the extension, alteration, repair or renovation of existing buildings. This affords Contract Works, Public Liability and Employers' Liability cover and automatically includes the interest of any Mortgagee. Cover is provided in one policy document, summarised as follows. This description of insurance must be regarded only as an outline. The policy is a legal document and as such defines the insurance in precise terms. A specimen copy of the policy form is available on request.

CONTRACT WORKS

Cover	"All Risks" of loss or damage to:
	(a) the new building whilst under construction and materials for incorporation therein
	(b) plant, tools, equipment, temporary buildings and caravans.
Sum insured	The full rebuilding cost of the property, excluding the value of the land.
Including	(a) your own and hired plant, tools and equipment used in connection with the work up to a total sum insured of £2000 (can be increased if required).
	(b) Employees personal effects and tools whilst on the site up to a sum insured of £330 any one employee in accordance with standard Building Industry/Union agreements.
	(c) Architects, Surveyors and other fees necessarily incurred in rebuilding following loss or damage.
	(d) the cost of removing debris following any claim.
Excluding	(a) the first £50 of each and every claim for loss or damage to employees personal effects or tools.
	(b) the first £500 of each and every other loss.

EMPLOYERS LIABILITY (compulsory by law)

Cover	Your legal liability for death or bodily injury to employees, including labour only sub-contractors, arising out of the building work.
Limit	£10,000,000 each occurrence.
Including	Legal costs and expenses in defending any claim.
Note	A Certificate of Insurance will be provided, and must by law be displayed on site.

PUBLIC LIABILITY

Cover	Your legal liability to members of the public, (including sub-contractors working on the site not classed as employees) for death, bodily injury or damage to property, arising out of the building work.
Limit	£1,000,000 any one loss. (Can be increased if required)
Including	Legal costs and expenses in defending any claim.
Excluding	The first £250 of any claim for damage to property.

PERIOD	From the commencement date you specify (which should be no earlier than the date you complete the proposal form) up to completion of the building work, subject to a maximum of 24 months. Extensions to this period may be available on payment of an additional premium. There is no refund for early completion.
THE POLICY	Will be sent direct to you by DMS Services Ltd. on behalf of the Insurance Company.
THE PREMIUM	£6.24 per 1,000 on the rebuilding cost of the property. (Minimum £80,000). This is a total rate for all the cover set out above, subject to submission of the completed proposal form overleaf. and includes insurance premium tax at 2½%. Proposal forms should be accompanied by cheques for the relevant premium made out to DMS Services Ltd.

Rebuilding Cost Up to £	Premium £	Rebuilding Cost Up to £	Premium £	Rebuilding Cost Up to £	Premium £
80,000	499.20	110,000	686.40	160,000	998.40
85,000	530.40	120,000	748.80	170,000	1060.80
90,000	561.60	130,000	811.20	180,000	1123.20
95,000	592.80	140,000	873.60	190,000	1185.60
100,000	624.00	150,000	936.00	200,000	1248.00

Over £200,000 @ £6.24 per £1000

TAX	The scale of premiums shown in this prospectus and proposal are inclusive of Insurance Premium Tax at 2½% and are only valid while the rate of tax remains at this level. DMS Services Ltd. will advise on revised premiums should the rate of tax change.

IMPORTANT The above terms only apply:
(a) up to 31st December 1996. Amended terms may be necessary for proposal forms completed after that date.
(b) to risks in Mainland Great Britain only. Proposals from N. Ireland are quoted individually and special excesses may apply. Phone 01909 591652 or fax 01909 591031 for a quotation. Proposals cannot be accepted from Eire.
(c) Where there is no abnormal exposure to risk of floods, storm damage or vandalism.

THE AGENCY

The Agency is DMS Services Ltd., a company which provides specialised insurance services to those building on their own. The proposal form overleaf should be completed and sent to the agency with a cheque for the premium payable to DMS Services Ltd.

D.M.S. Services Ltd., Orchard House, Blyth, Worksop, Notts. S81 8HF.
Phone 01909 591652   Fax 01909 591031

1996

Agency: DMS Services Ltd   Agency Reference: 50GA59   Policy No.

NORWICH UNION

Proposal — BUILDING OWN PRIVATE DWELLING
The Insurer: Norwich Union Fire Insurance Society Limited

Name of Proposer
MR/MRS/MISS

Phone No.

Full Postal Address
....
.... Postcode ....

Address of property to be erected
....
....

Name and address of any interested party — eg Bank or Building Society
....
....

Commencing date
of insurance
....

Important — Please give a definite answer to each question (block letters) or tick appropriate boxes

	Yes	No	If "Yes" please give details

1. Have you made any other proposal for insurance in respect of the risk proposed?   ☐ ☐

2. Has any company or underwriter declined your proposal?   ☐ ☐

3. Have you ever been convicted of (or charged but not yet tried with) arson or any offence involving dishonesty of any kind (eg fraud, theft, handling stolen goods)?   ☐ ☐

4. Will the property be

   (a) a completely new structure and not an extension, conversion or restoration of an existing building?   ☐ ☐   (If "No" please refer to DMS Services Ltd.) Phone 01909 591652

   (b) of conventional construction, either in loadbearing masonry, or with a timber frame, and built to drawings approved under the requirements of the Building Regulations as meeting the requirements of the regulations in full?   ☐ ☐

   (c) occupied as your permanent residence on completion?   ☐ ☐   (If "No" please refer to DMS Services Ltd.) Phone 01909 591652

5. (a) Will the total value of plant, tools, equipment and temporary buildings, whether hired or owned on site at any one time exceed £2,000. If so see page 6 of the prospectus for the additional premium required (cover for plant on site can be altered at any time while the policy is in force)   ☐ ☐   Contractors plant hired in with operators, such as excavators, need not be included if proposers are wholly satisfied the hirers insurances cover all risks. However if cover is required on such machines phone DMS Services on 01909 591652

6. Is there any abnormal exposure to risk of flooding, storm damage or vandalism?   ☐ ☐

7. State estimated value of building work on completion at builder price for reinstatement.   £ _____   It is important that this sum is the cost of a professional building firm rebuilding the entire dwelling should it be completely destroyed just prior to completion. This will be the limit of indemnity for item (A) of the Contract Works section, and payments of premium on a lesser figure will result in any contracts works claim being proportionately reduced. Please discuss with DMS Services Ltd if in any doubt.

8. Material facts — state any other material facts here. Failure to do so could invalidate the policy. A material fact is one which is likely to influence an insurer in the assessment and acceptance of the proposal. If you are in any doubt as to whether a fact is material it should be disclosed to the insurer.   If work on site has started certify here that there have been no incidents on site which would have given rise to a claim

Note: 1. You should keep a record (including copies of letters) of all information supplied to the insurer for the purpose of entering into the contract.
2. A copy of this proposal form will be supplied by the Insurer on request.
3. Please note that the details you are asked to supply may be used to provide you with information about other products and services which the Norwich Union Group can offer.

Declaration To be completed in all cases
I desire to insure with the Insurer in the terms of the Policy used in this class of Insurance. I declare that the above statements and particulars are true to the best of my knowledge and belief and that I have not withheld any material information. I agree to give immediate notice to the insurer of any alteration to the circumstances described herein and that this proposal shall form the basis of the contract between us.

Proposer's signature

Date

Send completed form to DMS Services Ltd., Orchard House, Blyth, Worksop, Notts. S81 8HF, together with a cheque made payable to DMS Services Ltd. Any queries to DMS Services. Phone 01909 591652.

Norwich Union Fire Insurance Society Limited. Registered in England No. 99122. Registered Office: Surrey Street, Norwich NR1 3NS. Member of the Association of British Insurers. Member of the Insurance Ombudsman Bureau.

HOME PLUS COVER
Quotations will be provided for household insurances when the building approaches completion. If this is not required please tick box — ☐

1996